Communication Systems

Communication Systems

N.T. Markad

Assistant Professor, Bharati Vidyapeeth's College of Engineering; GGSIPU,
New Delhi - 110063

Dr. M.S. Sutaone

Professor and Head of Department
Government College of Engineering
Shivajinagar, University of Pune (M.S.), Pune - 5

A.G. Raut

I/C H.O.D. & Lecturer
Government Polytechnic, Khamgaon (M.S.)

Ravi Prakash

Lecturer
University of Bombay, Bombay (M.S.)

TECHSAR™

NEW DELHI

Published and Printed by
Techsar Pvt. Ltd.
4435-36/7, Ansari Road, Daryaganj
New Delhi-110 002 (India)
E-mail: tech_info@techsarworld.com
Website: www.techsarworld.com

ISBN 978-93-81141-13-7

Reprint 2025

Dedicated …

To our family for their love and patience

&

To our students whose queries motivated us for more research

Foreword

It is my pleasure to write this foreword to the book *Communication Systems* by Prof. N.T. Markad and co-authored by Prof. M.S. Sutaone, Prof. A.G. Raut and Prof. Ravi Prakash. My former colleague Prof N.T. Markad and Prof M.S. Sutaone have more than two decades of experience in teaching subjects related to communication systems. They have clearly brought their teaching and research into effect in writing this book which emphasizes the learning of fundamental concept as well as the application of these concepts to develop efficient practical skills. I believe that it is a worthy presentation of their accomplishments and a valuable contribution to engineering education.

This book should be very useful for students pursuing undergraduate and graduate courses in Electronics and Telecommunication Engineering area.

Wishing the authors "Best of Luck" in all their future ventures.

Dr. Rajendra D. Kanphade
Principal
Dhole Patil College of Engineering
Wagholi, Pune – 412 207

Preface

First, I remember God Ganesh since God has given me strength and potential for writing this book which will be very beneficial for students. This book provides a comprehensive information to the modern study of electronics and telecommunication circuits. The book "Communication Systems" deals with electronics, and data communication. Authors have divided this book into five parts. First part deals with analog communication, second part digital communication, third part telecommunication network, fourth part data and computer communication and fifth and small part a few advances of telecommunication.

Readers can explore the book parts cum chapterwise like in first part, and its first chapter is methods of communication. In this chapter very preliminary concepts of telecommunication is given. Second chapter of first part is amplitude and angle modulation. In this chapter basic concept of modulation and its types treated in detail upto circuit level. Also detection is given upto circuit level. Third chapter of first part deals with AM, FM transmitters and receivers. Also noise is discussed in detail. Fourth chapter of first part related with pulse modulation and its types. Also digital pulse modulation like PCM is given. Fifth chapter of first part deals with propagation of electromagnetic waves. Types of propagation, microwave link and satellite communication are given in this chapter.

Chapter number 6, 8, 9 and 10 deals with second part. In second part chapter six gives you information about very preliminary concepts of digital communication. Chapter six also contain probability, information, entropy and coding & types of coding. Chapter eight of second part related with digital modulation system and its detection system Chapter nine of second part gives you information about types of digital modulation techniques and its error probability. Also line coding given in detail. Chapter ten of second part again gives you preliminary concepts of random process and its probability. Different functions and theorem also given in chapter ten.

Third part of this book related with telecommunication networks. Chapter seven of third part gives you information about telephone exchange and its architecture as well as its related different functions Also in chapter seven traffic engineering related with telephone exchange is given in detail along with problems.

Now, come to fourth part of book. Fourth part of this book related with advanced part of telecommunication like computers and data communication. In this part discussed chapter is chapter eleven. In chapter eleven simple data communication model, advanced switching system, protocol structure, advanced protocols, routing in advanced computer network and value-added services are written in such a manner students will understand easily.

This last that is fifth part of this book is related with advances in satellite communication and mobile communication in short. In fifth part of this book chapter discussed is chapter twelve. This chapter gives you clear understanding of a satellite based ALOHA Technique, mobile access technique and terrestrial packet radio system.

Now I have discussed preface in detail, but speciality of this book is that those who wants to learn telecommunication may be students or any other person, my opinion is that go through this book you will definitely learn telecommunication. Other than this so many unsolved problems are solved in this book. This book is written in such a manner students can understand very easily. Also a few unsolved problems and questions are given below each chapter. I would also like to mention that the book "communication systems" covers almost syllabus of GGSIPU, Delhi, UPTU, MDU, DU, Pune University, Bombay University, Shivaji University, Amravati University, Nagpur University, North Maharashtra University, Jalgaon, Marathwada University, Aurgangabad. For other universities this book will also play important role in the subject of telecommunication. I have written this book for benefit of students community. When students will benefited then that will be my pleasure, because the book has been written in students understandable language so that reader feel comfortable while exploring the book. This book also covers syllabus of diploma course in Electronics, Electronics and Telecommunication engineering. It also covers syllabus of AMIE courses in Electronics and Electronics and Telecommunication.

This book is useful to Instrumentation and Control, Electrical and Electronics, Computer science, Electronics, Electronics and Telecommunication, Electrical, Information Technology branches for subject communication system since syllabus of these branches covered in this book.

New Delhi — Prof. N.T. Markad

Acknowledgements

This book has taken three years to write and benefited from contributions of many individuals. We would like to say thank you to Dr. R.D. Kanphade for all his research oriented inputs. We are wholeheartedly grateful Dr. D.G. Wakade, Dr. P.P. Vitakar, Prof. N.B. Pasalkar, Prof. G.B. Dhanokar, Dr. Abhay Wagh, Dr. K.C. Tiwari, Dr. A. Basu, Dr. A.K. Sinha, Dr. N.D. Kaushika for their regular encouragement. A very special thank to Dr S.W. Dharmadhikari, Dr. G.V. Inamdar, Dr. D.S. Bilgi, Dr. A.A. Gathol, Dr. T.R. Sontakke, Prof. B.R. Jadhavar, Prof. M.D. Bhagat, Dr. P.S. Jagtap, Dr. M.A. Joshi, Dr. A.A. Sapkal, Dr. S.K. Mahajan, Dr. Barttke who has been always a source of inspiration for us since we could inculcate research methodology in our professional life. I would like to thank faculty S.K. Kundu, S.B. Kumar, J.C. Agarwal, and staff A.K. Srivasta, Bhagwan for their timely co-operation.

Thanks to all of our students over the years. Your questions and insight have shown challenged and strengthened us. Please come up and say hello if you see us at an event. Stay in touch!

Thanks to the folks. I.K. International Publishing House Pvt. Limited, New Delhi who have shown faith in us and come forward to publish this edition. We give special thanks to Krishan Makhijani, for his effort in publishing this book.

My deepest thanks to my wife, daughter Aishwarya and son Aum for their patience while preparing this book. Our deepest thanks go to our family and their continuous support as they have been always encouraging us while the project was dragged on.

Prof. N.T. Markad
Dr. M.S. Sutaone
Prof. A.G. Raut
Prof. Ravi Prakash

Contents

CHAPTER 1

Methods of Communication

1.1 INTRODUCTION

Communication refers to the sending, receiving and processing of information by electric means.

In the beginning communication took place with wire telegraphy, after that telephony and now by radio. And most recently, the use of satellites and fibre optics has made communication even more widespread, with increasing emphasis on computer and of data communications.

A modern communication system is first concerned with the sorting, processing and storing of information before its transmission. The actual transmission then follows, with the further processing and the filtering of noise. Finally, we have reception, which may include processing steps such as decoding, storage and interpretation.

1.2 BLOCK DIAGRAM OF BASIC COMMUNICATION SYSTEM

Description:

The origin of the message is from some information source. The message may be

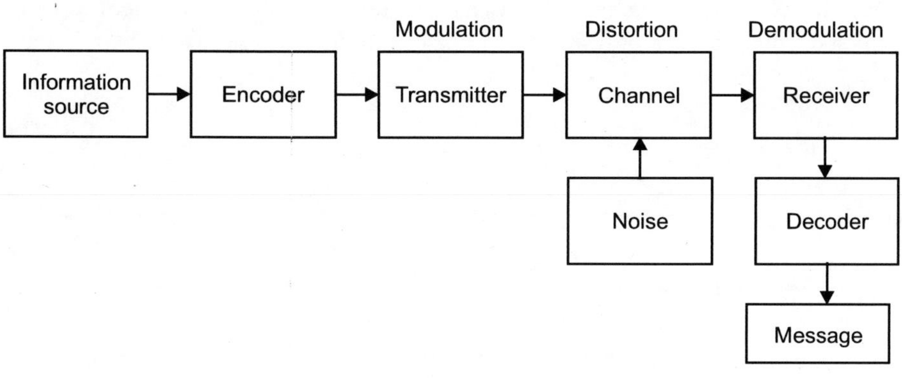

Fig. 1.1: Block Diagram of Basic Communication System

in the form of words, code symbols, etc. The amount of information conveyed in any message is measured in bits.

The encoder converts the message into electrical signals and feeds them to the transmitter. The signals are processed at the transmitter and transmitted to the receiver. In fact, the signal gets modulated in the transmitter.

The channel which may be a pair of wires in case of telephone and telegraph. In case of radio communication, the channel is open wire or electromagnetic signal.

When signal passed through pair of wire or coaxial cable in case of telephone and telegraph signal get distorted. In case of radio communication signal passed through open wire get distorted by external noise or internal noise.

Then modulated signal affected by noise covers long distance by electromagnetic propagation, then signal get received by receiver. Within the receiver demodulation takes place, i.e. carrier is removed and audio is given to decoder. Decoder converts analog signal to digital signal. Original message is derived from digital signal.

1.3 ANALOG AND DIGITAL SIGNAL

1.3.1 Analog Signal

A voltage waveform that is a sinusoidal function of time may be written as $e(t) = E_{max} \sin 2\pi f_0 t$. Here the constant are E_{max}; the peak value of the voltage, and f_0, the frequency of wave. The periodic time is $T_0 = \frac{1}{f_0}$.

1.3.2 Digital Signals

Digital signal comprise pulse occurring at discrete interval of time. Simplest type of digital signals are the telegraph and teleprinter signals but sometimes analog messages are also converted into digital signal form before being transmitted.

1.4 SPECTRUM OF A SIGNAL

Waveforms of different signals are drawn by plotting amplitudes of the concerned signal at different instant of time. Such representation is generally termed as Time Domain representation.

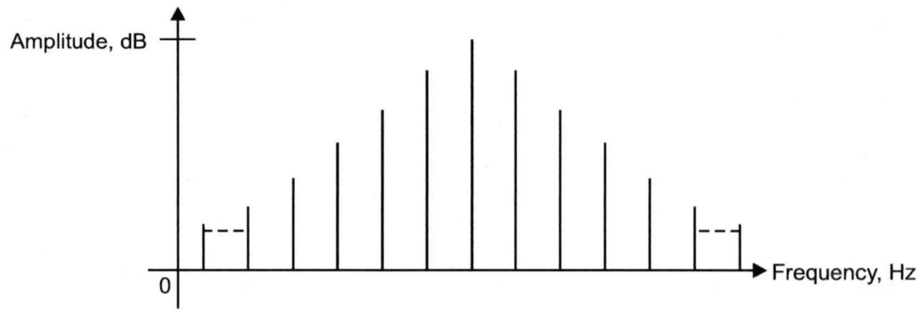

Fig. 1.2: Spectrum

Sometimes it is required that the frequencies present in the signal and relative amplitudes of different frequencies be known for this, the signal may be analyzed and amplitudes of different frequency components may be plotted. This plot is termed as Frequency Domain representation and gives spectra component amplitudes against frequency.

The analog signal when analyzed are found to comprise certain fundamental frequencies and their harmonics and occupy only a small portion of the frequency spectrum. Such a spectrum is termed *Discrete Spectrum.* The analysis of the digital signals, on the other hand, yields an infinite number of frequencies. Such a spectrum is termed *Continuous Spectrum.*

1.5 (TRANSMISSION LINES) OPEN WIRE LINES

Electrical energy may, in general, be transmitted from one point to another by any one of the following two methods:

By radiation of electromagnetic waves through free space or by the use of electrical conductors in one or the other form. A transmission line may be considered a guide that guides the electrical energy to the desired destination.

Consider a situation when an audio generator is connected to a load through a pair of wires as shown in Fig. 1.3a.

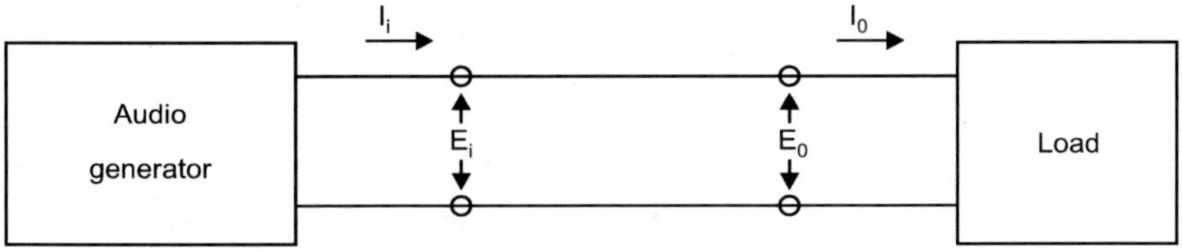

Fig. 1.3(a): Audio Generator Connected to a Load by a Pair of Wires

If lead resistances are considered to be negligible, then the current $I_i = I_0$ as also $E_i = E_0$.

The situation is not that simple while using the radio frequencies. When the lead length is an appreciable portion of the wavelength of the working frequency, neither of the above-mentioned equalities of voltages and currents are realized; E_0 differs from E_i and I_0 differs from I_i in both magnitude and phase., Lines which are greater than about 10% of the wavelength of frequency of operation, current and voltage undergoes a change in phase as well as in amplitude along the length of line. Such transmission lines cannot be analyzed by just considering their resistance and inductance. Here, shunt capacitance and shunt conductance also play an important role. Thus, to ascertain the exact nature or behaviour which the transmission line will exhibit at radio-frequencies, we analyze its equivalent circuit containing four parameters, viz. resistance, inductance, shunt capacitance and shunt conductance.

Primary Constants of Line: Consider a long line consisting of two parallel uniform conductors carrying current, there is a magnetic field around the conductors and voltage drop along them. The existence of magnetic field indicates that the lines have series inductance (L) and the voltage drop indicates the presence of series resistance (R).

Voltage applied across the conductor produces an electric field between the conductors and charges them. This shows that lines have shunt capacitance (C) and since capacitance is never loss-less or perfect, it will have some shunt conductance (G) as well. These four constants R, G, L and C are called primary constants of line. A unit section (of 1 km. length) of line may be represented as shown in Fig. 1.3b.

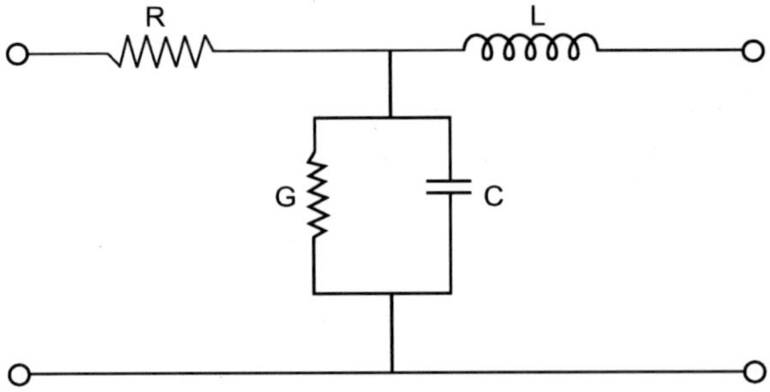

Fig. 1.3(b): Equivalent Circuit of Transmission Line.

These constants (R, G, L & C) are uniformly distributed along the entire length of transmission line, it is called uniform-transmission line. These constants depend upon the frequency of operation.

Series impedance (Z) and shunt admittance (Y) of the line per unit length are

$$Z = R + j\omega L$$

$$Y = G + j\omega c$$

Characteristic-Impedance (Z_0): When an alternating voltage is applied to the sending end of an infinite line, a finite current will flow due to the capacitance and the leakage conductance between the two wires constituting the line. The ratio of the voltage applied to the current flowing, will give the input impedance. This input impedance is called "characteristic impedance" of the line and is denoted by Z_0.

Z_0 of a line is defined as the impedance looking into an infinite length of line. The magnitude and angle of characteristic impedance will in general vary with frequency; and frequency at which the impedance has been measured should be stated.

$$\text{Characteristic impedance, } Z_0 = \sqrt{\frac{Z}{Y}}$$

Propagation Constant (γ): Propagation constant is normally a complex quantity.

The real part is called the *attenuation constant* and is denoted by 'α'. Imaginary part is called *phase constant* and is represented by 'β'.

Thus,

$$\gamma = \sqrt{ZY} = \alpha + j\,\beta$$

or

$$\alpha + jB = \sqrt{(R + j\omega L)(G + j\omega C)}\,.$$

Example 1.1: An open wire telephone line has R = 10 ohm per km.
L = 0.0037 henry per km, C = 0.0083 microfarad per km and
G = 0.4 × 10^{-6} mho - per km. Determine Z_0, α and β at 1000 Hz.

Solution: As we know,

Series impedance $= Z = R + j\omega L$

Since, $\omega = 2\,\pi\,f$

$$Z = 10 + j \times 2 \times 3.14 \times 1000 \times 0.0037$$

$$= 10 + j \times 23.2$$

$$= 25.3 \angle 66.8°$$

and shunt admittance

$$Y = G + j\omega C$$

$$Y = 0.4 \times 10^{-6} + j \times 2 \times 3.14 \times 1000 \times 0.0083 \times 10^{-6}$$

$$= (0.4 + j\,52.1) \times 10^{-6}$$

$$Y = 52.1 \times 10^{-6} \angle 89.6°$$

Now

$$Z_0 = \sqrt{\frac{Z}{Y}} = \frac{\sqrt{25.3\angle 66.8°}}{\sqrt{52.1 \times 10^{-6} \angle 89.6}} = 69.7\angle -11.4°$$

$$Z_0 = (683 - j\,38) \text{ ohms.}$$

The propagation constant

$$\gamma = \sqrt{Z \times Y}$$

$$= \sqrt{25.3 \times 52.1 \times 10^{-6}} \angle \frac{66.8 + 89.6}{2}$$

$$= 0.0363 \angle 78.2°$$

$$r = (0.0074 + j\,0.0356)$$

But $\gamma = \alpha + j\,\beta$

hence α = 0.0074 weber/km

β = 0.0356 radians/km

Advantages of open wire lines:

(1) They are simple to construct and have lowest cost.

(2) Insulation between the line conductors is completely air. As a result, the dielectric loss is extremely small.

(3) It is balanced with respect to ground.

Disadvantages:

(1) There is a significant energy loss due to radiation that makes these lines unsuitable for use at frequencies above 100 MHz.

1.5.1 Coaxial Cable: (High Frequency Lines)

To avoid radiation losses taking place in open-wire lines at high frequencies, a closed field configuration must be used by surrounding the inner conductor with an outer cylindrical hollow conductor and the arrangement is called as "coaxial cable".

With an increase in the frequencies transmitted, the parallel open wire line suffers from three great handicaps:

(1) Symmetrical balance pair does not radiate any appreciable power at low frequencies, but as the frequency is increased, the separation of the conductors becomes an appreciable fraction of the wavelength, the transmission lines becomes an antenna and energy is lost through radiation of E-H waves.

(2) Conductor heating of I^2R loss is proportional to current and therefore inversely proportional to characteristic impedance. It also increases with frequency because of skin-effect.

(3) Dielectric heating is proportional to the voltage across the dielectric and hence inversely proportional to characteristic impedance Z_0. It again increases with frequency because of gradually worsening properties with increasing frequencies for any given dielectric medium. These problems are overcome by the use of coaxial cable which consists of inner cylindrical conductor placed along the axis of an outer hollow cylindrical conductor.

The chief advantage of the system is that the fields are confined within the outer conductor thereby eliminating radiation losses totally. The system also provides effective shielding from outer interfering signals.

Figure 1.4 shows the constructional details of coaxial cable.

In practice the central-conductor is held in position accurately by insulating material which consists of a solid core or discs strung along the axis of the conductor.

At high frequencies if the dielectric loss is negligible, the coaxial pair has half the inductance and double the capacitance value of two parallel wire line of the same diameter as the coaxial pair, resulting in low attenuation at high frequencies.

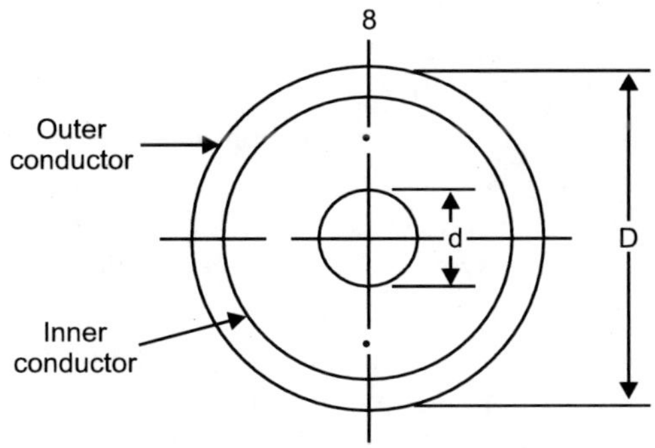

Fig. 1.4: Coaxial Cable

The presence of outer conductor provides shielding effect and minimizes radiation losses and external interference.

The characteristic impedance Z_0 of any transmission line is expressed as;

$$Z_0 = \sqrt{\frac{R + j\omega L}{G + j\omega C}}$$

At high frequencies

$$Z_0 = \sqrt{L/C}$$

This shows that this characteristic impedance is resistive at high frequencies. Physically, Z_0 is determined by the size and spacing of conductors and by the dielectric constant of the insulator separating them. In the case of coaxial cable, Z_0 is given by,

$$Z_0 = \frac{138}{\sqrt{k}} \log \frac{D}{d}$$

where, K = dielectric constant of insulation. The usual ranges of Z_0 from 40 to 150 Ω for coaxial lines. The coaxial cables can be conveniently employed in the frequency range extending upto about 1 GHz for transmission of signals but they are quite costlier in comparison to the open wire lines, losses in the dielectric increase with frequency and beyond 1 GHz, these cables cannot be used.

1.6 WAVEGUIDES

Waveguides are extensively used to minimize losses and for high power transmission at microwave frequencies. Waveguides consist of a single hollow metallic conductor in the shape of a rectangular box or cylinders through which electromagnetic waves are propagated. Propagation through waveguides is different than taking place through open wire coaxial lines and takes place in the form of transverse, electric (TE) or transverse magnetic (TM) waves. The first one is commonly termed E-mode while the latter is termed H-mode.

Waveguides are usually made of copper, brass or aluminium in various standard sizes corresponding to the signal frequencies to be propagated and their inner surface is plated with thin layers of silver or gold to reduce losses.

Waveguides are usually coupled together by flanges bolted together and supported on metal stands at different intervals. Waveguide is a hollow conducting metallic tube of uniform cross-section for transmitting electromagnetic waves by successive reflection from the inner walls of the tube. These waveguides are used in UHF range and microwave range as an alternative to coaxial cables. Waveguides are usually air filled and are superior to the coaxial cables and can handle greater power and possess less resistance. There is no need of return conductor in waveguide as energy propagates in the form of electric and magnetic fields. Waveguides may have many shapes such as rectangular, circular or elliptical. Waveguides are simpler to manufacture than coaxial lines. The power handling capacity of waveguides is about ten times as high as for coaxial air dielectric rigid cable of similar size.

1.6.1 Rectangular Waveguides

A pipe of any cross-section can be used as a waveguide but the simplest cross-section is preferred. Thus, waveguides with constant rectangular or circular cross-sections are normally preferred although other shapes may be used from time to time for some specialized purposes.

Figure 1.5 a & b shows rectangular and circular waveguides.

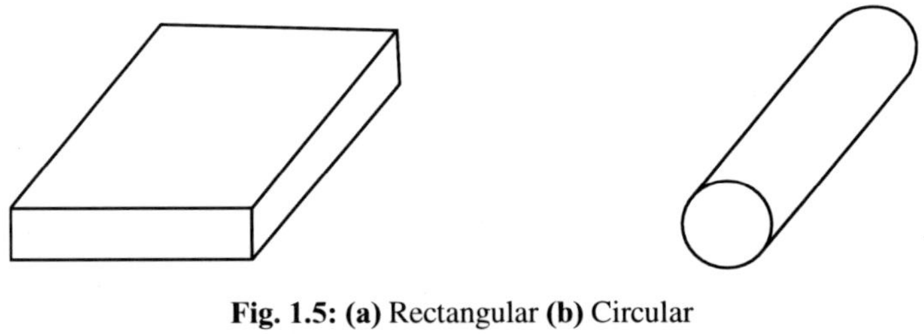

Fig. 1.5: (a) Rectangular **(b)** Circular

In a typical setup, a waveguide may be connected at one end to an antenna and other end to a transmitter or a receiver. The waveguide guides waves from the antenna to the receiver or from transmitter to antenna. The walls of a waveguide are made of conductor and therefore waves get reflected from them. The conduction of energy is through the dielectric filled in hollow waveguide which in most cases is air. The reflection from a conducting wall has two velocities, in a direction parallel to the wall, namely the group velocity and phase velocity.

1.6.2 Circular Waveguides

The behaviour of waves in circular waveguide is the same as in rectangular guides. Circular waveguides have a different geometry and some different applications. Circular waveguides are easier to manufacture than rectangular ones. They are also easier to join together, in the usual plumbing fashion. The circular waveguide is used whenever a rotating element such as a radar antenna has to be connected to a microwave system.

1.6.3 Matching

In communication networks, the elements of the network should be designed such that maximum power transfer takes place between the source and the load. This is governed by maximum power transfer theorem which states that the maximum power is absorbed by one network from another joined to it at two terminals when the impedance of one is a complex conjugate of other. This means that for maximum power transfer the reactance of the load should be equal to that of the source but opposite in sign, i.e. if the source is inductive the load should be capacitive. When this condition is achieved, it is called "impedance-matching" and the devices employed to achieve this matching are called impedance matching devices. The transmission lines can also be used as matching devices.

1.6.4 Balun

A balun or balance to unbalance transformer is a circuit element used to connect a balanced line such as a parallel wire to an unbalanced line such as coaxial cable or an antenna. The balun is a transmission line and is also called as choke, sleeve or bazooks. A $\frac{\lambda}{4}$ sleeve is placed around the outer conductor of the coaxial line and is connected to it at X, as shown in Fig. 1.6.

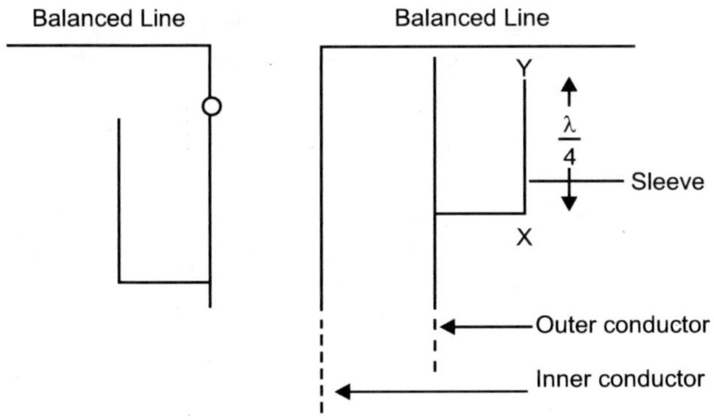

Fig. 1.6: Balun

At point Y, therefore, λ/4 away from X, the impedance seen when looking down into the transmission line formed of sleeve and the outer conductor of the coaxial line is infinite. In other words, the outer conductor of the coaxial line no longer has zero impedance to ground at Y. Thus, one of the line wire of the balanced line may be connected to it without fear of being short circuited to ground. The outer balanced wire is connected to the inner conductor of coaxial line. A balance load such as the dipole antenna may now be placed upon it.

1.7 OPTICAL FIBRES

Fibre optics is the technology related with the transfer of optical energy through guided media, like a glass fibre. One of the important media of application of fibre optics is telecommunication.

Conventional methods of communication like telephone lines, coaxial cable system, radio/T.V. broadcast or microwave communication have already saturated due to the increased demand of communication. Conventional communication system requires the use of repeaters, after a few km distance, because of the signal loss. Repeaters are used to amplify the signal. The number of channels provided by any communication system depends on the bandwidth, which is related to the working frequency, examples: (1) Coaxial cables works upto 60 MHz and can support around 10,800 channels. (2) Radio waves work upto MHz and microwaves upto megahertz (10^{10} Hz) and provides a few thousands of channels. Since optical frequencies are extremely large ($\approx 10^{15}$) Hz as compared to conventional radio waves ($\approx 10^6$ Hz) and microwaves ($\approx 10^{10}$ Hz), a light beam acting as a carrier wave is capable of carrying far more information in comparison to radio waves and microwaves.

The technique of optical modulation was in use earlier, like smoke signals or lamps flashed with morse code, but it was a slow process and was dependent on weather. Soon after the discovery of the laser, some preliminary experiments on the propagation of information carrying light waves through the open atmosphere were carried out, but it was realized that because of the vagaries of the terrestrial atmosphere (example: rain, fog, etc.), in order to have an efficient and dependable communication system, one would require a guiding medium in which the information carrying light waves could be transmitted. This guiding medium is the optical fibre which is hair thin and guides the light beam from one place to another.

In addition to the capability of carrying a huge amount of information, fibres fabricated with recently developed technology are characterized by extremely low losses ($\approx$ 0.2 dB/km) as a consequence of which the distance between two consecutive repeaters can be as large as 250 km. Today fibre-optic cable is the most promising signal transmitting media.

Advantages:

(1) Large signal width: Since the frequency of light used as a carrier is of the order of 10^{15} Hz and even 1% of modulation contains all information, the bandwidth is of the order of 10^{13} Hz/carrier, which can accommodate 6 millions of channels/carrier.

(2) The fibre-optic cable gives constant attenuation over a large bandwidth, as against the copper cables having increasing attenuation.

(3) The optical fibres are made of dielectric material, which offers electrical isolation between input and output parts of the circuit.

(4) The transmission is due to internal reflection, therefore, there is less loss. The cable is immune to electric, magnetic or RF fields in atmosphere because it is covered with cladding.

(5) The fibres are light and occupy less space.

(6) Due to the recent developments of source, detector and large distances between repeaters the system cost has lowered.

The attenuation of the signal in the fibre is because of Rayleigh scattering, absorption in the material, bending loss and imperfection at the bounding between core and clad. Light travels in straight lines in a homogenous medium. If it strikes the interface between two media, it suffers reflection and refraction. At shallow angle to perpendicular, most of the light travelling through one transparent medium will cross interface into another surface medium.

1.8 BANDWIDTH

Fundamental limitations in communication system:

The two major constraints on the design of communication system are:

(1) *Technological Constraints:* These arise due to engineering facts and are related to problems of feasibility; such diverse considerations as equipment availability, interaction and compatibility with the existing systems and economic factors, etc. These problems can be solved theoretically but may not be feasible in practice.

(2) *Fundamental Physical Limitations:* These are imposed by the system itself, the laws of nature that govern the task. These limitations just & cannot be resolved even in theory. Apart from technological constraints, these limitations ultimately dictate what can or cannot be accomplished.

1.8.1 Bandwidth Constraint: Bandwidth and S/N-Ratio Relationship

In most of information transmission, a message must be sent in real time, the output signal keeping pace with the input as it comes from the source. Clearly the rate at which the source is producing the message and hence the signalling speed governs the design of the system. Even if the source rate is adjustable, it is necessary to ensure efficient system utilization, i.e. to minimize the time during which the information is sent.

Channel bandwidth: It is the range of frequencies that a system can transmit with reasonable fidelity. For example, if a channel can transmit with reasonable fidelity a signal whose frequency components occupy a range from dc upto a maximum of 5 kHz the channel bandwidth B is 5 kHz.

Again, consider the possibility of increasing the speed of information transmission by time compression of the signal. If a signal is compressed in time by a factor of two, it can be transmitted in half the time, and the speed of transmission is doubled. But this implies that the frequency of its components are doubled. To transmit this compressed signal without distortion, the channel bandwidth must also be doubled. Thus, the rate of information transmission is directly proportional to bandwidth (B).

If a channel of bandwidth B can transmit N-pulses per second, then to transmit KN pulses per second we need a channel of bandwidth KB. We observe:

(1) The number of pulses/second that can be transmitted over a channel is directly proportional to its bandwidth.

(2) The signal speed is conceived in terms of bandwidth—the width of the signal spectrum in frequency domain. In other words, the rate at which a system can change stored energy is reflected by its usable frequency response, measured in terms of system bandwidth. It means that if large amount of information is to be transmitted in a small amount of time, it requires wideband signal to contain the information and in turn, wideband system would be necessary to accommodate these signals. Therefore, bandwidth is the fundamental limitation. If it is not possible to have sufficient bandwidth, it may become necessary to decrease signalling speed and as a result increase transmission time.

(3) The task of equipment design is governed largely not by absolute bandwidth, but fractional bandwidth, which means absolute bandwidth divided by the centre frequency.

According to Shannon-Hartley's law

$$C = B \log (1 + S/N) \text{ bits/sec.}$$

where, C – channel capacity, B – bandwidth, S/N – signal to noise ratio.

Thus, for a given C, we may increase B and reduce S/N and vice versa. In other words, it is possible to transmit weak signals over a given channel by increasing bandwidth of channel whereas reduction in bandwidth requires strong signals for accurate transmission.

Thus, bandwidth and S/N ratio parameters represents the ultimate limitation on a rate of communication.

1.9 BASEBANDS

The source originates the message, such as human voice, television picture, a teletype message or data. If the data is non-electric (human voice, teletype-message, television picture) it must be converted by an input transducer into an electrical waveform referred to as the baseband signal or the message signal.

Baseband communication corresponds to transmission of the band of frequencies of signal delivered by the source or the input transducer. In telephones, the baseband is the audio band (band of voice signals) of 0 to 3.4 kHz. In T.V. the baseband is the video band occupying 0 to 4.3 MHz. For digital data using bipolar signalling at a rate of f_0 pulses/second, the baseband is 0 to f_0 Hz.

In baseband communication, baseband signals are transmitted without modulation, that is, without any shift in the range of frequencies of the signal. Because the baseband signals have sizeable power at low frequencies they cannot be transmitted over a radio link but are suitable for transmission over a pair of wires or coaxial cables. Local telephone communication and short hour PCM (between two exchanges) use baseband communication. Because baseband communication uses only baseband frequencies, its uses are rather restricted.

1.9.1. Data Transfer Rate

The speed of data transmission can be expressed as data transfer rate. It is defined as:

It is the average of bits, characters or blocks per unit time passing between corresponding equipment in a data transmission system.

In practical systems, errors are introduced and the information bits accepted by a terminal are less than the number that are transmitted. In the case of band or code the start/stop elements do not contribute to the information received. For error-free transmission 2500 bits would be received for 500 five-unit band of characters transmitted on an average of $\frac{2500}{(500 \times 0.1)} = 50$ bits/sec.

1.9.2 Bands and Bits

Signalling Speed:

A band is used to represent the rate at which charges occur in a signal over a given period of time. It is the unit of signalling speed "S" or modulation rate and is equal to the reciprocal of the shortest signaling element. For example, the signalling speed of Bandot code shown in Fig. 1.7 is $\frac{1}{(0.0135)} = 74$ bands.

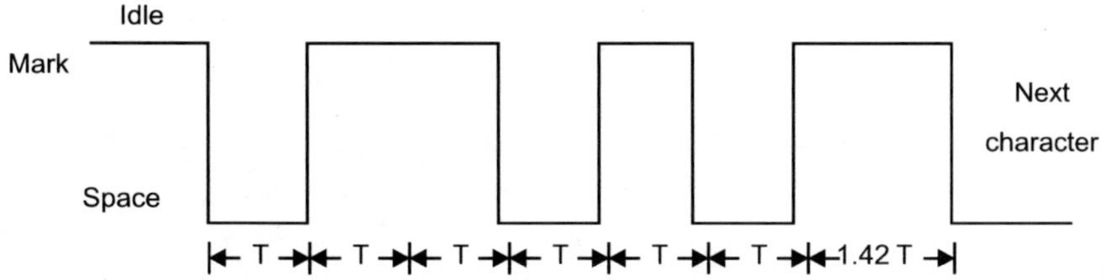

Fig. 1.7: Bandot Code

A signal which can take on n-signalling conditions such as a multi-frequency or multilevel signal having m-unique frequencies or levels, can be represented in its various conditions by $\log_2$ n bits.

To express the rate at which information can be transmitted, we indicate the data signalling rate in its bits/sec. For serial transmission, this is defined as:

$$\frac{1}{T}\log_2 n$$

where, T = Length of unit signal element.

T = Number of signalling conditions.

1.10 MULTIPLEXING

It is often desirable to transmit several messages on the transmission channel in order to make full use of channel capacity. This is made possible through a process called multiplexing.

Multiplexing is the sending of number of separate signals together, over the same cable simultaneously without interference. There are two basic multiplexing techniques:

1.10.1 Time Division Multiplexing (TDM)

TDM is a method of transmission of more than one signal by a method of interleaving pulses or by dividing the time domain into slots, one slot for each message. In TDM, use is made of the fact that narrow pulses with side spaces between them are sent in any of the forms of pulse modulation, so that the spaces can be used by signals from other sources. To understand working of TDM system, assume a 24-channel TDM system, transmitting at the rate of 8000 samples per second. This means that each of the 24 signals is sampled 8000 times every second. Thus, time period of samples corresponding to different channels equal $\frac{1}{8000}$ or 125 micoseconds. The magnitude of each sample is represented by 128 sampling levels and contains an additional bit for synchronization. Thus, for the 125 microseconds period available sample from every signal is made to occupy 5-microsecond and the total period comes out to be 24 × 5 = 120 microseconds. The remaining 5-microsecond period available is utilized for signalling and synchronization. The pulse width occupied by each bit in the 8-bit sample therefore equals $\frac{5}{8}$ microseconds or 0.625 microseconds.

The information contents of any bandlimited signal are completely specified by its samples spaced at least $\frac{f_m}{2}$ seconds apart where, f_m is maximum frequency of the signal. In other words, the time duration between the samples T should be less than $\frac{1}{2}f_m$ seconds, i.e.

$$T \le \frac{1}{2f_m}$$

or $$\frac{1}{f_s} \le \frac{1}{2f_m}$$

or $$f_s \ge 2f_m$$

where $f_s = \frac{1}{T}$ is called sampling frequency.

The time interval in between samples is utilized by the samples of the second signal as shown in Fig. 1.8a.

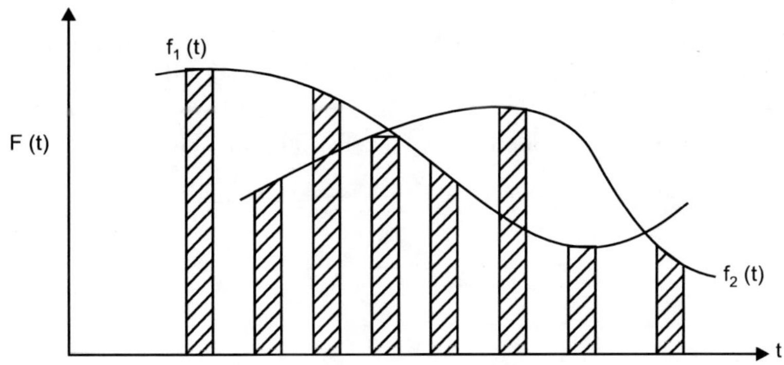

Fig. 1.8(a)

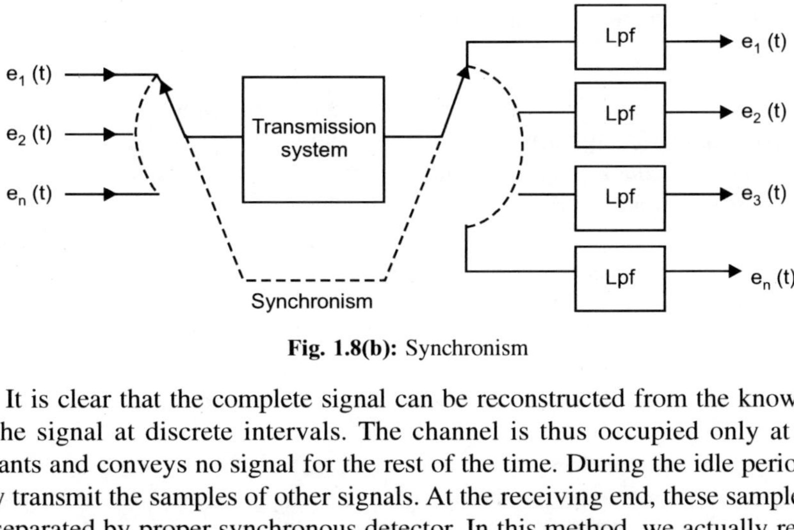

Fig. 1.8(b): Synchronism

It is clear that the complete signal can be reconstructed from the knowledge of the signal at discrete intervals. The channel is thus occupied only at these instants and conveys no signal for the rest of the time. During the idle period, we may transmit the samples of other signals. At the receiving end, these samples can be separated by proper synchronous detector. In this method, we actually recover the various signals in time domain. This approach is called TDM.

The essentials of TDM are shown in Fig. 1.8b.

The several input signals, all bandlimited in "f_m" are sequentially sampled at the transmitter by a rotary switch or commutator. The switch makes one complete revolution in $T \le \frac{1}{2f_m}$ seconds.

Extracting one sample from each input. Hence, the commutator output is a PAM waveform containing the individual message samples periodically interlaced in time. If there are n-inputs, the pulse to pulse spacing is $\frac{T}{n} = \frac{1}{nf_s}$, while spacing between successive samples from any one input is of course T. A set of pulses consisting of one sample from each input is called a frame.

At the receiver a similar rotary switch, the commutator or distributor, separates the samples and distributes them to a band of low-pass filters, which in turn reconstruct the original messages. The switching action is usually electronic and synchronising signals are provided to keep the distribution in step with commutator. The pulse duration (t) in the multiplexed TDM system should be less than $\frac{1}{n f_s}$, i.e.

$$t \leq \frac{1}{n f_s} \leq \frac{T}{n} \leq \frac{1}{2 n f_m}$$

or
$$\frac{1}{2t} \geq n f_m$$

The bandwidth required for transmitting n TDM signals is $B \geq \frac{1}{2t} \geq n f_m$ or $B = nf_m$.

However, the direct transmission of PAM signals by radiation is impractical, since energy in such a signal is concentrated at lower frequencies, necessitating unreasonable size radiating antennas. In such cases, the whole spectrum of PAM signals is shifted to a higher frequency by amplitude modulation giving rise to upper and lower sidebands and hence the bandwidth required for transmission of PAM/AM signal of time division multiplexed continuous signals bandlimited to f_m Hz is 2 nf_m Hz, i.e.

$$B = 2\, nf_m$$

for noise reduction purposes PPM - AM, PCM - FM are particularly attractive.

1.10.2 Frequency Division Multiplexing (FDM)

It is often required to send a large number of independent telephone or telegraph channels from one point to another. In developed country, there may be requirements of hundreds or even thousands of simultaneous telephone, telex or data transmissions between any two major cities. Obviously, it would be highly expensive to devote a separate cable or radio link to each transmission and thus some kind of combination of channels without mutual interference, is visualized. This is done in FDM by taking a bandwidth adequate for the number of channels required and allocating a frequency slot to every channel.

In FDM there is frequency translation, by the process of modulation, it is possible to shift or translate the frequency spectra of signals to various frequency ranges so that when they are transmitted over a single channel (or path), they do not interfere with each other. At the receiving end, various signals are separated by using appropriate filters and then demodulated to get the desired signals. In this method, all of the signals are mixed in time domain but their spectra are so separated that they occupy different frequency bands. This process is called “stacking”.

Suppose we wish to transmit simultaneously 960 signals on channel making use of FDM technique. These 960 signals are not modulated directly and transmitted but are divided into groups and super groups as follows:

First twelve signals are modulated using SSB technique and then all fed to the adder to form a group as shown in Fig. 1.9a.

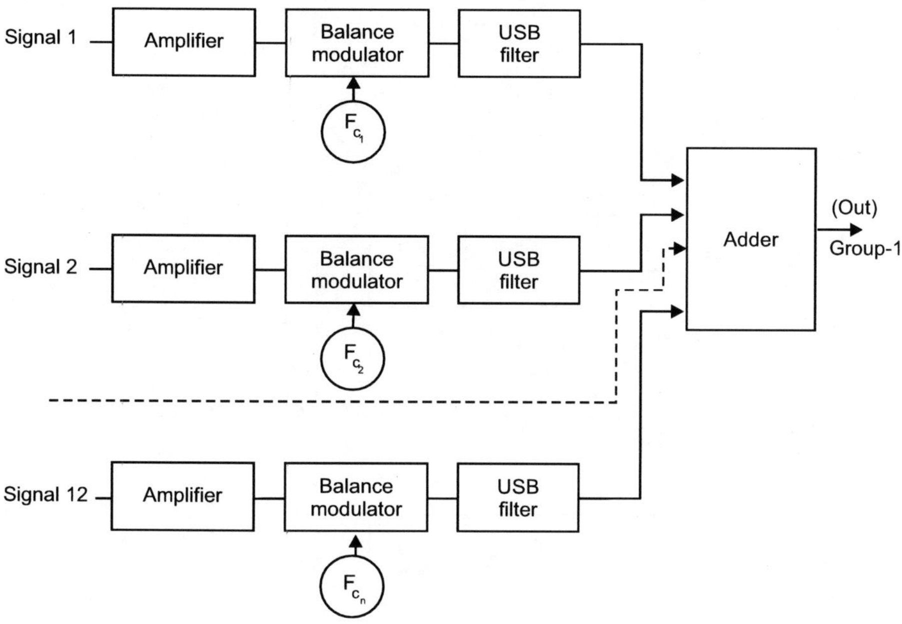

Fig. 1.9(a): FDM (Group Formation)

Exactly the same procedure is adopted for the second group, taking in signals (13 to 24), and so on for 5 groups accommodating 60 circuits or signals. These 5 groups are now combined to form a super group. This time SSB technique with lower side band is employed. Exactly the same procedure is adopted for the second super group and also the third and fourth upto sixteen, accommodating 960 signals.

These 16 super groups are now used to form the baseband signal.

Figure 1.9b shows the formation of baseband signal.

Lower side band is again used to stack the super groups. It is possible to generate the baseband directly by staking each circuits above the other until the 960 signals have been stacked. The same result would have been obtained in the end, but the method shown above is preferred in practice because, it is simpler. If each signal were translated up individually, a total 960 oscillators would be required with the method as it stands, 33 are sufficient (12 for signals, 5 for groups and 16 for super groups). Finally, the main carrier f_c may be frequency modulated with the baseband then transmitted.

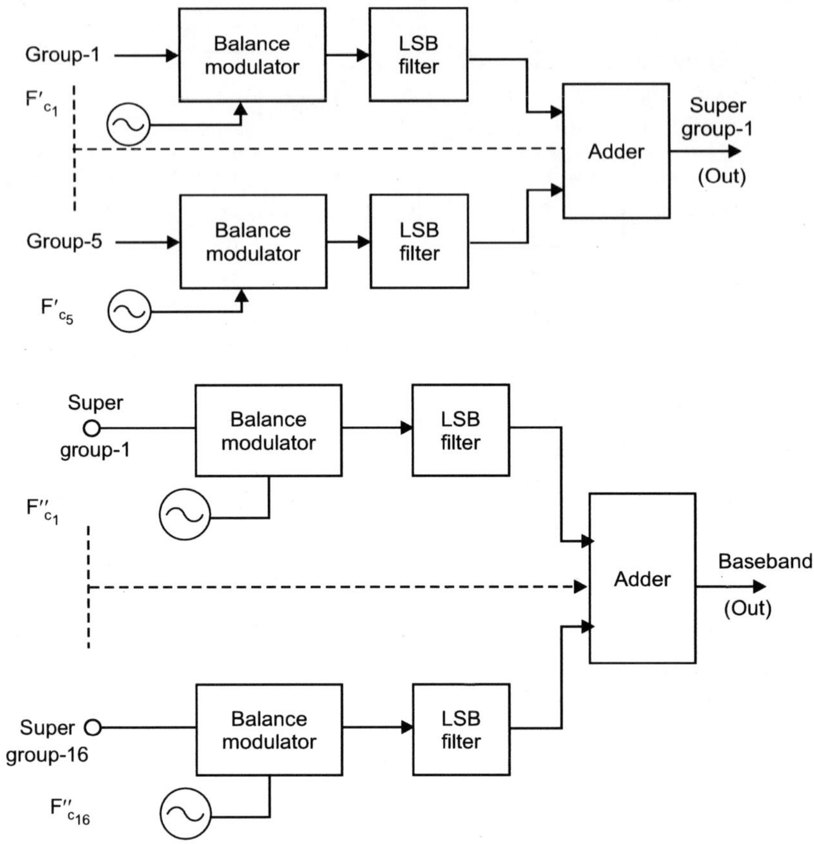

Fig. 1.9(b): FDM Technique

1.10.3 Comparison of FDM and TDM

Clearly, TDM and FDM accomplish the same goals, though the means are different. They can be visualized as dual techniques, for in TDM the signals are separated in time domain but jumbled together in frequency, whereas in FDM the signals are separated in the frequency domain but jumbled together in time.

From practical viewpoint TDM can be superior in two respects.

First TDM instrumentation is somewhat simpler. FDM requires modulators, filters and demodulators, these are all replaced by commutator and distributor of TDM. And TDM synchronization is but slightly more demanding than that of FDM with suppressed carrier modulation.

Second and equally important, TDM if invulnerable to the usual sources of FDM interchannel crosstalk.

There is no crosstalk in TDM if the pulses are completely isolated and non - overlapping since message separation is achieved by decommutation or gating in time rather than by filtering.

Practical TDM system will have guard times and guard bands, the former to suppress crosstalk, the latter to facilitate message reconstruction with practical filters.

One more point regarding the bandwidth requirement. TDM system for multiplexing "n" signals, each band, limited to f_m requires a bandwidth of n f_m Hz and if modulated by carrier the bandwidth becomes 2 n f_m. Now if the "n" signals are multiplexed in FDM, using single sideband (SSB) technique the bandwidth is "n f_m". Thus, TDM system using PAM signals has the same bandwidth as AM - SSB system. Similarly, if AM-SSB technique is used for sending n-signals, the bandwidth is 2 n f_m which is the same as required for PAM/AM systems. For the advantages mentioned above, TDM systems are being used more commonly in such applications as long distance telephone communications. TDM and FDM techniques are together used in radio telemetry.

Numericals

Characteristics impedance:

Example 1.2: A coaxial cable has a 75 Ω characteristic impedance of free space and nominal capacitance of 69 pF/m. What is the inductance per metre? If the diameter of the inner conductor is 0.584 mm and dielectric constant of the insulation is 2.23, what is the outer conductor diameter?

Solution:

$$Z_0 = \sqrt{\frac{L}{C}}$$

$$L = Z_0^2 C = 75^2 \times 69 \times 10^{-12}$$

$$= 3.88 \times 10^{-7} = 0.388\ \mu H/m$$

$$Z_0 = \frac{138}{\sqrt{k}} \log \frac{D}{d}$$

$$\log \frac{D}{d} = \frac{Z_0}{138/\sqrt{k}} = \frac{75}{138/\sqrt{2.23}} = 0.81$$

D = antilog 0.81 = 0.584 × 6.457 = 3.77 mm.

Example 1.3: What is the minimum value that the characteristic impedance of an air dielectric parallel wire line could have?

Solution: Minimum impedance will occur when 2s/d is also minimum, and this is reached when the two wires of parallel wire just touch. It is then seen that s = d, so that we have

$$Z_0\text{, min} = 276 \log 2 \times 1 = 276 \times 0.3010 = 83\ \Omega$$

Example 1.4: A coaxial cable, having an inner diameter of 0.025 mm and using an insulator with a dielectric constant of 2.56, is to have a characteristic impedance of 2000 Ω. What must be the outer conductor diameter?

Solution:

$$\log\frac{D}{d} = \frac{20}{138/\sqrt{k}} = \frac{2000}{138/\sqrt{2.56}}$$

$$= 2000\times\frac{1.6}{138}$$

$$= 23.1884$$

$$D = d \times \text{antilog } 23.1884 = 0.025 \times 10^{23} \times 1.543$$

$$= 3.86 \times 10^{21} \text{ mm}$$

$$= 3.86 \times 10^{15} \text{ km}$$

$$= \frac{3.86\times10^{15}}{9.44\times10^{12}} = 409 \text{ light years.}$$

(*Note:* From this solution it is clear that since the outer diameter is so large means such a high value of characteristic impedance is not possible).

Example 1.5: A lossless transmission line has a shunt capacitance of 100 pF/m and a series inductance of 4 μ H/m. What is its characteristic impedance?

Solution:

$$Z_0 = \sqrt{\frac{L}{C}}$$

$$Z_0 = \sqrt{\frac{4\times10^{-6}}{100\times10^{-12}}}$$

$$= \sqrt{40000}$$

$$= 200\ \Omega$$

Example 1.6: A coaxial line with an outer diameter of 6 mm has 50 Ω characteristic impedance. If the dielectric constant of the insulation is 1.60, calculate the inner diameter.

Solution:

$$Z_0 = \frac{138}{\sqrt{k}}\log\frac{D}{d}$$

$$50 = \frac{138}{\sqrt{1.6}} \log \frac{D}{d}$$

$$\log \frac{D}{d} = \frac{50 \times \sqrt{1.6}}{138}$$

$$\therefore \log \frac{D}{d} = 0.4583$$

$$D = d \times \text{antilog } 0.4583$$

$$6 = d \times 2.8727$$

$$\therefore d = 2.0885 \text{ mm.}$$

QUESTIONS

1.1. Briefly discuss the difference between analog and digital signals.
1.2. Define and explain the terms binits, bit and baud.
1.3. Explain what is meant by the characteristic impedance of transmission line.
1.4. Explain how standing waves can be produced in a waveguide.

PROBLEMS

1.1. A 50 Ω short-circuited lines is 0.1 λ in length, at a frequency of 500 MHz. Calculate (a) the equivalent inductive reactance and (b) the equivalent inductance.
1.2. The attenuation coefficient of a line is 0.0006 N/m. Determine the attenuation coefficient in (a) dB/m (b) dB/mile.
1.3. Determine λ_c, λ_g, λ, v_p and v_g for standing waveguide WR62 for which the outside dimensions are 0.702 × 0.391 inch and the wall thickness is 0.040 inch when the exciting frequency is 10 GHz.

CHAPTER 2

Amplitude and Angle Modulation

2.1 INTRODUCTION

Modulation is the process or result of the process by which a message is changed into information. Modulation plays vital role in the field of communication. Communication involves the transmission, reception and processing of information by electrical means. For the propagation of electric signals, the media used is electromagnetic field and when this field changes with time it takes the form of wave. Modulation is also the process whereby in response to the received wave either the original message or information pertaining the original message is made available in the desired form and is delivered when it is wanted. "Demodulation and detection" are the terms often observed to denote the recovery of the wanted message from a modulated signal. Modulation is fundamental to communication and it implies the bandwidth occupancy. In the chapter, we will basically deal with the fundamental concepts of Amplitude Modulation and Phase Modulation.

2.2 NEED FOR MODULATION

If the signal is send directly, i.e. without modulation, i.e. unmodulated carrier several difficulties arise which are listed below:

(1) *Antenna height:* Theory of antenna tells that for the efficient radiation of electromagnetic waves the height of antenna must be comparable to the quarter wavelength of the frequency which we are using.

Now suppose you want to transmit the audio frequency, i.e. 20 kHz, we know that

$$C = f\lambda$$

$$\therefore \lambda = \frac{C}{f} \qquad (1)$$

where, λ = wavelength, C = velocity of light, f = frequency

$$\lambda = \frac{3\times 10^8}{20 \text{ kHz}} = 15,000$$

$$\frac{\lambda}{4} = \text{Height of antenna} = 3750 \text{ m.}$$

Such a large height of antenna is practically impossible.

(2) *Mixing of signals:* When we transmit the signal frequencies directly all the sound is concentrated within the audio range, i.e. 20 Hz to 20 kHz. Hence, all the signals from the different sources would be mixed up and it is impossible to recover desired signal at receiver.

(3) *Poor Radiation:* At the lower frequencies, the radiation power is very low. Such signals die out over long distances.

It is also impossible to transmit high frequency on modulated carrier because it is having

(a) constant maximum amplitude.

(b) constant frequency.

(c) constant phase relationship with respect to some reference.

A message consists of rapid and unpredictable variations in amplitude and frequency.

It is impossible to represent these two variables by an unmodulated carrier and hence cannot be used to convey information.

The transmission of one signal at a time on channel is highly wasteful situation as given channel is being operated very much below its capacity to transmit information. If on the other hand, more than one signal on a single channel, all signals would be mixed up and no useful transmission is possible. Now, if the frequency spectra of various signals are shifted by means of modulation so that they occupy different frequency ranges without overlapping, then it is possible to transmit a large number of signals at the same time on one channel. This technique of transmitting more than one signal simultaneously on a single channel after these have been translated to different frequency ranges by the process of modulation is called, "Frequency Division Multiplexing".

At low frequencies, the size of coupling capacitors become so large that they introduce a drift in the operating point of amplifier. This drift varies the output signal. By means of modulation, frequency spectrum of low frequency input signal is shifted to suitable higher frequencies where it can be easily amplified.

The amplified signal is now demodulated to get back the amplified form of the original low frequency signal. Thus, modulation makes possible amplification of low frequency signals and simultaneous transmission of more than one message using practicably small antennas in all radio systems.

In the process of modulation low frequency bandlimited signal is mixed with high frequency wave called, "carrier wave". Such a carrier wave may be represented by the equation

$$e = E_m \sin(\omega t + \phi) \qquad (2)$$

where, e = instantaneous value of sine wave,
E_m = maximum amplitude,
ω = angular frequency,
ϕ = phase relation with respect to some reference.

Any of these last three characteristics or parameters of the carrier may be varied by the low frequency modulating signal during the process of modulation. Thus, in the process of modulation, some characteristics of high frequency sinusoidal wave is varied in accordance with the instantaneous value of modulating signal since there are three parameters E_m, ω and ϕ of a carrier wave, therefore, any of these parameters can be varied in proportion to the instantaneous value of the modulating signal, giving rise to amplitude, frequency or phase modulation respectively.

Frequency and phase modulation are together named as "angle modulation", as variation of any of the two varies the angle of the carrier wave.

Amplitude Modulation:

In this type of modulation, the amplitude of carrier signal is varied by the modulating voltage, whose frequency is less than that of carrier.

Let $V_c = V_c \sin \omega_c t$

and $V_m = V_m \sin \omega_m t$

in the above expressions, the phase angle has been ignored.

2.3 FREQUENCY SPECTRUM OF AM WAVE

Amplitude modulation is a system of modulation in which the amplitude of the carrier is made proportional to the instantaneous amplitude of modulating signal.

Sideband frequency is defined as

$$f_{SB} = f_c \pm n\, f_m \qquad (3)$$

and in the first pair n = 1.

When a carrier is amplitude modulated, the proportionality constant is made equal to unity and the instantaneous modulating voltage variations are superimposed on to the carrier amplitude. Hence, when there is no modulation, the amplitude of carrier is equal to its unmodulated value. When modulation is present the amplitude of the carrier is varied by its instantaneous value. This is shown in Fig. 2.1. In the figure it is clear that what will happen if E_m is greater than E_c.

Modulation index (m) is given by

$$m = \frac{E_m}{E_c} \qquad (4)$$

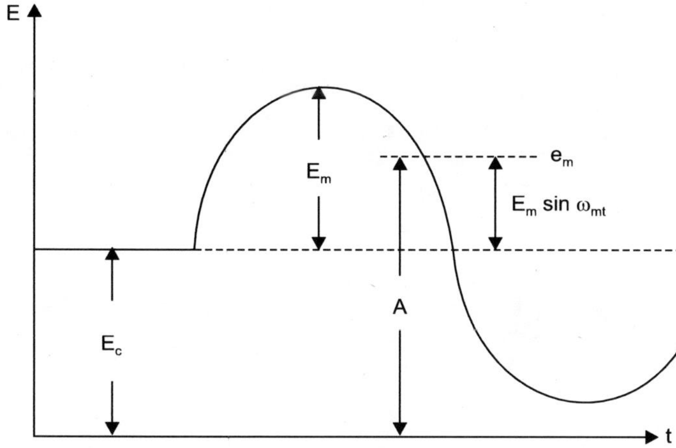

Fig. 2.1

The modulation index is a number lying between 0 and 1 and often expressed as percentage modulation.

Now, it is possible to write an equation for the amplitude of the amplitude modulated voltage. Thus,

$$A = E_c + e_m$$
$$= E_c + E_m \sin \omega_m t$$
$$= E_c \left(1 + \frac{E_m}{E_c} \sin \omega_m t\right)$$

since $$m = \frac{E_m}{E_c} \qquad (5)$$

$$\therefore \qquad A = E_c [1 + m \sin \omega_m t]$$

This instantaneous voltage of the resulting amplitude modulated wave is

$$E = A \sin \theta = A \sin \omega_c t$$
$$E = E_c [1 + m \sin \omega_m t] \sin \omega_c t \qquad (6)$$

This equation may be expanded by means of trigonometrical relation.

As $$\sin x \sin y = \frac{1}{2} [\cos (x - y) - \cos (x + y)]$$

Therefore, instantaneous voltage of AM is

$$E = E_c \sin \omega_c t + m E_c \sin \omega_m t \sin \omega_c t$$

$$E = E_c \sin \omega_c t + m E_c \left(\frac{1}{2} \cos(\omega_c - \omega_m)t - \cos(\omega_c + \omega_m)t\right)$$

$$E = E_c \sin \omega_c t + \frac{mE_c}{2}[\cos(\omega_c - \omega_m)t - \cos(\omega_c + \omega_m)t]$$

$$E = E_c \sin \omega_c t + \frac{mE_c}{2}\cos(\omega_c - \omega_m)t - \frac{mE_c}{2}\cos(\omega_c + \omega_m)t \quad (7)$$

where, $E_c \sin \omega_c t$ = Unmodulated carrier

$\frac{mE_c}{2}\cos(\omega_c - \omega_m)t$ = Lower sideband

$-\frac{mE_c}{2}\cos(\omega_c + \omega_m)t$ = Upper sideband

Total additional terms produced are the two sidebands.

$f_c - f_m$ = Lower sideband (LSB)

$f_c + f_m$ = Upper sideband (USB)

This important conclusion is that the bandwidth required for amplitude modulation is twice the frequency of the modulating signal.

2.4 REPRESENTATION OF AM WAVE

Figure 2.2 shows frequency spectrum of AM wave.

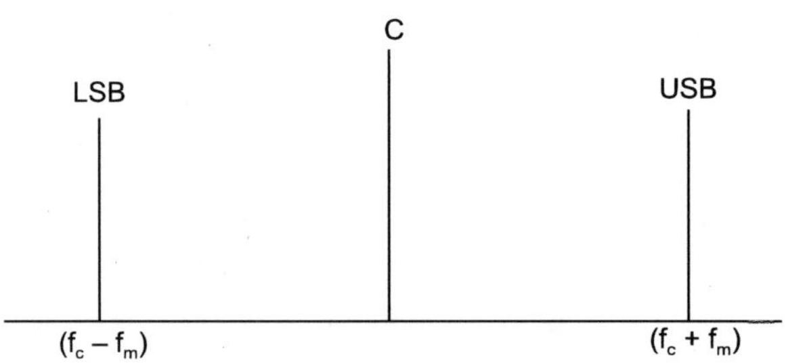

Fig. 2.2: Spectrum of AM Wave

Here in Fig. 2.2, AM is simply shown comprising of three different frequencies. The central frequency, i.e. carrier is having the highest amplitude and other two are placed symmetrical about it. They are having equal amplitudes. They never exceed half the carrier amplitude.

Amplitude modulated wave is shown in Fig. 2.3. Area of top envelope is given by

$$E_c + E_m \sin \omega_m t$$

Area of the bottom envelope is given by

$$-A = -(E_c + E_m \sin \omega_m t)$$

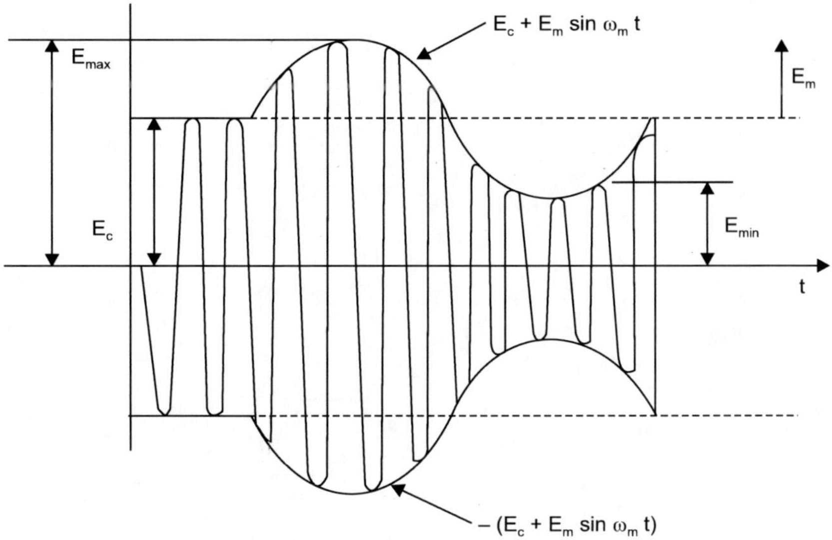

Fig. 2.3: Amplitude Modulated Wave

Modulated wave extends between these two limits and has repetition rate equal to modulated carrier frequency.

To Find Modulation Index:

Amplitude of modulated carrier varies as,

$$A = E_c\,[1 + m \sin \omega_m t]$$

The maximum and minimum values of A are,

E_c (1 + ma) & E_c (1 – ma) and are denoted as E_{max} & E_{min} respectively. Hence,

$$E_{max} = E_c\,(1 + ma)$$

$$E_{min} = E_c\,(1 - ma)$$

This gives

$$ma = \frac{E_{max} - E_{min}}{E_{max} + E_{min}} \qquad (8)$$

This equation is the standard method of evaluating the modulation index when calculating from a waveform. Such as may be seen on an oscilloscope.

Oscilloscope Display of AM

Pattern shown in Fig. 2.3 can be obtained directly on oscilloscope and modulation index can be measured directly from this modulated waveform.

(a) Modulated wave is applied to the vertical deflection circuit of oscilloscope and modulating signal to horizontal deflection circuit.

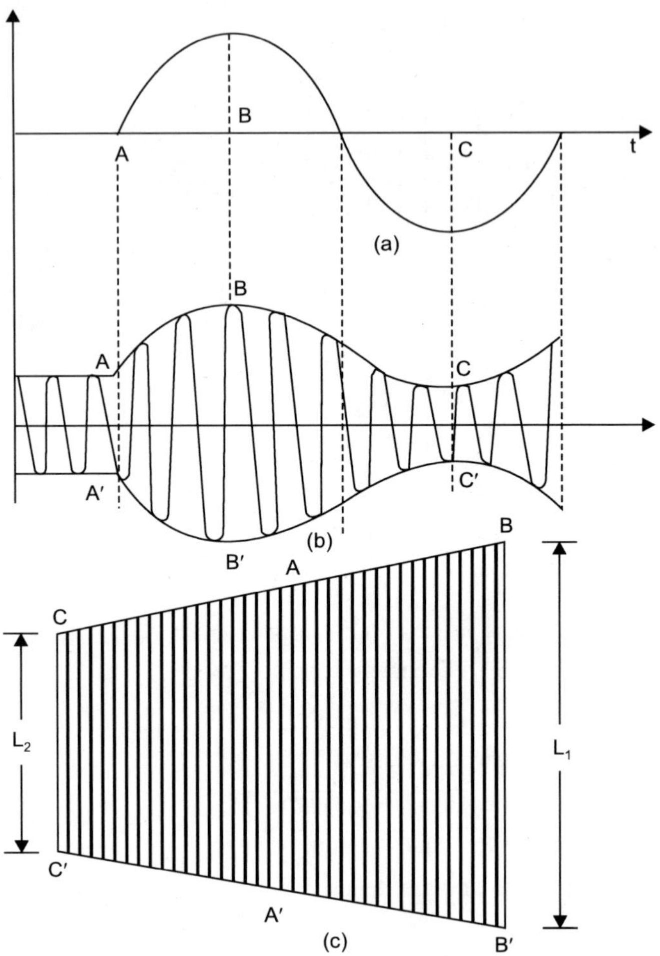

Fig. 2.3(a), (b) and (c)

Modulation index is given by

$$m = \frac{L_1 - L_2}{L_1 + L_2}$$

If the over modulation occurs, the trapezoid will become a triangle.

2.5 BLOCK DIAGRAM OF AM TRANSMITTER (HIGH LEVEL AND LOW LEVEL MODULATION)

Figure 2.4 shows the block diagram of a typical AM transmitter. It can be either high level modulated or low level modulated.

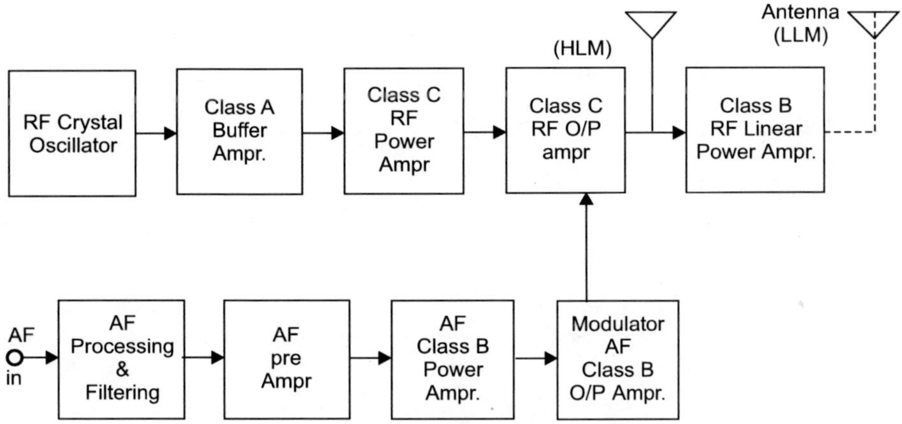

Fig. 2.4: AM Transmission Block Diagram

2.6 POWER RELATION IN AM WAVE

Modulated wave contains more power than the carrier had before modulation took place. Since the amplitude of the side bands depends on the modulation index, it is anticipated that the total power in the modulated wave will depend on the modulation index also.

The total power in the modulated wave will be

$$P_t = \frac{E^2_{carrier}}{R} + \frac{E^2_{LSB}}{R} + \frac{E^2_{USB}}{R} \tag{9}$$

where, all three voltages are rms values & R is the resistance in which power is dissipated.

$$\therefore P_c = \frac{E^2_{carrier}}{R} = \frac{\left(\frac{E_{carrier}}{\sqrt{2}}\right)^2}{R} = \frac{E_c^2}{2R} \tag{10}$$

$$P_{LSB} = P_{USB} = \left(\frac{\frac{mE_c}{\sqrt{2}}}{2}\right)^2 / R = \frac{m^2 E_c^2}{8R}$$

$$= \frac{m^2 E_c^2}{8R} = \frac{E_c^2}{2R}\frac{m^2}{4} \tag{11}$$

$\therefore$ Putting values of equations (11) & (10) in equation (9)

$$P_t = \frac{E_{carr}^2}{R} + \frac{E_{LSB}^2}{R} + \frac{E_{USB}^2}{R}$$

$$\therefore P_t = \frac{E_c^2}{2R} + \frac{E_c^2}{2R}\frac{m^2}{4} + \frac{E_c^2}{2R}\frac{m^2}{4}$$

$$= \frac{E_c^2}{2R} + \frac{E_c^2}{2R}\left[\frac{m^2}{4} + \frac{m^2}{4}\right] = \frac{E_c^2}{2R} + \frac{E_c^2}{2R}\left[\frac{m^2}{2}\right]$$

$$P_t = \frac{E_c^2}{2R}\left[1 + \frac{m^2}{2}\right]$$

but P_c = carrier power = $\dfrac{E_c^2}{2R}$

$$\therefore P_t = P_c\left[1 + \frac{m^2}{2}\right] \tag{12}$$

Example 2.1: A broadcast AM transmitter radiates 50 kW of carrier power what will be radiated power at 85% modulation?

Given

P_c = 50 kW

Percentage modulation = 85

To find : P_t = total power radiated.

Solution:

$$P_t = P_c\left[1 + \frac{m^2}{2}\right]$$

$$= 50 \text{ kW}\left[1 + \frac{(0.85)^2}{2}\right]$$

= 68.06 kW.

Example 2.2: A broadcast radio transmitter radiates 10 kW when the modulation percentage is 60. How much of this is carrier power?

Solution: $P_t = 10 \text{ kW}, m_a = 60\%$

$$P_t = P_c\left(1+\frac{m_a^2}{2}\right)$$

$$\therefore \quad P_c = \frac{P_t}{1+\frac{m_a^2}{2}}$$

$$P_c = \frac{P_t}{1+\frac{m_a^2}{2}} = \frac{10}{1+\frac{0.6^2}{2}} = 8.47 \text{ kW}$$

$$P_c = 8.47 \text{ kW}$$

Example 2.3: A transmitter radiates 9 kW without modulation and 10.125 kW after modulation. Determine depth of modulation.

Solution: Here, $P_c = 9\,\text{kW},\ P_t = 10.125 \text{ kW}$

$$\therefore \quad P_t = P_c\left[1+\frac{m_a^2}{2}\right]$$

$$10.125 = 9\left[1+\frac{m_a^2}{2}\right]$$

$$1+\frac{m_a^2}{2} = 1.125$$

$$m_a = 0.5$$

Example 2.4: The output voltage of a transmitter is given by 500(1 + 0.4 sin 3140 t) sin (6.28 × 10^7 t). This voltage is fed to a load of 600 Ω resistance. Determine (a) Carrier frequency (b) Modulating frequency (c) Carrier power (d) Mean power output (e) Peak power.

Solution: We have,

$$e_{AM} = 500\,(1 + 0.4 \sin 3140\, t) \sin 6.28 \times 10^7\, t$$

comparing with

$$e_{AM} = E_c\,(1 + m \sin \omega_m t) \sin \omega_c t$$

So we have,

$$E_c = 500,\ m = 0.4,\ \omega_m = 3140,\ \omega_c = 6.28 \times 10^7$$

(a) Carrier frequency:

$$\omega_c = 2\,\pi\, f_c$$

$$\therefore \qquad f_c = \frac{\omega_c}{2\pi} = \frac{6.28\times 10^7}{2\pi} = 10 \text{ MHz}$$

$$f_c = 10 \text{ MHz}$$

(b) Modulating frequency: $f_m = \frac{\omega_m}{2\pi} = \frac{3140}{2\pi} = 500 \text{ Hz}$

(c) Carrier power: $P_c = \frac{E_c^2}{2R} = \frac{500^2}{2\times 600} = 208.33 \text{ W}$

(d) Mean power output:

$$P_t = P_c\left(1+\frac{m_a^2}{2}\right) = \frac{2500}{12}\left[1+\frac{0.4^2}{2}\right] = \frac{2500}{12}\times 1.08$$

$$P_t = 225 \text{ watts}$$

(e) Peak power output results when the positive half cycle of the modulating signal occurs. The peak output voltage is given by the sum of E_c & E_m:

Peak output voltage = $E_c + m\,E_c = 500 + 0.4 \times 500$

Peak output voltage = 700 V

$$\text{Peak power} = P_{tm} = \frac{700}{\sqrt{2}}\times\frac{700}{\sqrt{2}}\times\frac{1}{600}$$

$$P_{tm} = 408.3 \text{ watts}$$

2.7 CURRENT CALCULATIONS FOR AM WAVE

Sometimes it is easy to measure RF currents instead of voltages. We analyze such a situation in this section.

Let I_c be rms unmodulated current.

I_t = Total rms modulated current of AM transmitter.

R = Resistance in which the above two currents flow. Then,

$$\frac{P_t}{P_c} = \frac{I_t^2 R}{I_c^2 R} = \frac{I_t^2}{I_c^2}$$

but

$$\frac{P_t}{P_c} = \left(1+\frac{m_a^2}{2}\right)$$

$$\therefore \qquad \frac{I_t^2}{I_c^2} = \left(1 + \frac{m_a^2}{2}\right)$$

$$\frac{I_t}{I_c} = \sqrt{1 + \frac{ma^2}{2}}$$

$$I_t = I_c\sqrt{1 + \frac{ma^2}{2}} \qquad (13)$$

Also

$$m_a = \sqrt{2\left[\left(\frac{I_t}{I_c}\right)^2 - 1\right]}$$

2.8 MODULATION BY SEVERAL SINE WAVES SIMULTANEOUSLY

Modulation of a carrier by a single sinusoidal wave is rarely done, instead, several sinusoidal waves are simultaneously used to modulate the carrier. To calculate the modulated power, we have first to calculate the modulation index which can be evaluated by any one of the two methods given below:

First Method:
We know that

$$P_t = P_c\left(1 + \frac{m_a^2}{2}\right) = P_c + \frac{P_c m_a^2}{2}$$

$$P_t = P_c + P_{SB}$$

If several sine waves simultaneous modulate the carrier, the carrier power will remain unaffected but the total sideband power will become the sum of the individual sideband powers.

Then, $\qquad P_{SB_t} = P_{SB_1} + P_{SB_2} + P_{SB_3} + \ldots$

or $\qquad \frac{P_c m_{a_t}^2}{2} = \frac{P_c m_{a_1}^2}{2} + \frac{P_c m_{a_2}^2}{2} + \frac{P_c m_{a_3}^2}{2} + \ldots$

or $\qquad m_{a_t}^2 = m_{a_1}^2 + m_{a_2}^2 + m_{a_3}^2 + \ldots$

Second Method: Let E_1, E_2 … be the simultaneous modulating voltages. Then total modulating voltage E_t will be

$$E_t = \sqrt{E_1^2 + E_2^2 + E_3^2 + \ldots}$$

or $$\frac{E_t}{E_c} = \sqrt{\frac{E_1^2}{E_c^2} + \frac{E_2^2}{E_c^2} + \frac{E_3^2}{E_c^2} + \ldots}$$

but $$\frac{E_m}{E_c} = m_a ; \therefore \quad m_{a_t} = \sqrt{m_{a_1}^2 + m_{a_2}^2 + m_{a_3}^2 + \ldots}$$

A point to be noted here is that in both cases, the result is the same and so any one method can be used to find m_{a_t} . Also the total modulation index should not exceed unity otherwise distortion will result.

2.9 AM GENERATION

There is a variety of modulator circuits employing tubes or solid state devices to produce amplitude modulated waves. All these circuits may however be grouped as (a) Square law or non-linear modulator circuits (b) Linear modulator circuits.

Another type of grouping is also sometimes done. This is based upon the power level at which modulation is carried out and may be termed (a) low level modulation & (b) high level modulation.

High Level Modulation:
Modulation is done at higher power levels. In this output stage in transmitter, is plate modulated.

Advantages:

(1) High efficiency.
(2) Low distortion.
(3) High power handling capacities.

Disadvantages:

Complex means of AM generation.

Low Level Modulation:

Modulation is applied at any other point including some other electrode of the output amplifier, then AM modulation is called low level modulation. Modulation is carried out at lower levels.

Advantages:

(1) Lower amount of power to be handled.
(2) Simple design.

Disadvantages:

(1) Low efficiency.
(2) High distortion.

2.10 NON-LINEAR MODULATION

In general, any device operated in non-linear region of its output characteristics is capable of producing amplitude modulated waves when the carrier and modulating signals are fed at the input. Thus, a transistor, a triode tube, a diode, etc. may be used as Square Law modulator. In such a modulator circuit, the output current flowing through the load is given by the power series

$$i = a_0 + a_1 e_1 + a_2 e_1^2 + \ldots$$

where, a_0, a_1, a_2, etc. are constant and e_1 is the input voltage to the device. Considering the modulator circuit of Fig. 2.5

$$e_1 = E_c \sin \omega_c t + E_m \sin \omega_m t$$

$$i = a_0 + a_1 (E_c \sin \omega_c t + E_m \sin \omega_m t) +$$

$$a_2 (E_c \sin \omega_c t + E_m \sin \omega_m t)^2$$

$$= a_0 + a_1 E_c \sin \omega_c t + a_1 E_m \sin \omega_m t + a_2 E_c^2 \sin^2 \omega_c t$$

$$+ a_2 E_m^2 \sin^2 \omega_m t + 2 E_c E_m \sin \omega_c t \sin \omega_m t$$

$$= a_0 + a_1 E_c \sin \omega_c t + a_1 E_m \sin \omega_m t + a_2 E_m^2 \sin^2 \omega_m t + a_2 E_c^2 \sin^2 \omega_c t$$

$$+ \frac{2 E_c a_2 E_m}{2} [\cos (\omega_c - \omega_m) t - \cos (\omega + \omega_m) t]$$

$$i = a_0 + a_1 E_c \sin \omega_c t + a_1 E_m \sin \omega_m t + a_2 E_c^2 \sin^2 \omega_c t$$

$$+ a_2 E_m^2 \sin^2 \omega_m t + a_2 E_c E_m [\cos (\omega_c - \omega_m) t - \cos (\omega_c + \omega_m) t]$$

The last term of this equation (underlined) gives the upper and lower sidebands while the second term gives the carrier. If the load is resonant circuit, sidebands and carrier may be selected giving the AM output. When all unwanted frequencies are rejected, the modulated component present at the output is represented by

$$i = a_1 E_c \sin \omega_c t + a_2 E_c E_m [\cos (\omega_c - \omega_m) t - \cos (\omega_c + \omega_m) t]$$

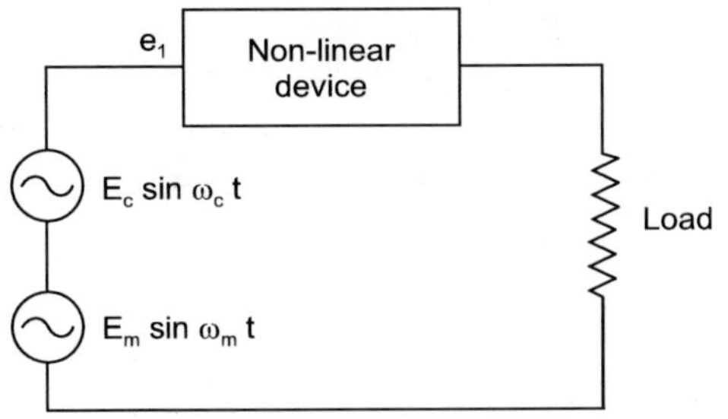

Fig. 2.5

As "a_1" is considerably larger than a_2, the depth of modulation that is available without distortion is low.

2.11 GRID MODULATED CLASS C AMPLIFIER

Figure 2.6a shows circuit diagram of grid modulated class C amplifier. A class C amplifier may be modulated by the introduction of modulating voltage in series with the grid bias.

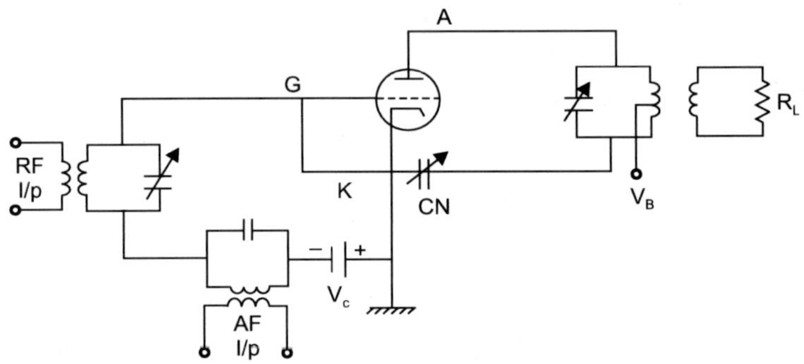

Fig. 2.6(a): Grid Modulated Class C Amplifier

The modulating voltage is superimposed on fixed negative grid bias. Hence, the amplitude of the total bias is proportional to the amplitude of the modulating signal and varies at the rate equal to the modulating frequency. The resulting plate current flows in pulses. The amplitude of each pulse is proportional to the instantaneous bias and therefore to the instantaneous modulating voltage. So the application of current pulses to the tuned tank circuit gives AM signal. The operation can be shown by waveforms as shown in Fig. 2.6b.

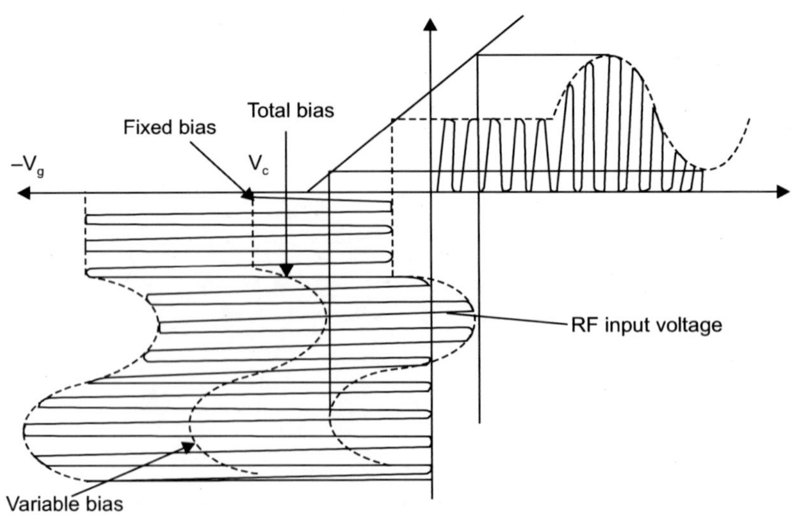

Fig. 2.6(b): Grid Voltage Plate Current Waveforms

This system will operate without distortion or have undistorted output only if the transfer characteristic of triode is perfectly linear. Because this can never be so, the output must be somewhat distorted.

Due to such bias conditions, the maximum output power from a grid modulated amplifier is very less than that would be obtained from the same tube if it is unmodulated. The disadvantages of grid modulation are counterbalanced by the lower modulating power needed in comparison with plate modulation. Harmonics generated due to non-linearity of transfer characteristics are reduced by operating amplifier in push pull.

2.12 PLATE MODULATED CLASS C AMPLIFIER (Transformer Modulation Using Triode)

For radio broadcasting purpose and high power application, the plate modulated class C amplifier is most widely used and best. This is because of high powers available with good efficiency.

Plate modulated class C amplifier shown in Fig. 2.7(a) and its equivalent circuit shown in Fig. 2.7(b).

From Fig.2.7(a) it is seen that the audio voltage is applied in series with the plate supply voltage of a class C amplifier, whose plate current is varied in accordance with modulating voltage. The final power amplifier most frequently work as modulator of transmitter.

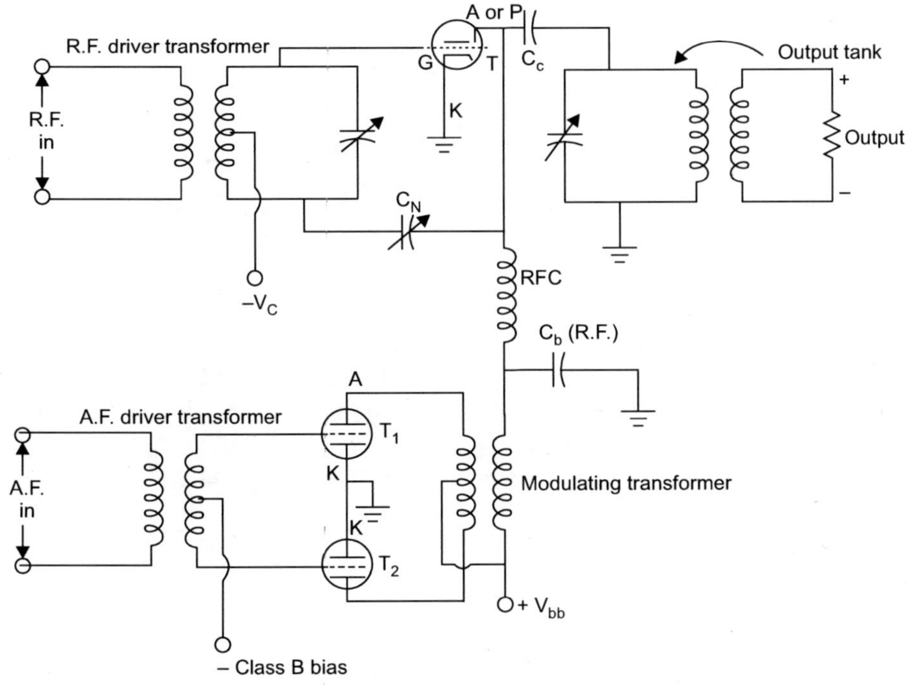

Fig. 2.7(a): Plate-modulated triode class C Amplifier

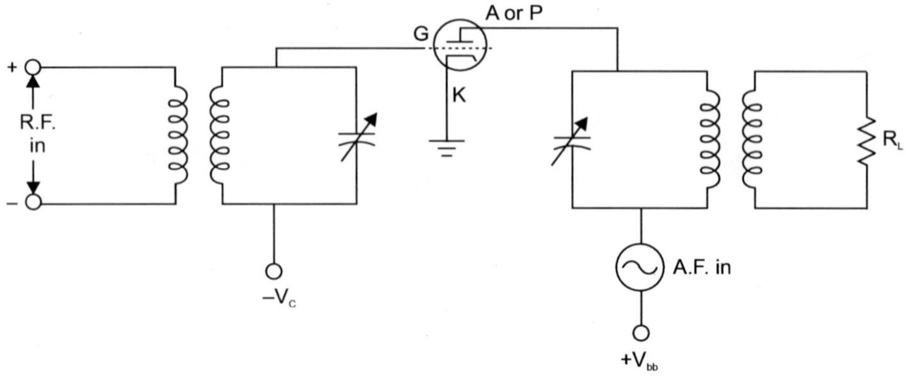

Fig. 2.7(b): Plate-modulation Equivalent Circuit

The output of modulating amplifier is applied to the modulator through modulating transformer. This system sometimes called anode-B modulation, i.e. anode modulation of output power amplifier and class B operation of class B modulator, giving good audio efficiency. It permits 100% modulation to be achieved, since output of modulator can take any value required. As a result of this consideration, this modulator system is employed in a vast majority of AM broadcasting transmitters.

Transformer Modulator Using Triode

The equivalent circuit of Fig. 2.7(b) has been transformed into practical circuit shown in Fig. 2.7(a) by introducing of the modulator with its transformer output. Neutralization capacitor C_N provides stability to circuit at higher frequencies. A choke or RFC is placed in series with the modulation transformer to protect it from R.F. damage.

Waveforms of Plate Modulator

For positive cycle of audio triode T_1 ON and triode T_2 OFF, positive cycle exist across modulating transformer. For negative cycle of audio T_1 OFF and T_2 ON, negative cycle exist across transformer. V_{bb} will come in series with modulating voltage; which forms total bias where V_{bb} is fixed bias and modulating voltage is variable bias. Triode T operate for beyond cut-off hence its output will be pulses. These pulses get amplitude modulated in accordance with total bias. At the output of tank circuit you will get swing in both directions of reference. Hence, resulting output will amplitude modulated as shown in Fig. 2.7(c), (d), (e), (f) and (g).

We have to check this position. If it is not checked, two things arise:

(1) There is possibility that driver may become overloaded when grid current rises, thereby giving distorted wave at the output.

(2) Grid of power amplifier due to excess heat will melt.

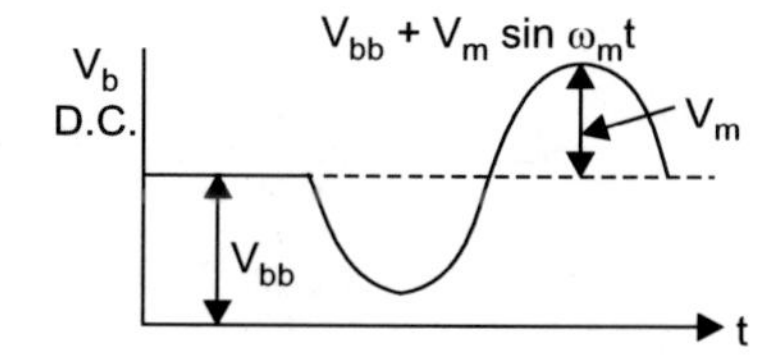

Fig. 2.7(c): Plate Supply Voltage

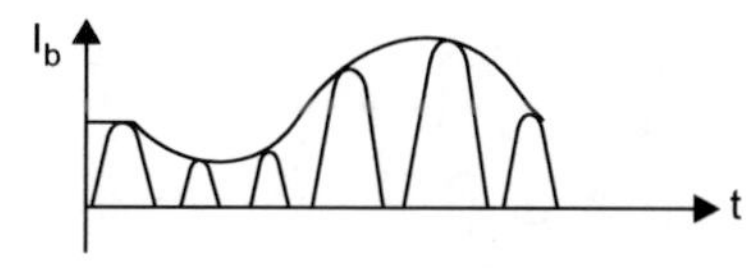

Fig. 2.7(d): Resulting Plate Current

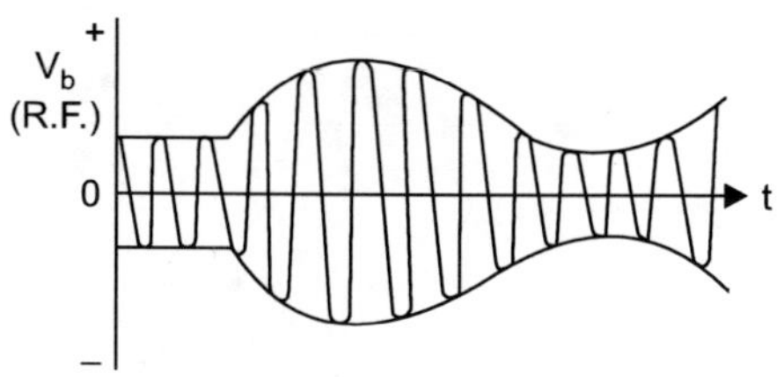

Fig. 2.7(e): Plate R.F. (Modulated Waveform)

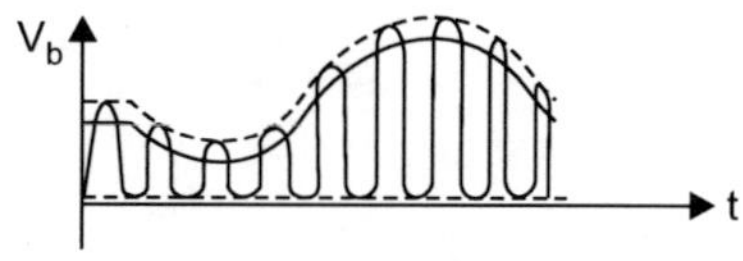

Fig. 2.7(f): Total Plate Voltage

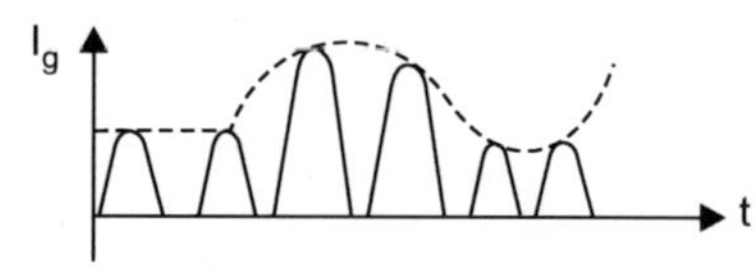

Fig. 2.7(g): Grid Current

Fig. 2.7(c–g): Plate Modulation Waveform

2.13 BIOPOLAR TRANSISTOR COLLECTOR MODULATOR

Especially tubes are used for high level modulation. Suppose we have used transistor for low level modulation and when signal is boosted by power amplifier, we will get high level modulation.

Here modulation used in class C amplifier collector modulation. Carrier signal is coupled by transformer T_1 to the base of transistor which is used in common emitter configuration. Tank circuit formed by L_1 C_1 is having appearance of parallel resonant tank circuit, but practically it will work as series resonant circuit as it is inductively coupled to the tank circuit. At resonant frequency or radio band frequency range maximum signal is fed to the base of transistor since at resonance circuit offers less impedance and impedance is totally resistive. Components C_B and R_B maintain the transistor in class C mode. We know that in class C mode transistor is kept far beyond the cutoff due to self biasing provided by R_B and C_B. As transistor is kept far beyond the cutoff it will not respond to negative half cycle as well as more than half positive cycle. This means that transistor will respond only for less than positive half cycle. Therefore, we will get only positive peak of applied signal whose nature is as shown in Fig. 2.8 and circuit diagram of collector modulator in Fig. 2.9.

As modulating signal comes in series with supply voltage V_{cc} then collector current is made proportional to total bias, i.e. fixed bias and modulating signal therefore resulting waveforms is as shown in Fig. 2.10.

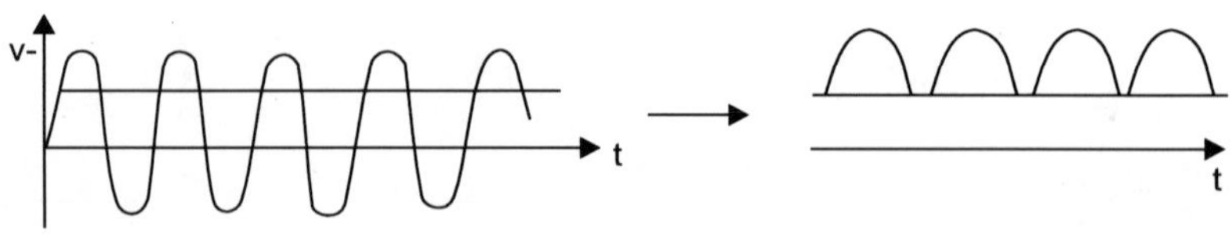

Fig. 2.8

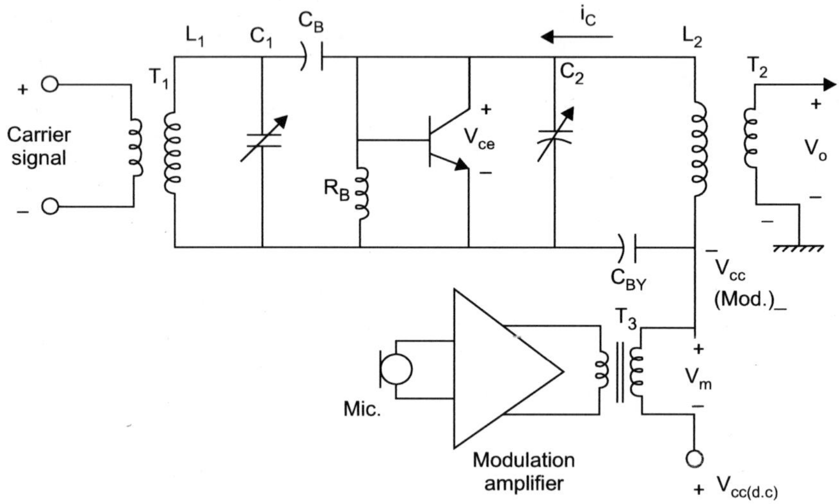

Fig. 2.9: Circuit diagram of collector modulator

2.14 MODULATED TRANSISTOR AMPLIFIERS

With stepping in of solid state technology, tubes are being replaced by transistors, wherever possible. As transistors are low power devices, these cannot be used for high power broadcast transmitters. However, transistor modulated amplifiers are successfully being used for AM generation with an output of few hundred watts. To achieve maximum modulated power output when using transistors, the output stage is made to work in push pull mode as shown in Fig. 2.11.

The modulation methods for transistors are analogous to tube methods example. collector and base modulation of a class C transistor amplifier in place of plate and control grid modulation respectively. Collector and base type of modulation have the same properties and advantages as their tube equivalents. Thus, collector modulation is preferred here.

Collector modulation has higher collector efficiency, better linearity and higher power output as compared to base modulation, except that collector modulation requires higher modulating power. A FET can also be used in place of transistors, in the drain and gate modulation mode.

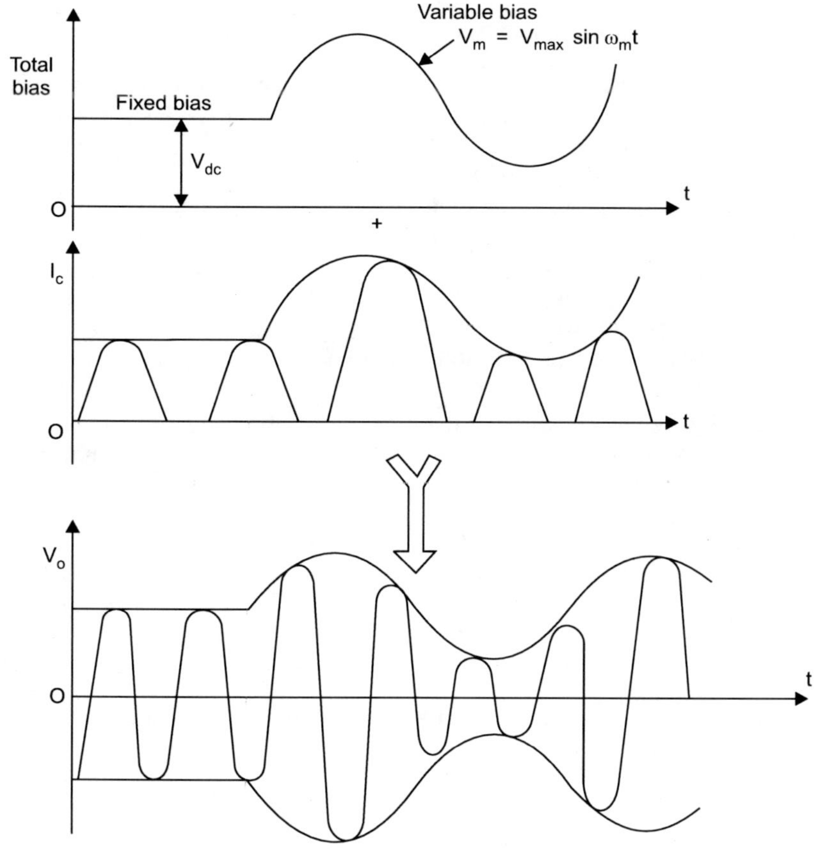

Fig. 2.10

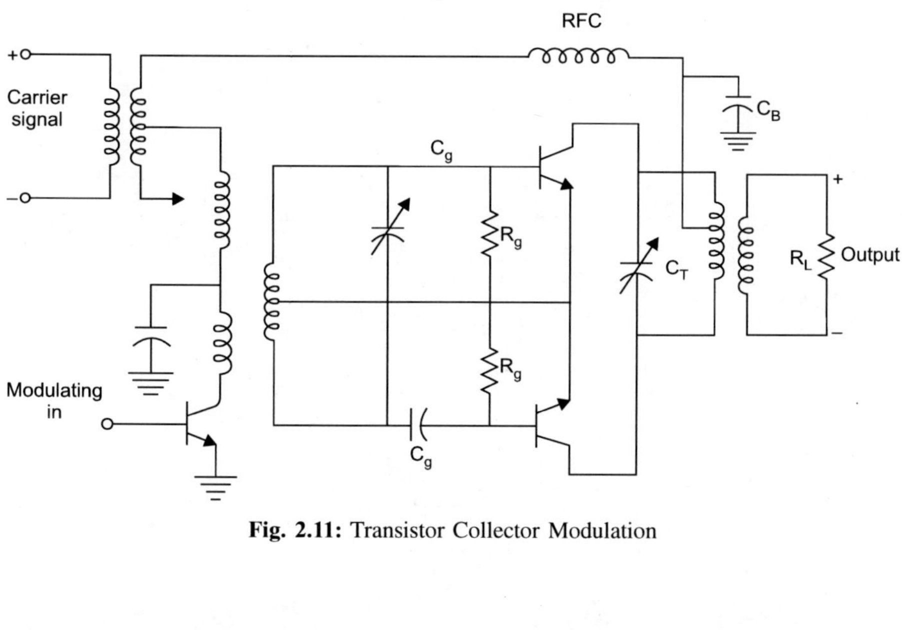

Fig. 2.11: Transistor Collector Modulation

2.15 FREQUENCY MODULATION (FM)

In FM the frequency of the carrier is varied in accordance with the instantaneous value of modulating signal. Here amplitude and phase is kept constant.

Mathematical Representation of FM

Figure 2.12 shows the modulating signal and frequency modulated signal. The instantaneous frequency of frequency modulated signal is given by

$$f = f_c\,(1 + K\,E_m \cos \omega_m t) \qquad (14)$$

where, K = Proportionality constant

E_m = Maximum value of modulating voltage.

f_c = Unmodulated carrier

ω_m = Angular frequency of the modulating signal.

Equation (14) is maximum when the cosine term $\cos \omega_m t = \pm 1$. In this case, the instantaneous frequency is

$$f = f_c\,(1 \pm K\,E_m) \qquad (15)$$

$$f = f_c \pm K\,E_m\,f_c$$

From equation (15), we see that maximum frequency deviation.

$$\delta = K\,E_m f_c$$

The instantaneous amplitude of a frequency modulated signal is given by

$$e = A \sin\,[F(\omega_m, \omega_c)]$$

$$e = A \sin \phi$$

Here, $F(\omega_m, \omega_c)$ is a function of the carrier and the modulating frequencies. This function can be represented by an angle ϕ. This is the angle traced out by a

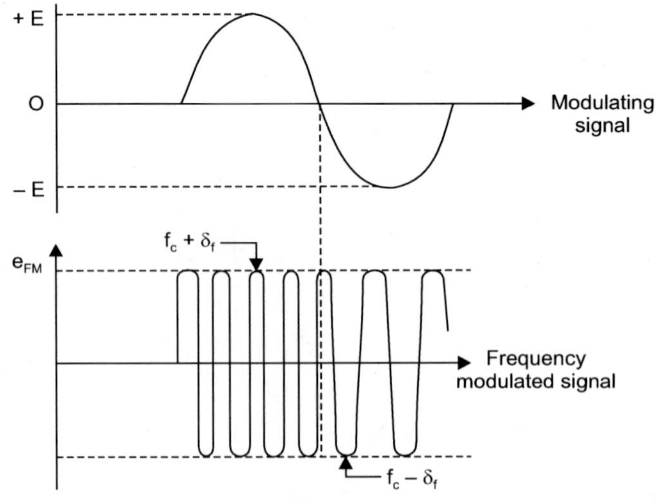

Fig. 2.12

vector A in time t. If we assume that this vector rotates at a constant speed of p, then angle ϕ would be pt radians. In fact, the speed of this vector is not constant and is dependent on equation (14), i.e.

$$\omega = \omega_c(1 + K E_m \cos \omega_m t)$$

To calculate the value of ϕ; we integrate ω with respect to time. Thus,

$$\phi = \int \omega dt = \int \omega_c (1 + K E_m \cos \omega_m t) dt$$

$$= \omega_c \int (1 + K E_m \cos \omega_m t) dt$$

$$= \omega_c \left[t + \frac{K E_m \sin \omega_m t}{\omega_m} \right]$$

$$= \omega_c t + \frac{K E_m \sin \omega_m t}{\omega_m}$$

$$= \omega_c t + \frac{K E_m f_c}{f_m} \sin \omega_m t$$

$$\phi = \omega_c t + \frac{\delta}{f_m} \sin \omega_m t$$

$$\phi = \omega_c t + m_f \sin \omega_m t$$

where m_f = modulation index

$$= \frac{\delta}{f_m} = \frac{\text{Maximum deviation}}{\text{Modulating frequency}}$$

∴ Instantaneous voltage of FM

$$e_{FM} = A \sin \phi$$

$$e_{FM} = A \sin [\omega_c t + m_f \sin \omega_m t]$$

A point to be noted here is that as the modulating frequency is decreased, while the modulating voltage amplitude is kept constant, the modulation index m_f increases. This is the basis of distinguishing frequency modulation from phase modulation.

Example 2.5: If a FM wave is represented by the equation

$$e = 10 \sin (8 \times 10^8 t + 4 \sin 1500 t)$$

Calculate the carrier frequency, modulating frequency, m_f, maximum δ. What power will this FM wave dissipate in 8 ohm resistor?

Solution:

We know, $$e_{FM} = A\sin\left[\omega_c t + \frac{\delta}{f_m}\sin\omega_m t\right]$$

$$e = 10\sin[8\times10^8 t + 4\sin1500t]$$

Here, $$\omega_c = 8\times10^8,\ f_c = \frac{8\times10^8}{2\pi} = 127.32\text{ MHz}$$

$$f_m = \frac{1500}{2\pi} = 238.732\text{ Hz}$$

$$m_f = 4$$

$$\delta = m_f \times f_m = 4 \times 238.732$$

$$\delta = 954.93\text{ Hz}$$

$$\text{Power} = P = \frac{E_{rms}^2}{R} = \frac{\left(\frac{10}{\sqrt{2}}\right)^2}{8} = 6.25\ \text{watts}$$

2.16 FREQUENCY SPECTRUM OF FM WAVE

Different frequency components in FM wave can be determined in the same wave. As followed for AM wave, i.e. by expanding the expression for the waves. Here Bessels function is used.

$$e_{modulated} = A\sin(\omega_c t + m_f \sin\omega_m t)$$

$$= A[\sin\omega_c t\cos(m_f\sin\omega_m t) + \cos\omega_c t\sin(m_f\sin\omega_m) t]$$

$$= A[J_0(m_f)\sin\omega_c t + J_1(m_f)\{\sin(\omega_c + \omega_m)t - \sin(\omega_c - \omega_m)t\}$$

$$+ J_2(m_f)\{\sin(\omega_c + 2\omega_m)t + \sin(\omega_c - 2\omega_m)t\}$$

$$+ J_3(m_f)\{\sin(\omega_c + 3\omega_m)t + \sin(\omega_c - 3\omega_m)t\}$$

$$+ J_4(m_f)\{\sin(\omega_c + 4\omega_m)t + \sin(\omega_c - 4\omega_m)t\} + -]$$

where J_0, J_1, J_2, J_3 ... are the coefficients of zero, first, second, third order of Bessel's function and m_f is the argument. Values of $J_n(m_f)$ may be found out from Bessel function chart. Typical plot of Bessel function is given in Fig. 2.13.

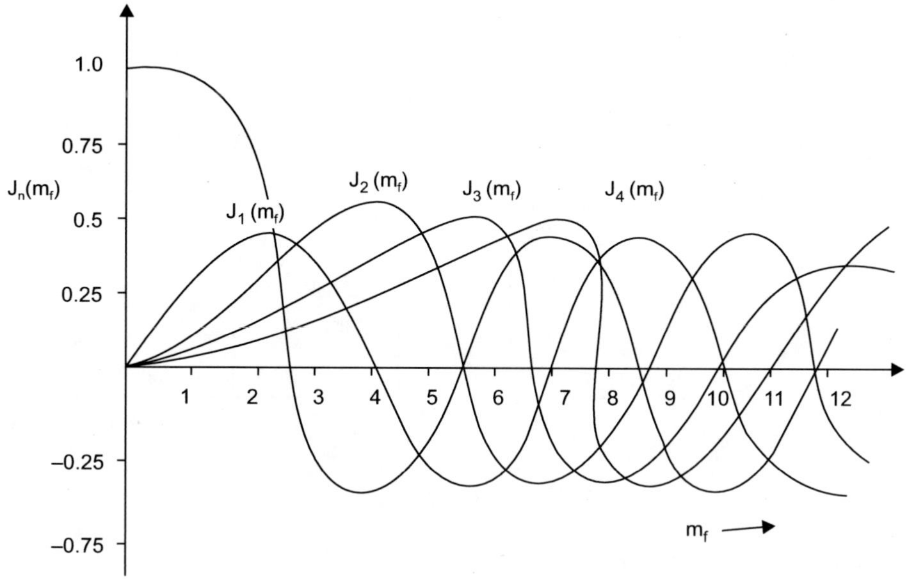

Fig. 2.13: Plot of Bessel Function

In order to evaluate the amplitude of any sideband, it is only necessary to find out the corresponding value of $J_n(m_f)$ and multiply it with E_c. Example for an FM wave with maximum deviation $\Delta f = \pm 75$ kHz and maximum audio frequency of 15 kHz has a modulation index $m_f = 5$. The wave has a total of 8 upper sidebands and an equal number of lower sidebands. The magnitude of 8th sideband is only 2% of the carrier amplitude. The FM wave as a matter of fact, contains an infinite number of sidebands, each sideband is separated from the next by f_m. Out of these, there are only few sidebands which carry significant power. The remaining sidebands have such a low power that they get attenuated during propagation and do not convey any message to the receiver. Following limits have been set by FCC and CCIR.

(1) Maximum permitted frequency deviation = ± 75 kHz

(2) Frequency stability of the carrier = ± 2 kHz

(3) Maximum allowed audio frequency = 15 kHz

(4) Guard bands = 50 kHz

(5) Maximum bandwidth allowed/channel = 200 kHz

Frequency spectrum of FM wave may be plotted in the usual way. Fig. 2.14 shows the plot of a frequency modulated wave for $m_f = 0.5$ and $m_f = 5$.

It can be seen from Fig. (2.14) that in the plot for $m_f < 1.0$, there are few sidebands of large magnitudes. As m_f becomes large, the number of sideband frequencies increase but their amplitude becomes relatively small.

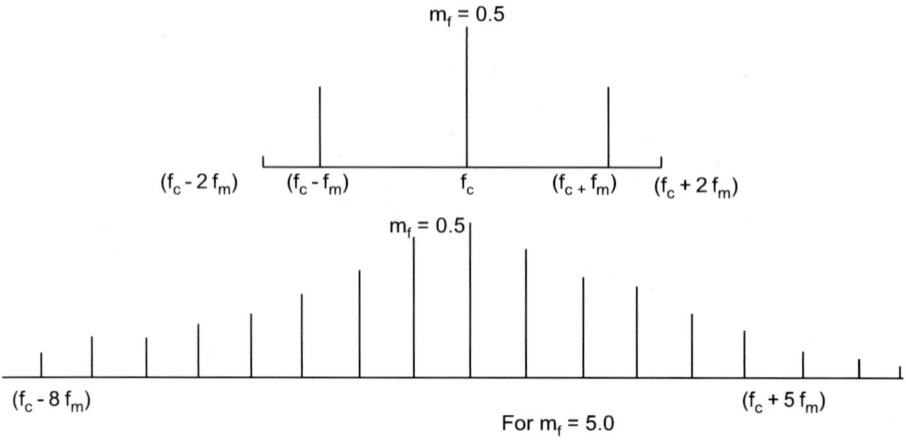

Fig. 2.14: Frequency Spectrum of FM

2.17 GENERATION OF FM

The frequency of an LC oscillator is determined by tuning capacitor and the inductor. If the system includes a reactive element whose reactance can be varied by the modulating signal, this results in production of a frequency modulated wave. A number of electrical and electronic parameters exist, which can provide a variation of capacitance with change in voltage. Similarly, inductance can also be varied by some suitable method. Generally, while using this type of method a voltage variable reactance is placed in parallel with the tank circuit and the tank circuit is tuned to desired carrier frequency (in the absence of any modulating signal). The capacitance or inductance of the variable element is varied with the modulating voltage which results in the variation of frequency about the carrier frequency. When the modulating voltage is zero, the variable reactance has its average value. Thus, at the carrier frequency the oscillator inductance is tuned by its own fixed capacitance in parallel with the average reactance which can give a variable reactance with application of modulating voltage. Example FET, transistor, vacuum tube, etc. These devices are biased in such a way that a proper reactance variation is achieved. Another method which does not employ variable reactance device can also be used to obtain FM and is called Indirect Method.

2.18 EFFECT OF NOISE ON CARRIER

Noise Triangle

Let the signal noise vector (V_n) is superimposed on the carrier vector V_c which is rotating constant angular velocity ω_c, then the noise vector will affect the amplitude of carrier by V_n and will rotate with constant angular velocity $\omega_n - \omega_c$ which is shown in figure and phase angle,

$$\phi = \sin^{-1}\left(\frac{V_n}{V_c}\right)$$

$$= \sin^{-1}\left(\frac{0.25}{1}\right)$$

$$= 14.5°$$

Let $V_n = 0.25$

and $V_c = 1$

Then $m_a = \left(\frac{V_n}{V_c}\right)$

$$= \frac{0.25}{1} = 0.25$$

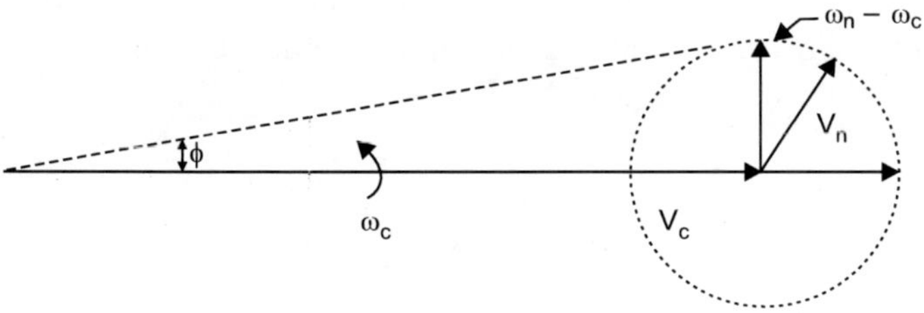

Fig. 2.15

Now, we know that in AM phase will not be affected by noise and in FM. Amplitude will not affected by noise which can be removed by limiter.

Now, we have to see effect of noise on the amplitude in AM and effect of noise on phase in FM.

For considering effect of noise on amplitude of AM and phase of FM. We will take worst case audio frequency to be maximum, i.e.

f_m = 15 kHz and take

$m_a = 1$

and m_f = 1 rad (57.3°)

Under such a condition $\left.\frac{N}{S}\right|_{AM} = \frac{0.25}{1} = 0.25$

and in case of FM $\left.\frac{N}{S}\right|_{FM} = \frac{14.5°}{57.3°} = 0.253$

just worst in FM than AM.

$\left.\frac{N}{S}\right|_{FM}$ just worst than $\left.\frac{N}{S}\right|_{AM}$

Now suppose audio frequency is reduced from 15 kHz to 30 Hz in audio band.

Then in case of AM there is no effect on $\frac{N}{S}$ ratio. In FM $\left(\frac{V_n}{V_c}\right)$ remains the constant but noise phase modulate the carrier signal. As audio frequency goes on reducing modulation index goes on increasing when noise and audio signal both having frequency equal to 30 Hz. Then N/S ratio is equal to

$$\frac{0.253}{15,000} \times 30 = 0.000505.$$

This means that N/S ratio which was 25.3% for f_m = 5 kHz is reduced to 0.05% for f_m = 30 Hz. This means we will say there is improvement in N/S ratio when audio frequency is reduced from 15 kHz to 30 Hz. In AM noise distribution is rectangular. While noise distribution in FM is triangular. As (N/S) ratio reduces than (S/N) power will become weak, when signal becomes weak as compared to noise. Then limiter present in FM receiver rejects our desired signal; so we have to adjust m_f in such a way that signal should not be weaker than noise.

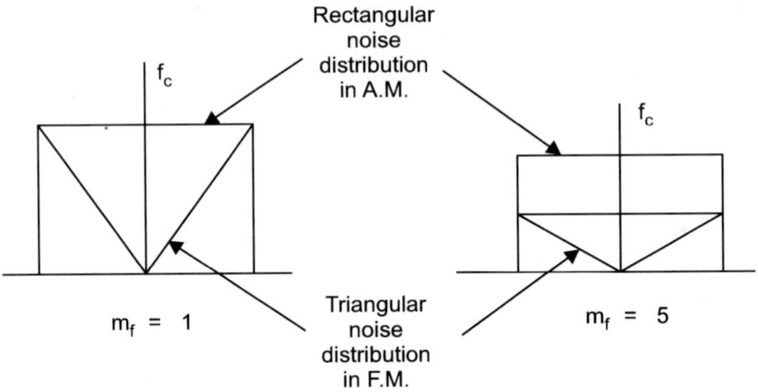

Fig. 2.16

For example, for wide band FM broadcasting system deviation frequency is kept at 75 kHz, f_m = 15, kHz and m_f = 5 and voltage ratio = 4.

2.19 PRE-EMPHASIS AND DE-EMPHASIS

From the noise triangle it is seen that noise existence is large at higher frequencies than lower frequencies. When signal present at higher frequencies is artificially boosted at transmitter and cut out at the receiver then we will be able to reduce, the noise at large extent. This boosting at higher frequencies, modulating signal voltage in accordance with pre-arranged curve is known as pre-emphasis and compensation of signal at receiver is known as De-emphasis.

Suppose we have taken two signals having same amplitude. Out of these two signals one has higher modulating frequencies and on the other hand other is lower modulating frequencies. Out of these two signals suppose higher frequency modulating signal is pre-emphasized signal has twice the frequency deviation as compared to un-emphasized lower modulating signal so that pre-emphasized signal is less affected by noise compared to lower modulating frequency signal. Out of these two signals ultimately higher modulating frequencies will be de-emphasised at the receiver therefore amplitude of this signal will be less than without emphasized signal therefore effect of noise on pre-emphasized signal will reduce.

RC time constant decided by this circuit is 75 μ sec. 3 dB frequency is given by

$$f = \frac{1}{2\pi RC}$$

and it is 2120 Hz for 75 μ sec RC time constant. Another important point is that signal should not be overemphasized.

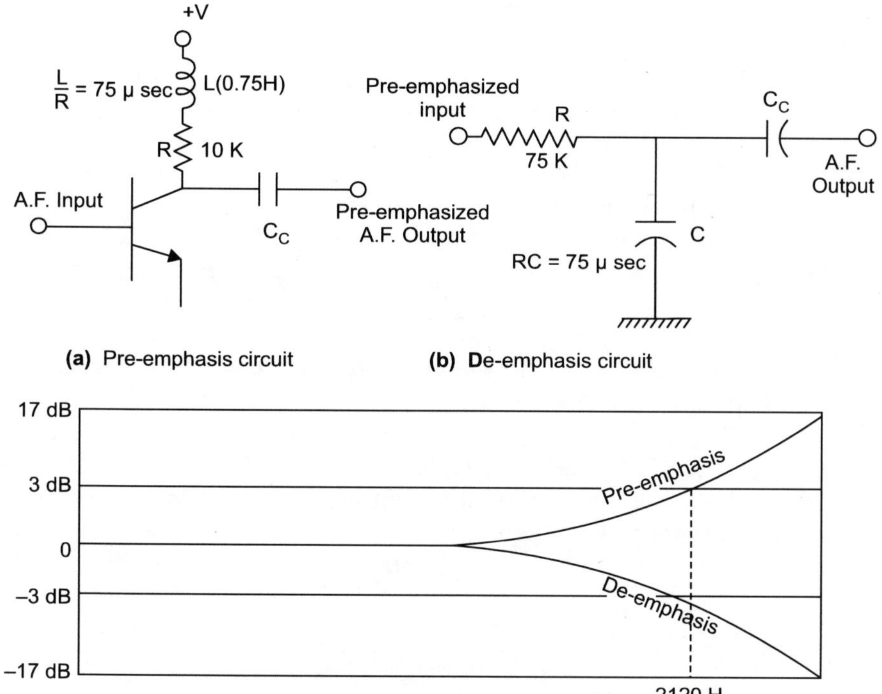

(a) Pre-emphasis circuit **(b)** De-emphasis circuit

(c) Pre-emphasis and de-emphasis response

Fig. 2.17

2.20 ADJACENT – CHANNEL INTERFERENCE

We know that frequency modulation improves (S/N) ratio. Also FM has better capability to reject the noise as well as FM has better capability to reject adjacent channel signal interference. FM having maximum frequency deviation 15 kHz and 75 μ sec pre-emphasis gives the noise rejection at least 24 dB better AM. Suppose AM has (S/N) ratio 60 and FM has (S/N) ratio 36 both circuits give same performance for perfect reception whether the unwanted signal may be noise or adjacent channel signal.

We know that in wide band FM channel, width required will be 200 kHz. Out of these 200 kHz, 180 kHz is used for transmitting desired signal and 20 kHz is used as a guard band therefore adjacent channel interference is avoided.

2.21 CO-CHANNEL INTERFERENCE OR CAPTURE EFFECT

Basically, we know that limiter eliminates weak signals and it passes stronger signal. When there are two signals, suppose one is desired signal and another signal is unwanted signal, i.e. noise. When desired signal is two times stronger than noise signal then noise is completely eliminated by the limiter.

In AM consider AM receiver and two AM transmitters.

Suppose AM transmitter (1) is nearer to AM receiver and AM transmitter (2) is longer to AM receiver. Then AM transmitter (1) predominates AM transmitter (2). This means that program of AM transmitter (1) is heard with the help of AM receiver but AM transmitter (2) will create interference on the program of AM transmitter (1). This is straightforward effect.

In case of FM this effect will be quite interesting

From Fig. 2.18(d) it is seen that FM receiver is nearer to FM transmitter (1) compared to FM transmitter (2). Then FM transmitter (2) program is completely eliminated and only FM transmitter (1) program is heard by FM receiver.

From Fig. 2.18(b) it is seen that FM receiver is at a centre of FM transmitter (1) and FM transmitter (2). Then FM receiver heard program of FM transmitter (1) and FM transmitter (2) alternatively.

From Fig. 2.18(c) it is seen that when FM receiver is moved towards FM transmitter (2) from FM transmitter (1) until $d_2 < d_1$. Then program of FM transmitter (1) is completely eliminated by FM receiver. While program of FM transmitter (2) is heard by FM receiver. This means that FM receiver is captured by FM transmitter (2). This co-channel interference phenomena is known as *capture effect*.

2.22 DIRECT METHODS OF FM GENERATION

Direct methods are the one which use a voltage variable reactance device. The most common types are the reactance modulator and varactor diode type.

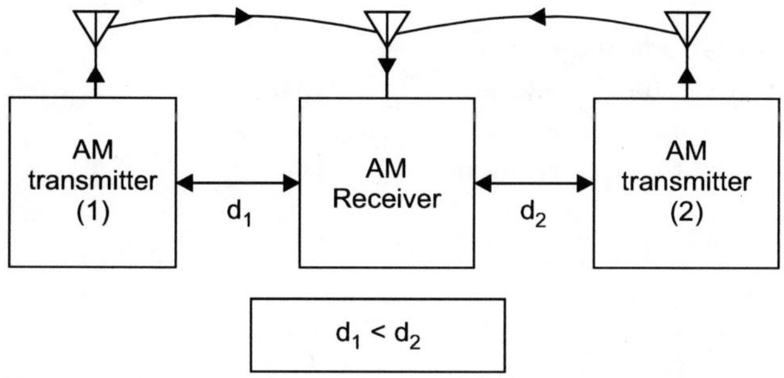

Fig. 2.18(a)

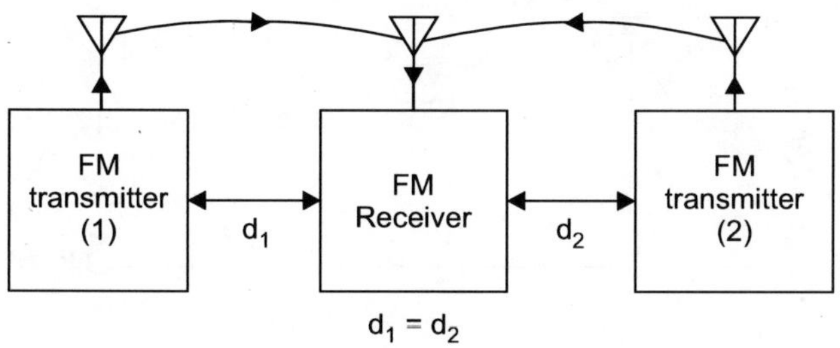

Fig. 2.18(b)

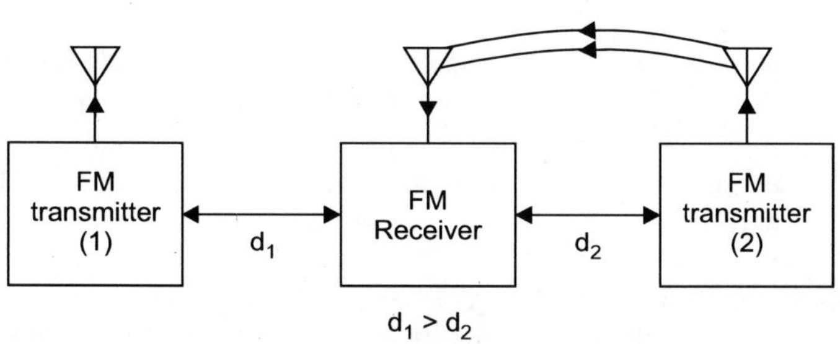

Fig. 2.18(c)

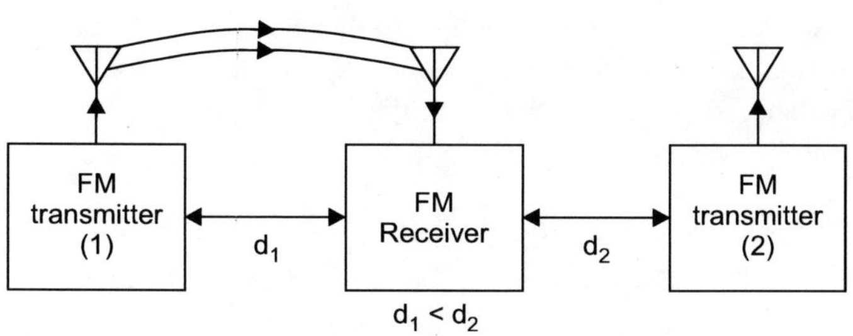

Fig. 2.18(d)

Fig. 2.18

2.22.1 Reactance Modulator

Consider a basic reactance modulator shown in Fig. (2.19). It uses FET and behaves as a three terminal reactance that may be connected a cross a tank circuit of the oscillator to be frequency modulated.

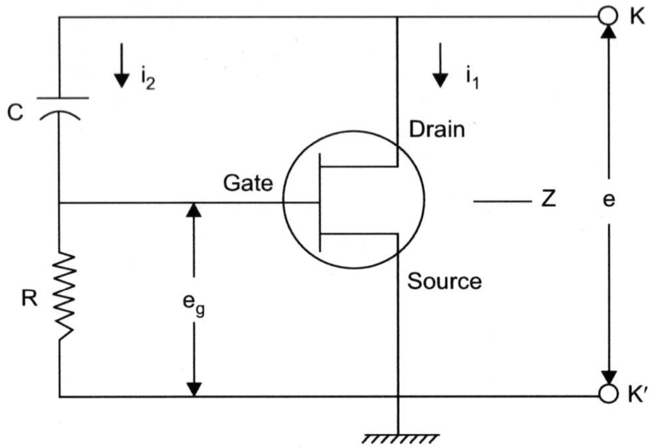

Fig. 2.19: Basic Reactance Modulator

Under certain conditions the impedance "Z" between terminals KK′ becomes entirely reactive. It can be capacitive or inductive by just interchanging one component. The value of the reactance Z is proportional to the transconductance of the device, which is turn depends upon the gate bias and its variation. The FET can be replaced by bipolar transistor or vacuum tube.

Theory: To evaluate the value of Z, a voltage "e" is applied at terminal KK′. The resulting current i_1 is calculated. In order for the impedance to be a pure reactance two requirements have to be fulfilled. First is that the bias networked current i_2 should be negligible as compared to drain current i_1. In other words, the impedance of the bias network should be large enough so that it can be ignored. The second requirement is that the drain to gate impedance X_c should be greater than the gate to source impedance R preferably by more than five times. (i.e. $X_C >> R$)

Then e_g can be written as

$$e_g = i_2 R = \frac{e}{R - jX_c} R = \frac{Re}{R - jX_c}$$

$$e_g = \frac{Re}{R - jX_c}$$

The FET drain current will be:

$$i_1 = g_m e_g = \frac{g_m Re}{R - jX_c}$$

∴ impedance seen at terminals KK′ will be

$$Z = \frac{e}{i_1} = e \div \frac{g_m Re}{R - jX_c}$$

$$= e \times \frac{R - jX_c}{g_m Re} = \frac{R - jX_c}{g_m R}$$

$$Z = \frac{1}{g_m}\left[1 - \frac{jX_c}{R}\right] \tag{16}$$

Assuming $X_c >> R$, equation (16) reduces to

$$Z = -j\frac{X_c}{g_m R} \tag{17}$$

From Equation (17) represents a reactance which is decidely capacitive in nature and therefore written as:

$$X_{eq} = \frac{X_c}{g_m R} = \frac{1}{2\pi fC \times g_m R}$$

$$X_{eq} = \frac{1}{2\pi f g_m CR}$$

$$X_{eq} = \frac{1}{2\pi f C_{eq}}$$

It can be seen that under conditions assumed that input impedance of the device at terminals KK′ results in a pure reactance given by

$$C_{eq.} = g_m RC \tag{18}$$

Thus, it can be seen that equivalent capacitance C_{eq} is dependent on the device transconductance which in turn can be varied by bias voltage. Initially the capacitance can be adjusted to any desired value by varying the component R and C. The gate to drain impedance should be much larger as compared to gate to source impedance.

If X_c/R had not been much greater than unity, Z would have a resistive component also. If R is not kept quite lower than X_c, the gate voltage in that case will no longer remain exactly 90° out of phase with the applied voltage e_i nor will be the drain current i_1. This results in the input impedance also not remaining purely reactive. The resistive component of input impedance is $\frac{1}{g_m}$; g_m is dependent on the applied modulating voltage. This variable resistance appears directly across the tank circuit, thus varying the Q, which further varies the output voltage. This results in some amplitude modulation of the output signal. The solve this problem, an amplitude limiter is used only in the cases when amplitude modulation becomes too excessive or it cannot be reduced otherwise.

Usually, X_c is made 5 to 10 times of R.

Let this factor be "n" i.e. X_c = n R for FET reactance modulator then

$$X_c = \frac{1}{\omega C} = n R$$

$$C = \frac{1}{\omega n R} = \frac{1}{2\pi f n R} \tag{19}$$

From equation (19) and (18)

$$C_{eq} = g_m R C$$

$$= g_m R \times \frac{1}{2\pi f n R}$$

$$C_{eq} = \frac{g_m}{2\pi f n}$$

Example 2.6: Evaluate the value of capacitance reactance from a reactance FET having g_m = 10 ms. The gate-to-source resistance is one sixth of reactance of the gate-to-drain. The frequency of operation is 4.6 MHz.

Solution:

We have g_m = 10 m sec, n = 6

$$C_{eq} = \frac{g_m}{2\pi f n}$$

$$\therefore 2\pi f C_{eq} = \frac{g_m}{n} = \frac{1}{X_{eq}}$$

$$X_{eq} = \frac{n}{g_m} = \frac{6}{10 \times 10^{-3}}$$

$$X_{eq} = 6000\ \Omega$$

2.22.2 Transistor Reactance Modulator

The general condition for all types of reactance modulator is that the drain current should always be much greater than the bias network current.

Figure (2.20) shows circuit diagram of transistor reactance modulator.

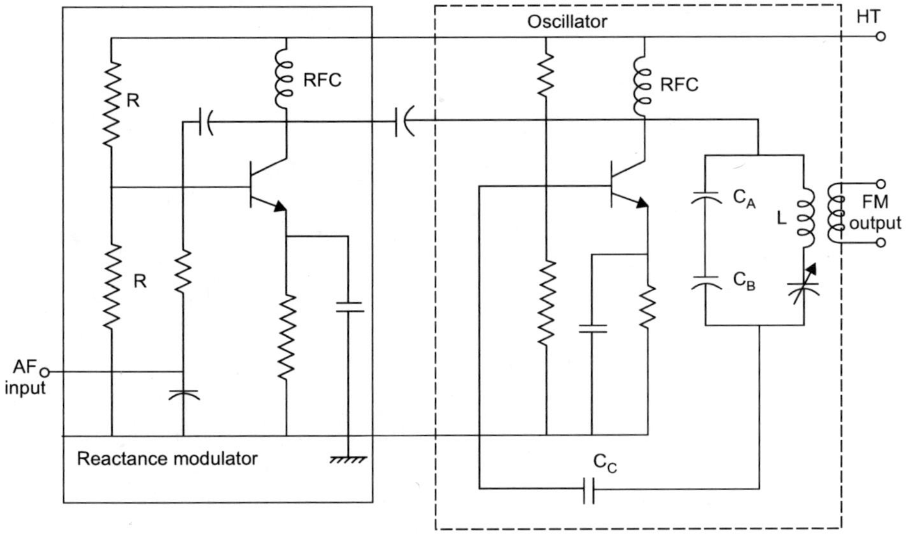

Fig. 2.20: Transistor Reactance Modulator

A RC-capacitive reactance modulator uses a transistor instead of FET and modulates a Clapp-Gouriet oscillator. Any modulator can be used to modulate any kind of oscillator provided correct component values and other parameters are used. Another condition is that the oscillator used must not be the one which requires more than one tuned circuit for its operation.

The Hartley and Colpitts (or Clapp-Gouriet) oscillators are most commonly used.

Proper isolation between the oscillator and modulator should be used and for that a buffer stage should be provided. The RF choke used in circuit allow to pass dc while isolating the various sections of this circuit for alternating current which may get coupled through the HT supply line.

2.23 VARACTOR DIODE MODULATOR

Varactor diode is a semiconductor diode having a variable junction capacitance. The variation is achieved by applying reverse bias. The junction capacitance of varactor diode varies linearly with the applied voltage. Varactor diodes are being extensively used these days to produce frequency modulation.

Figure 2.21 shows modulator using varactor diode.

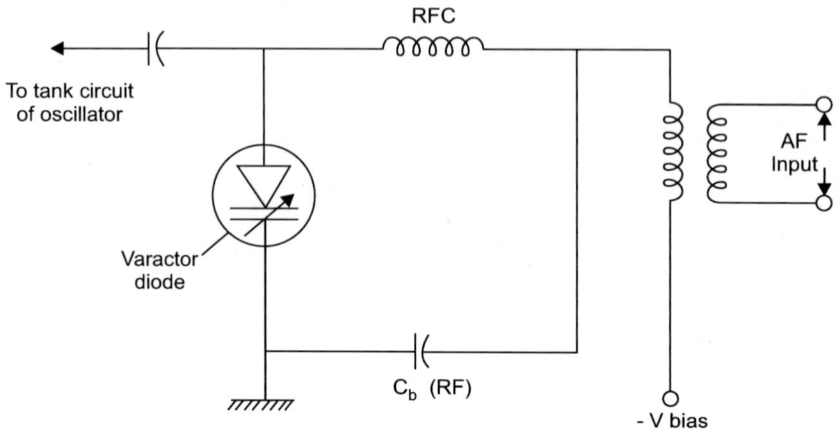

Fig. 2.21: Modulator Using Varactor Diode

The diode is back biased to obtain the junction capacitance effect, this reverse bias is achieved by the modulating voltage which in turn varies the junction capacitance causing the oscillator frequency to vary proportionally.

Disadvantage of varactor diode modulator is that it uses a two terminal device, thus it is used mostly in automatic frequency-control and remote tuning.

2.24 FREQUENCY STABILIZATION OF FREQUENCY MODULATOR

A prime requirement of any transmitter is the stabilization of the carrier frequency. It is necessary that the average or carrier frequency of a frequency modulator be maintained very nearly constant, even though the instantaneous frequency of frequency modulator varies with the modulating signal. When a reactance modulator is used, to modulate the carrier, the carrier cannot be crystal controlled. The average frequency depends to some extent on the temperature, the device characteristics and the operating potentials, slight drift in the operating characteristics is accompanied by an appreciable change in average frequency. This necessitates setting up of frequency stabilization system in the frequency modulator to maintain the carrier frequency within the 2 kHz deviation specified by FCC regulations.

Figure 2.22 shows block diagram of typical AFC system for FM transmitter.

It uses crystal oscillator and carrier frequency of the FM signal is compared with it. The reactance modulator works across the tank circuit of LC oscillator, whose output is isolated by a buffer stage. The output of buffer is fed to an amplitude limiter and subsequently to class-C power amplifiers. A small part of the signals is taken from the limiter output and fed to a mixer in which this signal is mixed with a signal from the crystal oscillator.

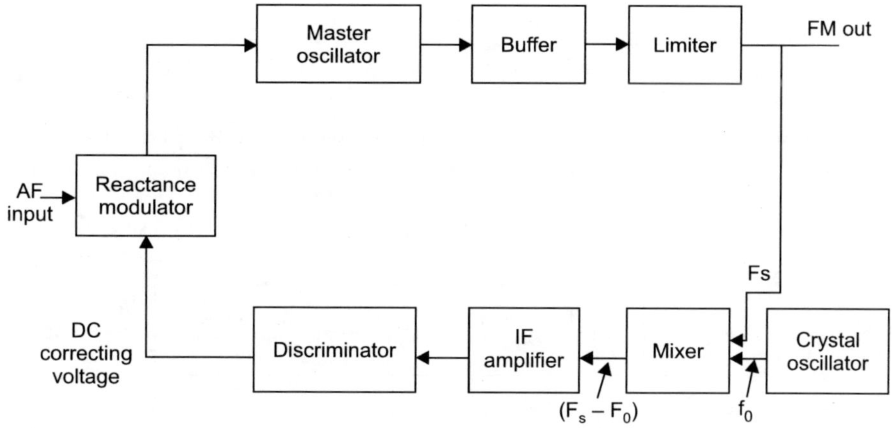

Fig. 2.22: AFC System for AM Transmitter

The difference signal which is usually about one tenth to one twentieth of the master oscillator frequency is amplified and applied to a phase discriminator. The output of the phase discriminator is a d.c. signal which is applied to the reactance modulator as a correcting voltage to counteract any drift in the average frequency of the master-oscillator.

The discriminator is designed in such a way that it reacts to slow changes in the incoming frequency, but not to the normal frequency variation due to frequency modulation. Also the discriminator is connected to give a positive output voltage if the input frequency exceeds the discriminator tuned frequency and a negative output voltage if it is lower.

Consider that the master oscillator frequency drifts on the higher side. This higher frequency is fed to a mixer where it is mixed with stable frequency obtained from a crystal-oscillator. This causes the difference frequency ($f_s - f_0$) to be fed to the discriminator through an IF amplifier stage. The discriminator is tuned to the correct frequency difference which should normally exist between the two oscillators, when the difference frequency input to the discriminator exceeds this tuned frequency a positive dc signal is developed at the output of the discriminator which is proportional to the increased frequency difference.

This type of frequency stabilization can be used for any type of oscillator which cannot for some reason be crystal controlled. The difference in this case in the circuit would be that there should be no reactance modulator and the discriminator load time constant will be low, so that it reacts to frequency changes quickly.

2.25 INDIRECT METHOD OF FM GENERATION

2.25.1 Armstrong Method

The direct method (modulation) have the disadvantage of being based on an oscillator which is not stable enough for communication and broadcast purposes. This requires stabilization of the reactance modulator, with attendant circuit complexity. It is possible to generate FM via phase modulation. This method is called "Armstrong method".

If we integrate the modulating signal first and then allow it to phase modulate, we obtain FM wave. This is the principle of Armstrong's method of generating FM wave.

Figure 2.23 gives the idea of generation of FM by Indirect Method.

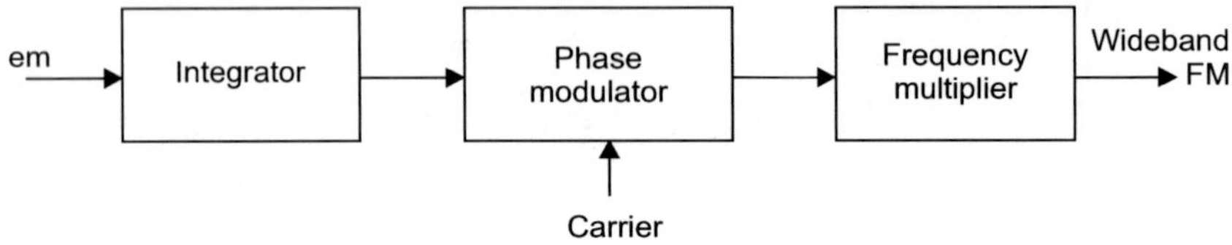

Fig. 2.23: Indirect Method

The frequency multiplier is a non-linear device which multiplies the frequency of the input. The equation for phase modulated wave is:

$$e = A\sin(\omega_c t + \phi_m \sin\omega_m t) \text{ or } e = A\sin(\omega_c t + m_p \sin\omega_m t)$$

m_p = Modulation index for phase modulation.

Figure 2.24 shows Armstrong method of obtaining frequency modulation.

A stabilized 200 kHz frequency oscillator is used to control, the carrier frequency of the radiated wave. The audio modulating signal is applied at the input of pre-distorter circuit. Such a that its amplitude is made to vary inversely with its frequency. The frequency distorted version of audio signal and a part of the 200 kHz carrier signal are mixed in balanced modulator. The output of balance modulator consists of two sideband components. The output of balance modulator consists of two sideband components with the carrier component completely suppressed (eliminated). The output of balance modulator is shifted through 90° in phase and then combined with the carrier (200 kHz) in the combining buffer amplifier. The resultant wave is the frequency modulated wave which has been obtained from the phase-modulated wave, the phase deviation, which has been made to vary inversely as the frequency of the modulating wave. The resultant frequency modulated wave is multiplied in frequency multiplier circuits until the frequency is brought to the desired frequency level. Then the wave is finally amplified and transmitted.

2.26 SINGLE SIDEBAND SYSTEM (SSB)

We know, AM signal consists of a carrier and two sidebands. In this chapter, it is not necessary to transmit all the three components for reconstructing the signal at the receiver. If carrier and one sideband are removed or attenuated, the transmitter

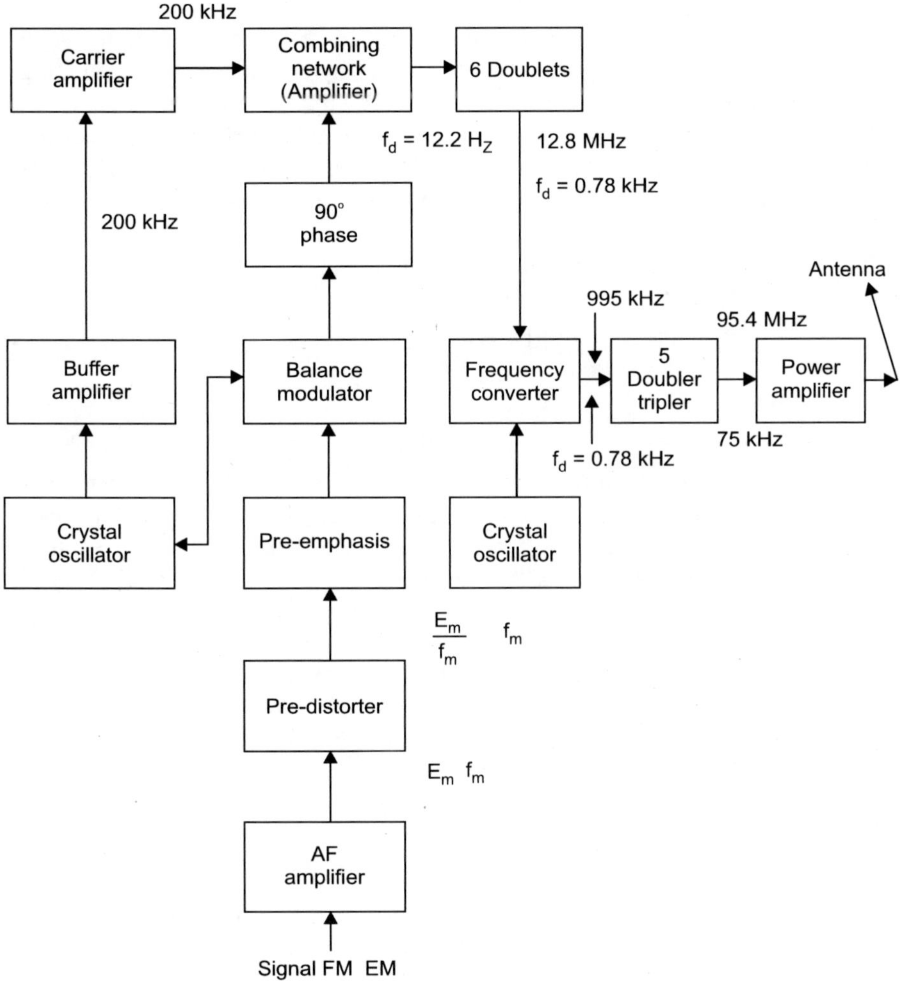

Fig. 2.24: Block Diagram of Armstrong Type of FM Transmitter

power and bandwidth requirements are reduced, still allowing acceptable communication possible. There are three standard methods of removing the carrier and unwanted sideband. Another type of sideband transmission called "Vestigial Sideband transmission".

An amplitude modulated signal consists of three different frequencies: The original carrier frequency, lower-sideband ($f_c - f_m$) and upper sideband ($f_c + f_m$). These three frequencies are automatically generated whenever amplitude modulation is done, unless and until additional steps are taken to remove any of these frequencies from the output. The ordinary AM is called DSBFC (Double Sideband Full Carrier). The carrier of DSBFC carries no information as it remains constant in amplitude and frequency irrespective of changes in modulating voltage.

Also each sideband is a mirror image of other and each carries the same information, since each is affected in a similar way by the changes in modulating voltage amplitude. Thus, each sideband will convey the same information.

In other words, we can say that only one sideband which can be either the lower or the upper can convey the complete information which in ordinary amplitude modulated signal conveys. The fact that an expensive modification is required for the receiver to receive an SSB signal.

The total transmitted power in case of AM signal, is

$$P_c\left(1+\frac{m_a^2}{2}\right)$$

where P_c carrier power and $\frac{P_c m_a^2}{2}$ is the sideband power. At 100% modulation, the sideband power is one-third of the total power. Thus, at 100% modulation, a two-thirds carrier power can be saved, which can even go higher at lower modulation levels. If now one sideband is also suppressed, the saving in power goes to a staggering value of 83.3% over DSBFC system.

Example 2.7: Calculate the percentage saving in power if only one sideband transmission is used over DSBFC system at:

(a) 100% modulation, (b) 80% modulation (c) 50% modulation.

Solution: *(a) Total power of AM, DSBFC signal is:*

$$P_t = P_c\left(1+\frac{m_a^2}{2}\right)$$

At $\quad m = 100\%$

$$P_t = P_c\left(1+\frac{1}{2}\right)$$

$$P_t = 1.5\ P_c$$

$$P_{SB} = P_c\frac{m^2}{4} = P_c\frac{1}{4}$$

$$P_{SB} = 0.25\ P_c$$

$$\therefore \text{ Saving Percentage } = \frac{1.5-0.25}{1.5} = \frac{1.25}{1.5}$$

Saving percentage = 83.3%

(b) 80% modulation:

$$P_t = P_c\left[1+\frac{0.8^2}{2}\right] = 1.32 P_c$$

$$P_t = 1.32\ P_c$$

$$P_{SB} = P_c \frac{m_a^2}{4} = P_c \frac{(0.8)^2}{4} = 0.16\ P_c$$

$$\text{Saving Percentage} = \frac{1.32 - 0.16}{1.32} = 0.879$$

Saving Percentage = 87.9%

(c) 50% Modulation:

$$P_t = P_c\left(1 + \frac{m_a^2}{2}\right) = P_c\left(1 + \frac{0.5^2}{2}\right)$$

$$P_t = 1.125\ P_c$$

$$P_{SB} = P_c\left(1 + \frac{m_a^2}{4}\right) = P_c\left(1 + \frac{0.5^2}{4}\right)$$

$$P_{SB} = 0.0625\ P_c$$

$$\text{Saving Percentage} = \frac{1.125 - 0.0625}{1.125} = 94.4\%$$

The above example shows, how wasteful it is to transmit the carrier and two sidebands where only one sideband can solve the purpose. Another advantage of SSB is that the bandwidth requirements are reduced by 50% as compared to DSBFC.

Figure 2.25 shows waveforms.

2.27 GENERATION OF SSB OR CARRIER SUPPRESSION

There are three different methods used for generation of SSB: (1) filter method, (2) the phase shift method, and (3) the Weavers method.

Each method has its one way of unwanted sideband suppression. For carrier suppression balance modulators are used. Thus, balanced modulator forms a very important part of SSB-system.

Effect of Non-linear Resistance on addition of signals:
The voltage and current relation for a linear resistance is

$$i = be$$

where, b = constant of proportionality, which can also be called as conductance of resistance. Similar voltage and current relation can also be found for active devices, such as a transistor. If the amplifier operates in class-A and a d.c. current component is also present in the collect or current which is independent of base voltage. Thus,

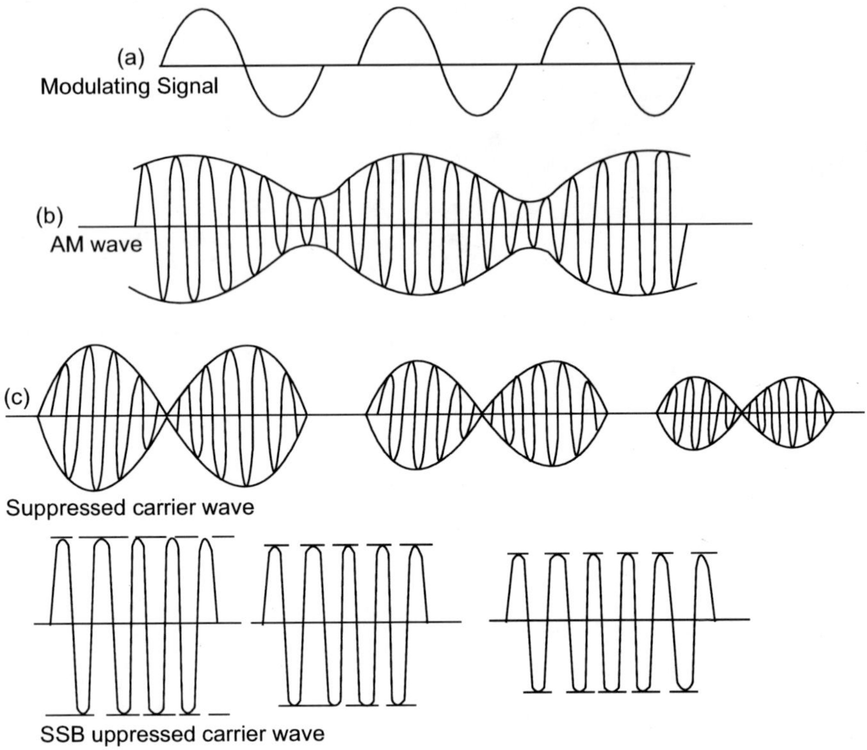

Fig. 2.25

$$i = a + be$$

a = dc current component of collector current and

b = transconductance of the transistor.

If we use a non-linear resistance, the current does not remain exactly proportional to the applied voltage as was in the case of linear resistance. Figure 2.26 shows current voltage characteristics for non-linear resistance.

The curve remains linear only upto a certain point and thereafter it has a sudden increase or a saturation.

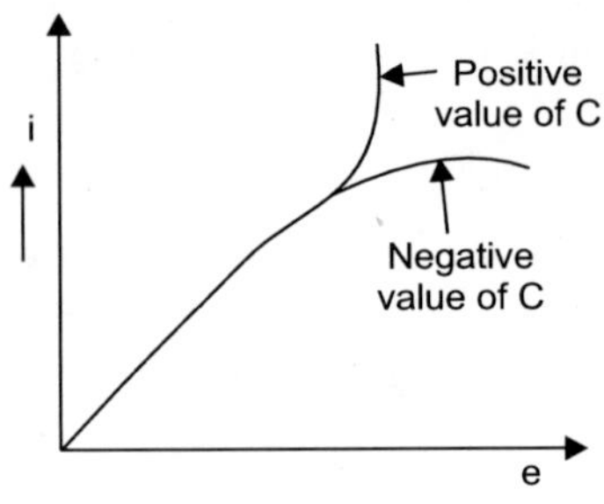

Fig. 2.26

Non-linear characteristic can be expressed as:

$i = a + be + ce^2 + de^3$ + higher order powers of e.

Current now becomes proportional not only to voltage but also to the square, cube and higher powers of voltage.

The constants a, b, c, ... determines the shapes of curve. Usually, only a and b dominate, while higher order constants have very low value. When e becomes large, then square term also attains a significant value and so cannot be neglected. This causes the curve to deviate from a straight line. The cube and higher order terms can be easily neglected. Thus, equation left would be

$$i = a + be + ce^2 \tag{20}$$

Here, the coefficient "c" is called "coefficient of non-linearity". Let apply equation (20) to a non-linear characteristic of FET

$$i = a + b(e_1 + e_2) + c(e_1 + e_2)^2$$

$$i = a + b(e_1 + e_2) + c(e_1{}^2 + e_2{}^2 + 2e_1 e_2) \tag{21}$$

Let the two input voltages be represented by equations:

$$e_1 = E_1 \sin \omega_1 t \tag{22}$$

and

$$e_2 = E_2 \sin \omega_2 t \tag{23}$$

where, ω_1 and ω_2 are angular velocities of two voltages.

Using equation (21), (22) and (23) we get

$$\begin{aligned} i &= a + b(E_1 \sin \omega_1 t + E_2 \sin \omega_2 t) + \\ &\quad c(E_1^2 \sin^2 \omega_1 t + E_2^2 \sin^2 \omega_2 t + 2E_1 E_2 \sin \omega_1 t \sin \omega_2 t) \\ &= a + bE_1 \sin \omega_1 t + bE_2 \sin \omega_2 t + \frac{1}{2}CE_1^2(1 - \cos 2\omega_1 t) \\ &\quad + \frac{1}{2}CE_2^2(1 + \cos 2\omega_2 t) + CE_1 E_2 \times \\ &\quad [\cos(\omega_1 - \omega_2)t - \cos(\omega_1 + \omega_2)t] \\ &= \underbrace{\left(a + \frac{1}{2}CE_1^2 + \frac{1}{2}CE_2^2\right)}_{(A)} + \underbrace{bE_1 \sin \omega_1 t}_{(B)} + \underbrace{bE_2 \sin \omega_2 t}_{(C)} \\ &\quad \underbrace{-\left(\frac{1}{2}CE_1^2 \cos 2\omega_1 t + \frac{1}{2}CE_2^2 \cos 2\omega_2 t\right)}_{(D)} \end{aligned}$$

$$+CE_1 E_2 \cos(\omega_1 - \omega_2)t - CE_1 E_2 \cos(\omega_1 + \omega_2)t \tag{24}$$

(E) (F)

Equation (24) a very important relation and will be used at number of places. If we assume ω_1 to be carrier angular frequency and ω_2 as the modulating angular frequency then term:

(A) represents dc current component,
(B) represents carrier,
(C) represents modulating signal,
(D) represents the harmonics of carrier and modulating signal,
(E) and (F) lower and upper sidebands respectively.

The equation proves that when the two frequencies are passed simultaneously through a non-linear resistance, amplitude modulation takes place. The output of modulation circuit can be tuned to the carrier frequency, having sufficient bandwidth so as to allow two sidebands and reject the other frequencies.

2.28 BALANCED MODULATOR (CARRIER SUPPRESSION)

(DSB - SC):

Balance modulator has a very important property of generating amplitude modulation and simultaneously suppressing the carrier. Figure 2.27 a and b shows the balance

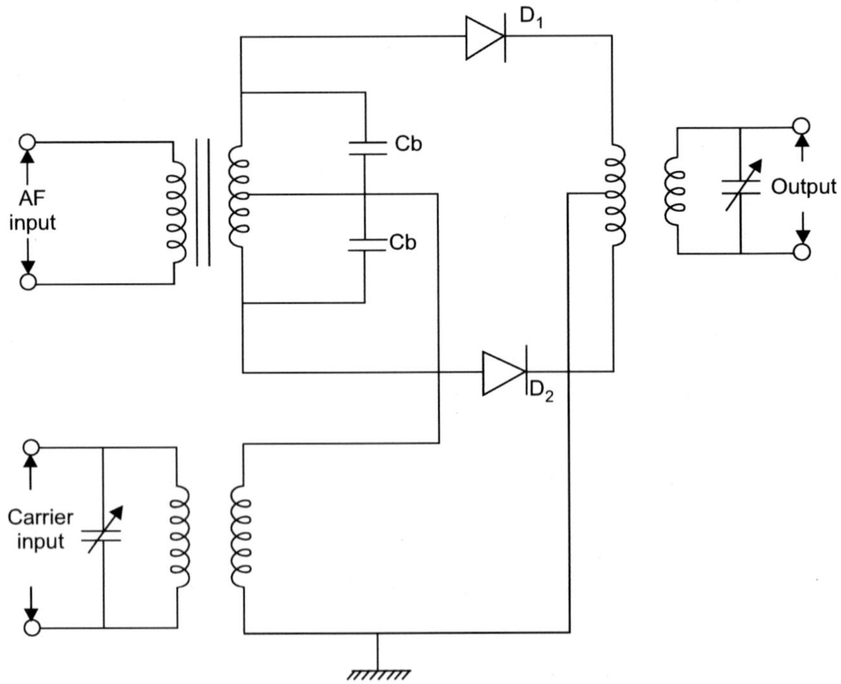

Fig. 2.27(a): Diode Type Balance Modulator

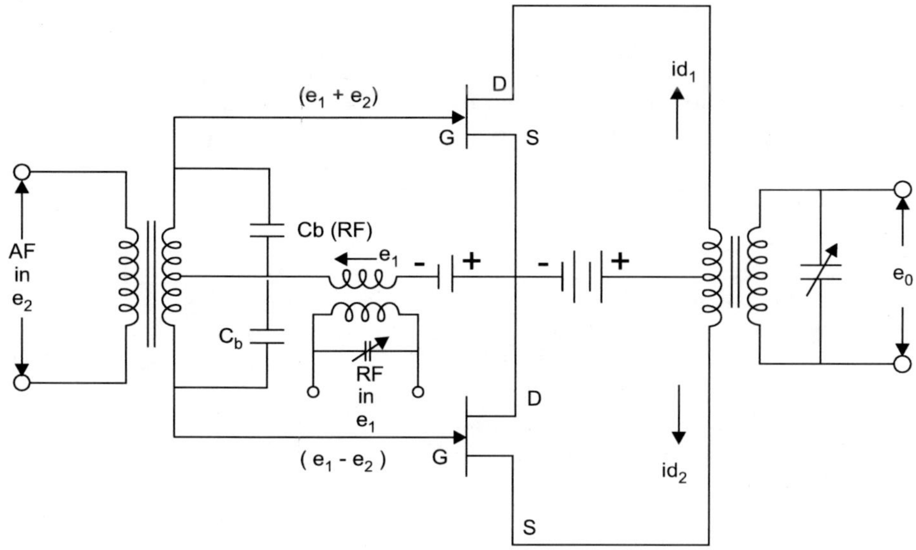

Fig. 2.27(b): FET Type Balance Modulators

modulator circuit using diodes and FETs. These two circuits utilize the principles of non-linear resistance.

The modulation voltage (let it be e_2) is fed in push pull mode and the carrier voltage (let it be e_1) is applied in parallel to a pair of exactly similar diodes (or transistor or FET acting as class - A amplifier).

In case of FET circuit the carrier voltage is applied in phase to the two gates and the modulating voltage is applied 180° out of phase at the gates. This is because these signals are applied at opposite ends of a centre tapped transformer. The modulated output currents of the two FETs are combined in primary of centre tapped push-pull output transformer. These two currents substract in the direction shown. If the system can be made completely symmetrical the carrier frequency can be eliminated. Exact similarity, however, cannot be achieved practically, so that carrier gets heavily suppressed instead of being completely eliminated. The harmonics can be easily eliminated at the output transformer stage by tuning the required frequency. Thus, the output consists of only the two sidebands. Let us make a mathematical analysis of FET type-modulator.

Assuming perfect symmetry, the two drain currents will be given as

$$i_{d_1} = a + b(e_1 + e_2) + c(e_1 + e_2)^2$$

$$= a + be_1 + be_2 + ce_1^2 + ce_2^2 + 2ce_1 e_2 \quad (25)$$

$$i_{d_2} = a + b(e_1 - e_2) + c(e_1 - e_2)^2$$

$$= a + be_1 - be_2 + ce_1^2 + ce_2^2 - 2ce_1 e_2 \quad (26)$$

The total primary current is given by the difference of the two individual currents. Using equations (25) and (26)

$$i_1 = i_{d_1} - i_{d_2}$$

$$= 2be_2 + 4ce_1 e_2 \quad (27)$$

Let modulating voltage is

$$e_2 = E_m \sin \omega_m t \text{ and}$$

carrier voltage

$$e_1 = E_c \sin \omega_c t$$

Thus, equation (27) can be written as

$$i_1 = 2bE_m \sin \omega_m t + 4cE_m E_c \sin \omega_m t \cdot \sin \omega_c t$$

$$i_1 = 2bE_m \sin \omega_m t + 4cE_m \times$$

$$E_c \times \frac{1}{2}[\cos(\omega_c - \omega_m)t - \cos(\omega_c + \omega_m)t]$$

The output voltage e_0 can be expressed as

$$e_0 = \alpha i$$

where, α = proportionality constant.

$$\therefore e_0 = 2 \alpha b E_m \sin \omega_m t + 2 \alpha c E_m E_c [\cos (\omega_c - \omega_m) t$$

$$- \cos (\omega + \omega_m) t]$$

$$= \underset{\text{(I)}}{2 \alpha b E_m \sin \omega_m t} + \underset{\text{(II)}}{2 \alpha c E_m E_c \cos (\omega_c - \omega_m) t}$$

$$\underset{\text{(III)}}{- 2 \alpha c E_m E_c \cos (\omega_c + \omega_m) t} \quad (28)$$

Equation (28) illustrates that under conditions of similarity the carrier is completely cancelled out, leaving behind the modulating frequency (term-I) and the two sidebands (term II and III).

The output transformer is used to tune out the modulating frequency, but if additional circuit is used the modulating voltage can be eliminated prior to the output transformer.

2.29 UNWANTED SIDEBAND SUPPRESSION

Balance modulator suppresses the carrier. To obtain the SSB signal still another frequency, i.e. the unwanted sideband has also to be removed. Three different methods are used. All three circuits have the capability to remove sidebands desired by little change in circuit arrangement.

(1) *Filter Method:*

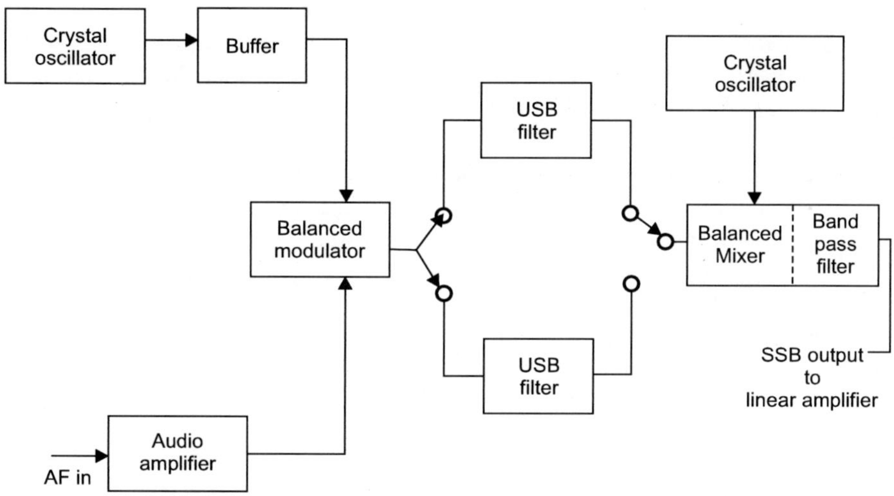

Fig. 2.28(a): Filter Method of Sideband Suppression

This is easiest of the three methods. In this, the unwanted sideband is heavily attenuated using a filter as shown in Fig. (2.28 a). This most important circuits in this block diagram are of the balance modulator and the sideband suppression filter.

Basically this filter should have a flat passband and very high attenuation outside the passband, higher the attenuation better will be the performance. In usual communication circuits the frequency range used for voice is about 300 to 3400 Hz. If it is required to suppress the LSB and if the transmitting frequency be f, then the lowest frequency that the filter should pass without attenuation is f + 300 Hz whereas the highest frequency that must be fully attenuated is f – 300 Hz. Therefore, the filters response must be change from zero attenuation to full attenuation within a range of only 600 Hz. Considering that the carrier frequency is few MHz, this sharp variation in filter characteristic is quite impossible. If lower modulating frequencies such as 50 Hz are employed, as in case of broadcasting, the situation becomes even worse. To achieve the required filter response as suggested above the tuned circuits are designed with very high Q.

As the frequency is increased, the requirement of Q for the tuned circuits also increases, till it becomes practically impossible to further increase the value of Q.

For upper frequency limit of the filters, it has been found that the multistage LC-filters cannot be used efficiently above 100 kHz, as above this frequency, the attenuation in the stop band becomes insufficient. Thus, crystal or mechanical filters are used for this purpose. Mechanical filters can be used upto 500 kHz usually, while crystal filters can go upto 20 MHz. Mechanical filters are the best as they offer best all-round properties example:

Small size, very good attenuation characteristics, good bandpass and sufficient upper frequency limit. The crystal filter, works out to be cheaper, but is preferred only at frequencies in excess of 1 MHz. The transmitting frequencies are still higher than operating frequencies of these filters. Thus, a balanced mixer is used. This mixer is similar to a balanced modulator, except that the sum frequency used here is much greater than the crystal oscillator frequency, than the upper sideband is from the carrier, so that it can be selected with use of tuned circuits.

In balanced mixer, the SSB - signal from the filter is added to crystal oscillator frequency, so as to achieve the desired transmission frequency. This gives an added advantage of changing the transmission frequency by just changing the crystal oscillator frequency to desired value.

The output of mixer is fed to a chain of class-BRF amplifiers. These amplifiers have a linear transfer characteristics because the amplitude of SSB signal is variable and so cannot be fed to a class-C amplifier, which will lead to distortion.

(2) *Phase-shift Method:*

This method uses two balanced modulators. The block diagram is shown in Fig. (2.28b).

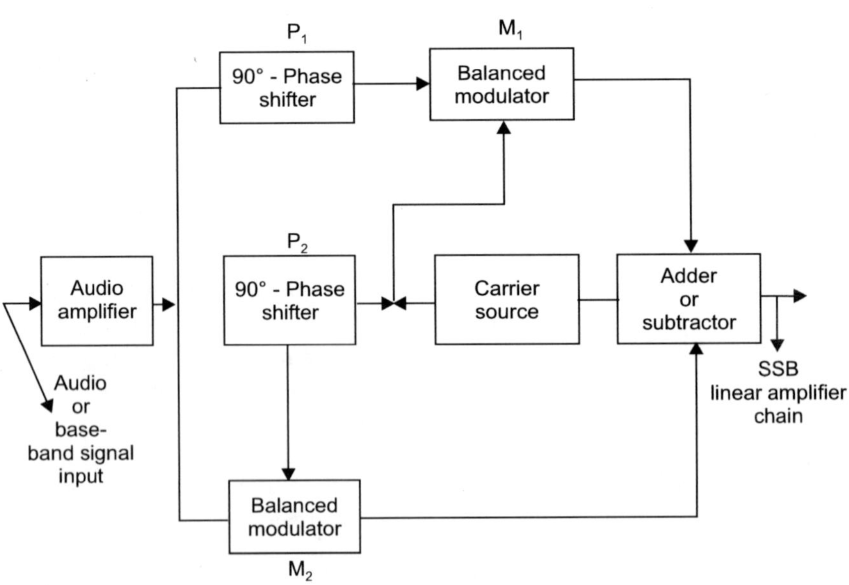

Fig. 2.28(b)

The non-use of filters makes this system free from the usual disadvantages encountered in filter use.

The baseband signal or the audio signal is fed first to an amplifier. The output of this amplifier is divided into the two parts. One is fed to an audio phase shifting network P_1 which gives it a 90° phase shift. The output of P_1 is fed to a balanced modulator M_1 along with the carrier. The second part of audio amplifier is directly fed to a balanced modulator M_2 along with the carrier which is phase shifted by 90° by P_2. Sometimes, a different audio frequency phase shift arrangement is done. One part is given a +45° phase shift and other is given –45° shift. However, the final result is the same in both cases. The output of both balanced modulators consists of two sidebands each, but whereas both USB's lead the input carrier voltage by 90° one of the USB leads the reference voltage by 90°, and other lags it by the same amount. Thus, the two LSB - signals when fed to an adder cancel each other.

The USB being in phase add in the adder, giving pure SSB. If the LSB is desired, we use subtractor instead of adder circuit.

Mathematical Analysis: We observe that the two balanced modulators are perfectly balanced with respect to each other. The output amplitudes of these modulators don't pose any problems as they do not affect the result, since both modulators are fed from same signal sources.

Let $\sin \omega_c t$ be the carrier and $\sin \omega_m t$ be the modulating signal. From the block diagram it can be seen that balanced modulator M_1 will receive $\sin (\omega_m t + 90°)$ and $\sin \omega_m t$ whereas M_2 mill get $\sin \omega_m t$ and $\sin (\omega_c t + 90°)$. The output of balanced modulator M_1 can be represented by e_1 and that of M_2 by e_2. Thus,

$$e_1 = \cos [\omega_c t - (\omega_m t + 90°)] - \cos \{\omega_c t + (\omega_m t + 90°)],$$

$$= \cos (\omega_c t - \omega_m t - 90°) - \cos (\omega_c t + \omega_m t + 90°)$$

$$\text{(LSB)} \qquad\qquad\qquad \text{(USB)}$$

Similarly,

$$e_2 = \cos [(\omega_c t + 90°) - \omega_m t] - \cos [(\omega_c t + 90°) + \omega_m t]$$

$$= \cos (\omega_c t - \omega_m t + 90°) - \cos (\omega_c t + \omega_m t + 90°)$$

$$\text{(LSB)} \qquad\qquad\qquad \text{(USB)}$$

Therefore, the adder output will be

$$e_0 = e_1 + e_2 = 2 \cos (\omega_c t + \omega_m t + 90°)$$

If subtractor is used instead of adder then

$$e_0 = 2 \cos (\omega_c t - \omega_m t + 90°)$$

Thus, it proved that SSB signal can be generated using the phase shift method.

(3) *Weaver Method:* This method is bit complicated and so its use is rejected for ordinary commercial systems. This method retains the advantage of using the

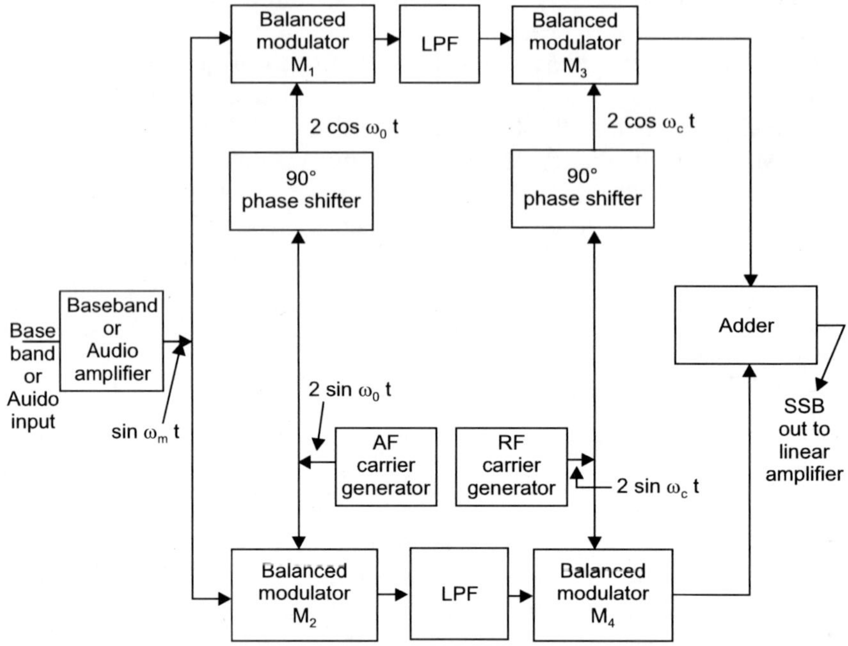

Fig. 2.28(c)

phase shift method without the inherent disadvantage of an AF phase shift network which is required to operate over a large range of audio frequencies. This method also has the ability to generate SSB at any frequency. Thus, this method falls in direct competition with the filter method.

Figure 2.28c shows block diagram of Weaver's method of SSB generator.

It looks quite complicated as compared to other two systems. Here, four balanced modulators are used. Instead of phase shifting the whole range of audio frequencies this method combines them with an AF carrier f_0, which is fixed audio frequency lying in the middle of audio frequency band used. A phase shift is then applied to this fixed frequency only. The resulting voltages at the output of the balanced modulators M_1 and M_2 are applied to low-pass filters whose cut off frequency is designed to be f_0 to ensure that the input to the last stage of balanced modulators, i.e. M_3 and M_4 result in a proper sideband suppression. All LSB signals will be cancelled regardless of whether the audio frequency lies above or below f_0. If a LSB is required at the final output the phase of the carrier voltage being fed to M_1 should be changed by 180°.

1. Pilot Carrier SSB Transmission System

In the pilot carrier SSB system a low level carrier signal is transmitted with signal sideband, in its proper place in the spectrum. This pilot carrier is used at the receiver to synchronise the local oscillator used for demodulator to the original carrier, thus greatly improving the demodulator operation.

Figure 2.29(a) shows a SSB pilot carrier radio transmitter, Fig. 2.29(b) shows the frequency spectra of signals at various points within the system. The audio signal (may be a telephone channel) in the frequency range 0–4 kHz (A) is fed to the balanced modulator to create upper and lower sidebands around 100 kHz carrier position (B)

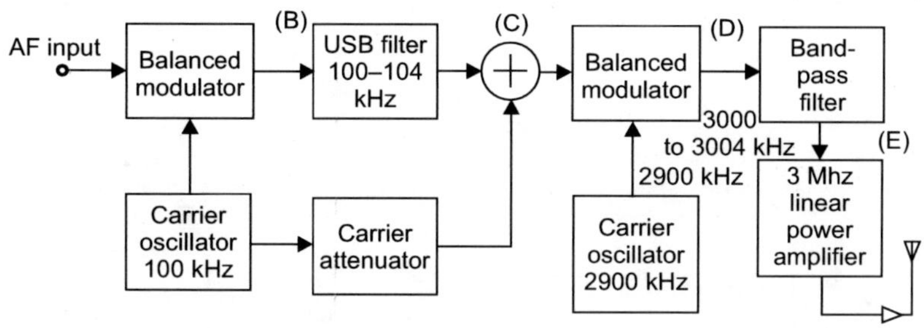

Fig. 2.29(a)

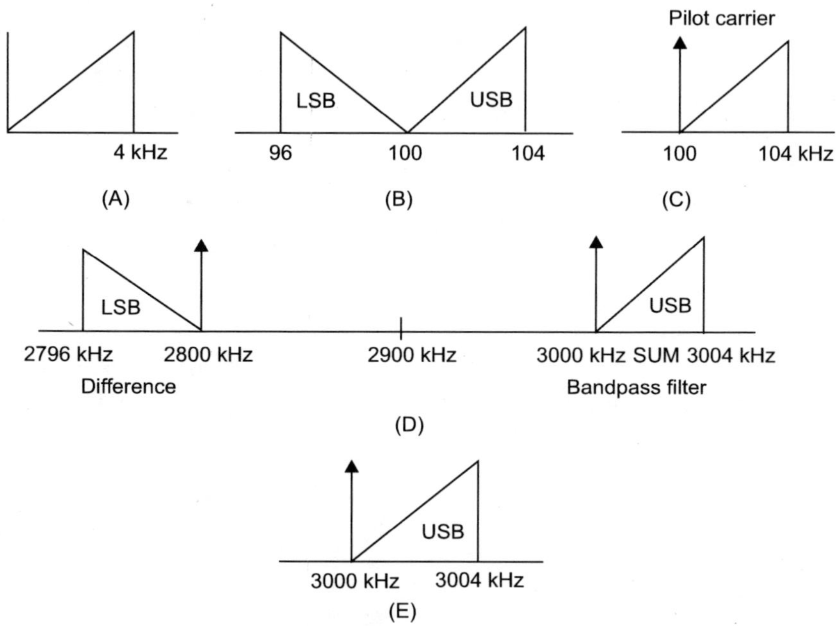

Fig. 2.29(b): Signal Spectra

An upper sideband filter passes the upper sideband between 100 and 104 kHz, to which the attenuated carrier is added to produce the signal (C). The reinserted carrier level is adjusted through the attenuator. This signal is now modulated on 2900 kHz carrier by a second balanced modulator which acts as mixer to produce an upper sideband of 3000 to 3004 kHz and lower sideband between 2800 and 2796 kHz (Signal (D)). A bandpass filter passes the upper sideband and rejects the lower sideband to produce the signal (E) which is then amplified and transmitted.

Independent Side Band System (I.S.B)

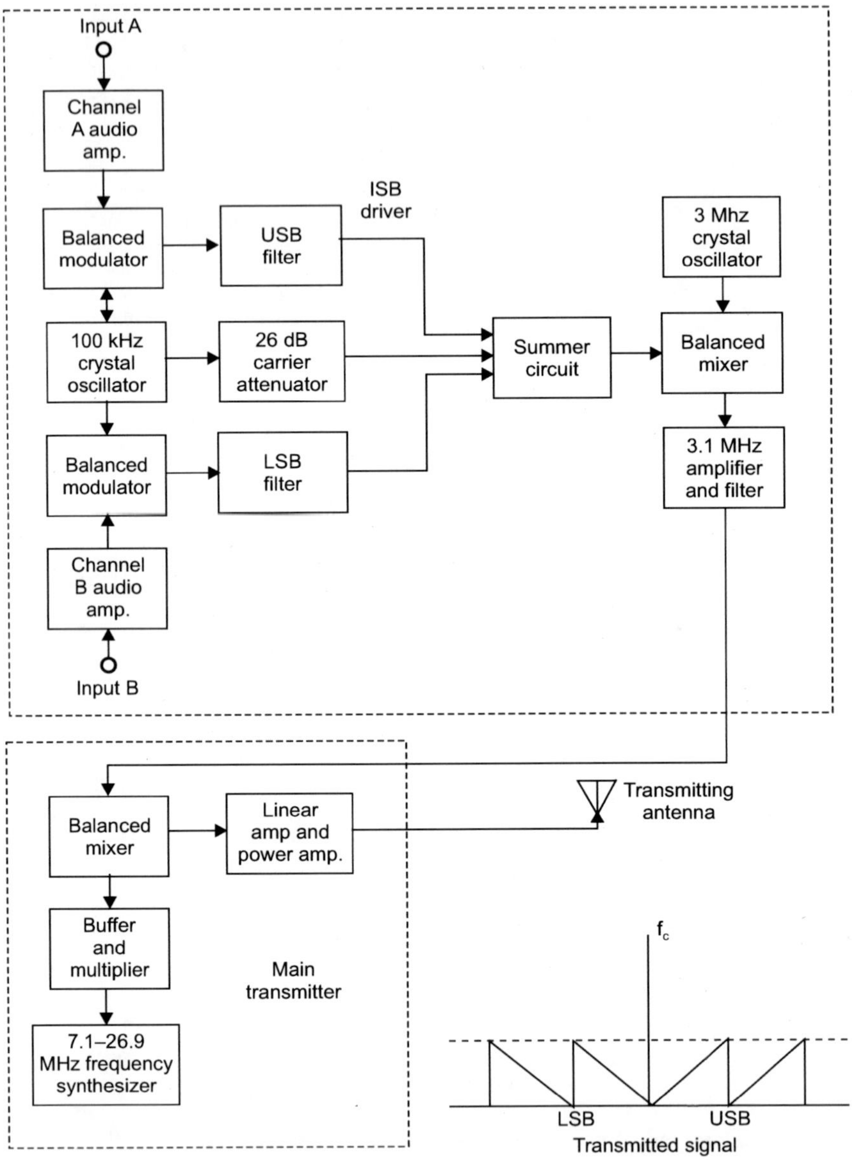

Fig. 2.30

Independent sideband transmission is used for:

1. High density point-to-point communication.
2. Medium density traffic control.
3. Ship to shore point-to-point communication.
4. ISB is also used in telephony and telegraphy.

1. Inputs: Here two inputs are applied, say input A and input B. These inputs are decided by international telecommunication union [I.T.U.]. According to I.T.U. applied inputs is of R3E modulation type. R3E modulation type means single side-band reduced carrier type. We know that receiver is tuned at different frequencies, for becoming tuning possible at different frequencies. For making demodulation easy single sideband reduced carrier type modulation is used. These inputs are given independently to channel A and channel B audio amplifiers respectively. These inputs are transmitted simultaneously since inputs are independent one sideband is used for telephony and other sideband is used for telegraphy at receiving end.

2. Audio channel A and channel B amp: These amplifiers are used for amplifying or raising voltage levels of SSB reduced carrier type input signals.

3. Balanced modulator: The 6 kHz channel A is fed to one balanced modulator; while another 6 kHz channel B is fed to another balanced modulator. 100 kHz crystal oscillator output is given to both balanced modulator. Carrier is attenuated by 45 dB or more by balanced modulator.

4. USB and LSB filter: USB filter will reject LSB and it will pass USB. On the other hand, LSB filter will reject USB and it will pass LSB.

5. Summer circuit: Outputs of USB filter and LSB filter and 26 dB attenuator are given to inputs of summer circuit. Output of summer circuit will be low frequency ISB, with pilot carrier present.

6. Crystal oscillator and Balanced mixer: 1 MHz crystal oscillator and balanced mixer will raise frequency level of signal from 100 kHz to 3.1 MHz. Balanced mixer will provide easier removal of unwanted frequencies with the help of output filter.

The signal is then given to main transmitter from ISB drive unit.

7. Frequency synthesizer, Buffer amplifier and multiplier and Balanced mixer: With the help of above 7.1 to 26.9 MHz frequency synthesizer, balanced mixer frequency of signal is again raised. This is done because this transmitter is used for HF band from 3 to 30 MHz.

8. Linear amplifier and power amplifier: Linear amplifier will raise voltage level of signal and power amplifier will raise power level of signal. Typical power level is generally 10 kW and 50 kW. This is fed to direction antenna for transmission.

Since width of channel is 6 kHz, it can carry two circuits of 3 kHz and hence four conversation is possible simultaneously.

2.30 DEMODULATION OF AM

Diode-detector: A diode is the most widely used device for demodulation. Figure (2.31) shows a simple circuit using a diode for detection. It has a parallel RC

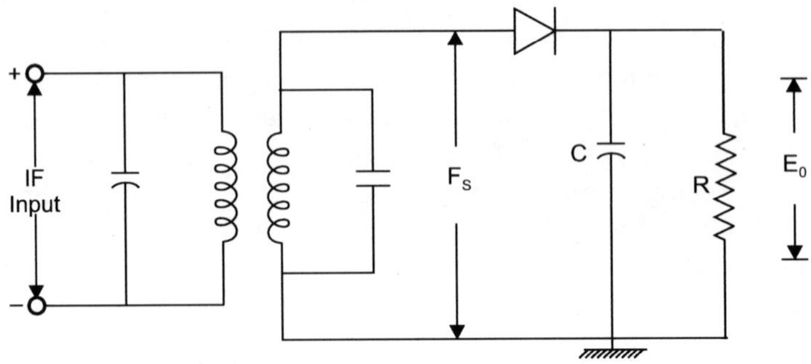

Fig. 2.31: Simple Diode Detector Circuit

combination, where C is small and R is large. This RC combination forms a load across which rectified output voltage E_0 is developed.

At each positive peak of RF-cycle, C charges up to a potential almost equal to the peak signal voltage E_s. The difference is due to the diode drop since the forward resistance of diode is very small. In between peaks, a little of the charge of C decays through R, to be recharged at the next positive peak. This causes a voltage E_0 across the RC-parallel combination which is nearly at exact reproduction of the modulating voltage, except for some RF ripple. The time constant of RF ripple small, but should be fast enough for the detector circuit to follow the fastest modulation variations. The simple diode detector suffers from disadvantages, e.g. dc component also is present in the output voltage E_0, besides the RF-ripple. The dc components represent the average envelope amplitude (i.e., carrier strength). A modified diode detector shown in Fig. 2.32 eliminates these defects to a great extent.

A practical diode detector has a number of resistances and capacitances shown. An important thing is a diode is reversed here, so that now negative envelope is demodulated. This is to derive negative AGC voltage to be applied to IF stage. The

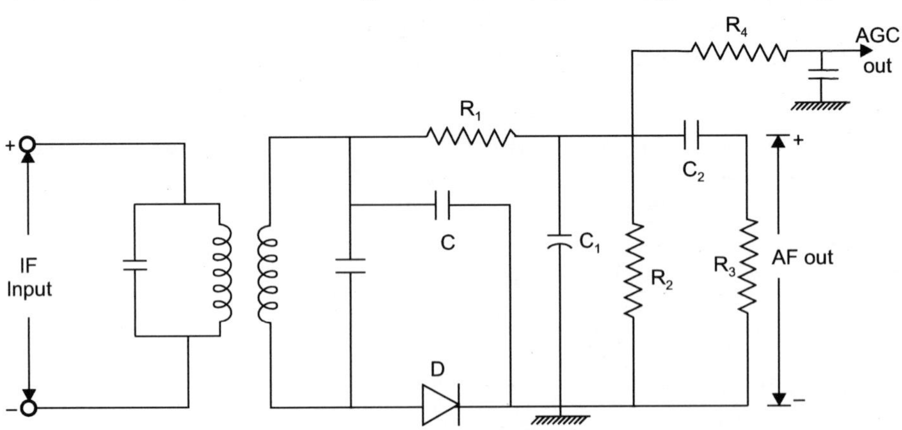

Fig. 2.32: A Practical Diode Detector

load resistance R of Fig. 2.31 is split here into two parts (R_1 and R_2) to obtain a series dc path to ground for the diode. A low-pass filter R_1 C_1 is added to remove any RF ripple still represent. C_2 is used to prevent any diode dc output reaching the volume control resistance (R_3). The R_4C_3 combination works as low-pass filter for removing AF components, thus providing dc voltage whose amplitude is proportional to the carrier strength which can be very easily used for AGC.

Distortion in diode detectors: Basically, there are two types of distortions encountered in diode detectors. The first one is caused due to unequal ac and dc diode load impedances, while the second results from the fact that the ac load impedance acquires a reactive components at the highest audio frequencies.

Figure 2.33 shows diode currents in two possible cases.

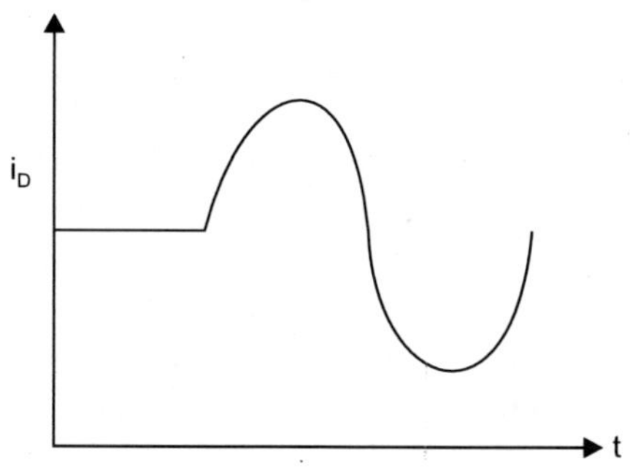

Fig. 2.33(a): Small Transmitted Modulation Index, No Clipping

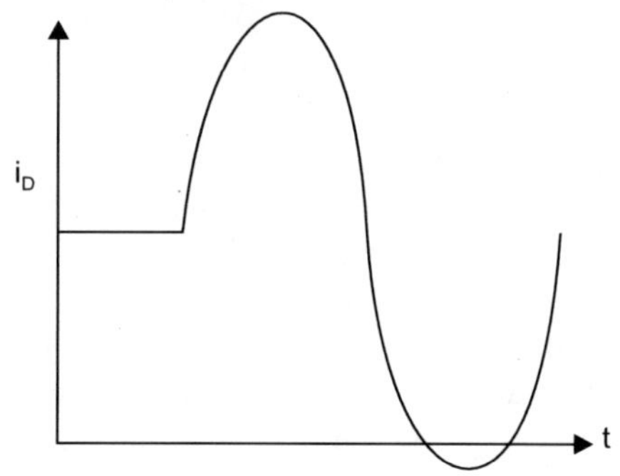

Fig. 2.33(b): Larger Transmitted Modulation Index, Negative Feed Clipping

The modulation index of a demodulated wave is defined as

$$m_d = \frac{I_m}{I_c}$$

Diode being a current operated device, m_d is defined in terms of currents. It has to be kept in mind that the current vlaues are peak values.

$$I_m = \frac{E_m}{Z_m}$$

and
$$I_c = \frac{E_c}{R_c}$$

where, R_c = dc diode load resistance
and Z_m = audio diode load impedance.

As Z_m is smaller than R_c, the AF current I_m will be larger in proportion to the d.c. current, than it would have been if both load resistance had been exactly the same. In other words, the modulation index in the demodulated wave is higher than the modulation index of the signal applied at the input of the diode detector. Thus, if we transmit a signal with 100% modulation there is every possibility of over modulation to exist at the output of the detector.

The modulation index in the demodulated wave will be

$$m_d = \frac{I_m}{I_c} = \frac{E_m / Z_m}{E_c / R_c} = m\frac{R_C}{Z_m}$$

Since the maximum allowable value of m_d is unity, the maximum value of m will be

$$m_{max} = m_{dmax}\frac{Z_m}{R_c} = \frac{Z_m}{R_c}$$

(since $m_{dmax} = 1$)

In AM broadcasting system modulation index very unlikely crosses 70%. The output of volume control resistance R_3 is usually connected to the base of the audio amplifier transistor. If the input impedance of this transistor is rather low, it will load the detector reducing the diode audio load impedance. To overcome this problem, the first stage, of an audio amplifier should have a FET instead of a bipolar transistor or a series resistance can be added in between the volume control resistance R_3 and the base of audio amplifier transistor.

But the latter solution reduces the volume fed to the audio amplifier, transistor thus reducing the output.

Diagonal clipping is the other form of trouble that may arise with diode detectors. At higher frequencies, the assumption that Z_m is purely resistive may not work resulting in a reactive component of Z_m due to C and C_1. At high modulation depths, current will be changing so quickly that the time constant of load may be too slow to follow the change. This results in a exponential decay of current instead of following the waveform as shown in Fig. 2.34a.

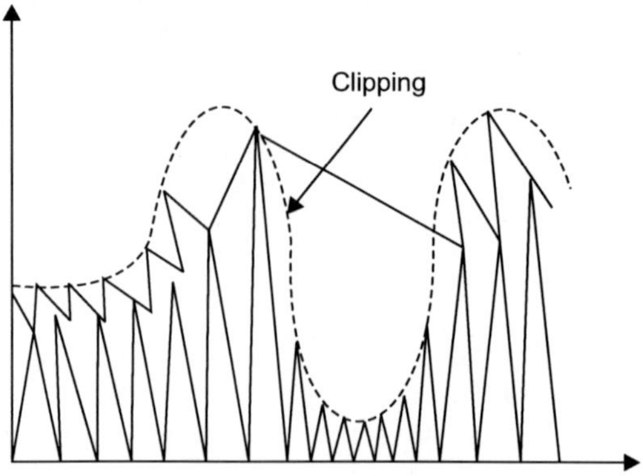

Fig. 2.34(a): Diagonal Clipping

Diagonal clipping does not normally occur, when modulation depth is below 60%. Thus, it is possible to design a diode detector that is free from this type of distortion. The RC time constant for diode detector has to be properly designed.

2.31 DEMODULATION OF FM

Basically, FM demodulator consists of a FM to AM converter. This conversion has to be done very efficiently and linearly. The detection circuit should be insensitive to any amplitude changes and should not be too critical in its adjustment and operation. A FM demodulation converts the frequency modulated IF of constant amplitude into a voltage which is both frequency and amplitude modulated. This voltage, is then applied to a detector system which detects the amplitude variations ignoring the frequency changes. Thus, a circuit has to be designed whose output voltage amplitude depends on the frequency deviation of the input voltage.

Slope-detector:

Consider a tuned circuit fed by a frequency modulated signal, the tuned circuit being resonant at the centre frequency of FM-signal. The amplitude of the output of the tuned circuit is found to vary in accordance with the frequency deviation of input signal.

Figure 2.34(b) shows that the circuit is detuned by an amount δf, to bring the carrier centre frequency to point 1 on the selectivity curve. The other side of the slope with point '1' would works equally well. The frequency variation at input produces an output voltage proportional to the frequency deviation of the carrier.

The output of this tuned circuit is applied to a diode detector with parallel combination of RC as load. The circuit is in fact similar to an ordinary AM detector

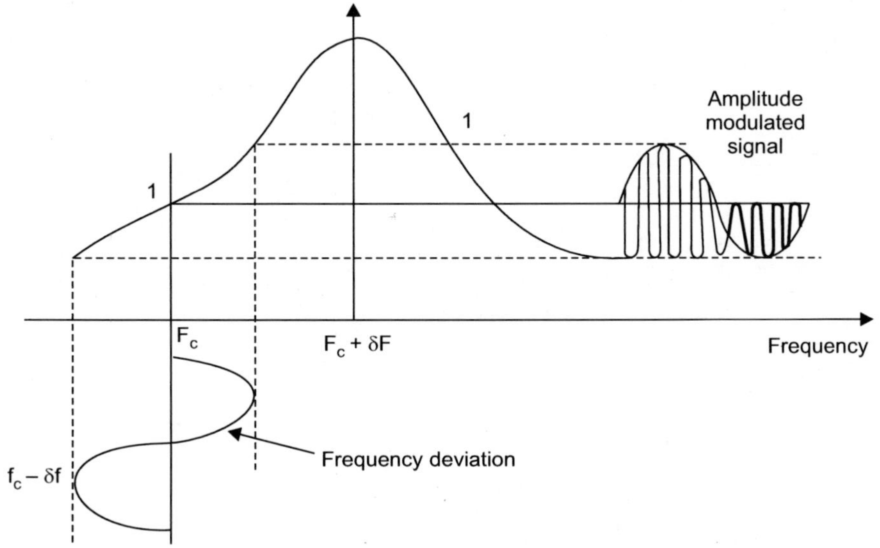

Fig. 2.34(b): Simple Slope Detector Characteristics Curve

except that the secondary winding of the IF transformer is detuned a little. In case of emergency it is thus possible to receive FM on an AM receiver by just detuning the IF transformer secondary. However, a simple slope detector discussed above has some discrepancies such as it is insufficient and is highly non-linear. The linear region is very small. Also, it is difficult to adjust the primary and secondary of a transformer to different frequencies. Simple diode detector however lays down the basis of operation of modified version called "Balanced Slope detector".

2.32 BALANCED SLOPE DETECTOR

A balanced slope detector uses two slope detectors connected back to back as shown in Fig. 2.35.

Each slope detector is connected to one end each of the centre-tapped secondary of the IF-transformer as shown. Thus, the two detectors are fed 180° out of phase. The usual FM receivers have a maximum frequency deviation of 75 kHz. The top secondary of the transformer is tuned above the IF by about 100 kHz, i.e. somewhat above 75 kHz. Similarly, the bottom circuit is tuned below IF by same amount. Each of the tuned circuit is connected through a diode to an RC-load. The RC loads for two diodes are connected as shown and the detected output is taken from across the series combination of these loads. This gives a sum of the two individual outputs.

Consider f_c to be the IF to which the primary of the transformer is tuned. The two secondary circuits are therefore tuned to $(f_c + \delta f)$ and $f_c - \delta f)$.

Let T_1 and T_2 be the upper and lower tuned circuits for reference. When the input frequency is f_c, a little voltage will be developed across T_1 because it is tuned little above f_c. This voltage will be available at the RC-load of diode D_1. Similarly, a voltage is developed across T_2 since it is tuned a little below f_c. This voltage is

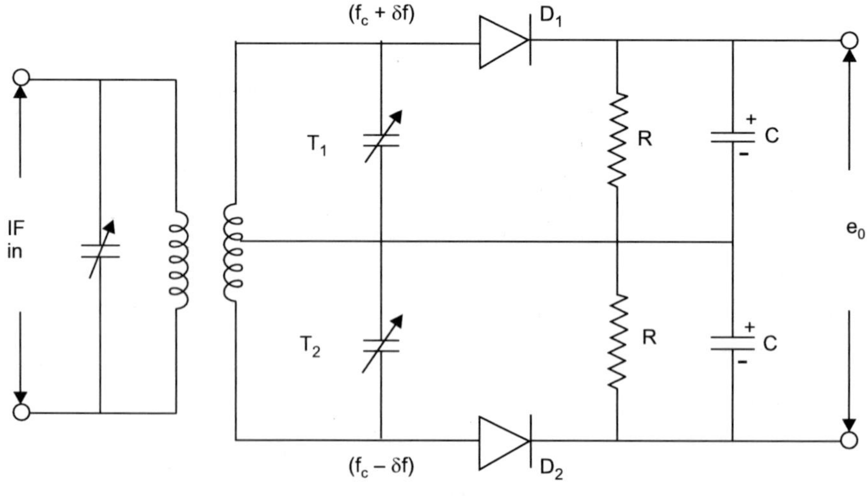

Fig. 2.35: Balanced Slope Detector

available at output of diode D_2 across the RC-load. If the two slope detectors are assume to be perfectly identical, the two voltages developed at the output will completely cancel each other as they are of opposite sign. The net output of the detector will thus be zero.

Consider now the case when the instantaneous frequency becomes ($f_c + \delta f$). As T_1 is tuned to this frequency, the output of D_1 will be quite large positive voltage, whereas the output from D_2 will be negligible as ($f_c + \delta f$) is away by 2 δf lower tuned circuit. Similarly, when the instantaneous frequency equals the other extreme, i.e. $f_c - \delta f$, the output from diode D_2 is very high negative voltage, whereas D_1 will give negligible output. Thus, in first case, the overall voltage will be maximum and positive, whereas in second case, it will be maximum and negative when instantaneous frequency lies between these two extremes, the output voltage will have some intermediate value between the two extremes. This output voltage will be positive or negative depending on the frequency, i.e. whether it lies above or below f_c. The required characteristic curve of this balanced slope detector is shown in Fig. 2.36.

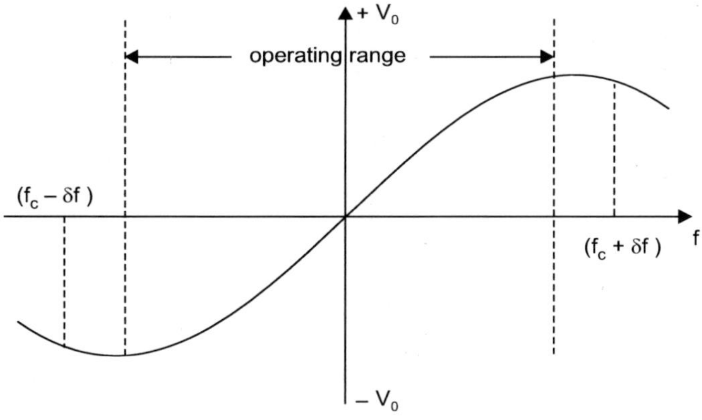

Fig. 2.36: Balanced Slope Detector Characteristics

A balanced slope detector is improvement over a simple slope detector, but still it has problems. Adjusting three frequencies is very difficult. Moreover, amplitude limiting has to be provided separately and linearly although it is better than simple slope detector but still is not good enough.

2.33 DISCRIMINATOR

It is also sometimes called a phase discriminator or Foster-Seeley discriminator or centre tuned discriminator. A phase discriminator solves the problem of tuning three resonant circuits. Besides solving this problem, a phase discriminator is still able to achieve the S-curve characteristics. The circuit of a phase discriminator is shown in Fig. 2.37. It has both its primary and secondary windings of the

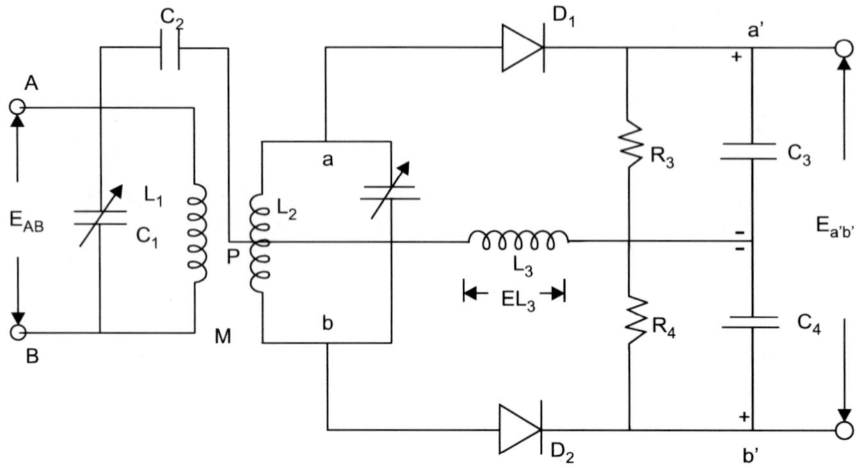

Fig. 2.37: Phase Discriminator Circuit

transformer tuned to the centre frequency. This simplifies to great extent the aligning of the circuit and also gives a better linearity. Circuit of a phase discriminator seems to be quite identical to that of a balanced slope detector, except for minor changes. The changes made in the circuit is to ensure that the voltages fed to be diodes vary linearly with the deviation in frequency of the input signal.

A mathematical analysis is now being given to show that the voltage at each diode is sum of the primary voltage and corresponding half secondary voltage. The following relations between primary and secondary voltages will also be proved.

(1) Primary and secondary voltages are exactly 90° out of phase at an input frequency f_c.

(2) This phase difference is less than 90°, if the input frequency is higher than f_c.

(3) Phase difference is more than 90°, if the input frequency is less than f_c.

In this case, with change in input frequency, there is only a phase change in the voltage fed to the diodes. The individual voltage components at the two diode inputs remain same at all frequencies, but their vector sums differ with the change in phase between the primary and secondary windings. Thus, the two individual voltages will be equal only at f_c and at all other frequencies the output of one diode will be greater than that of the other. Which diode has a large output will depend on whether input frequency is above or below f_c. The final output will be positive or negative depending on the input frequency. Thus, the magnitude of the output will depend upon the input frequency deviation, which is the desired objective.

The load resistance are made much more than the reactance offered by the load capacitors. Figure 2.38 shows a circuit depicting the primary windings and components camping across them.

From Fig. 2.38, it can be seen that C_2, L_3 and C_4 come effectively across primary of the circuit.

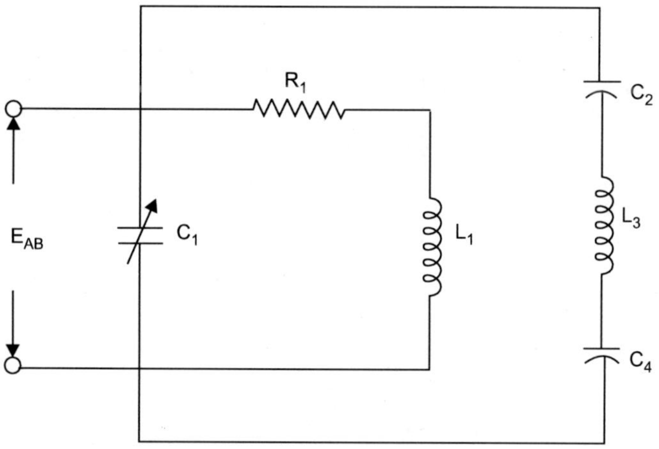

Fig. 2.38: Discriminator Primary Circuit

Thus, the voltage across L_3 will be given by

$$E_L = \frac{E_{AB} Z_{L3}}{Z_{C_2} + Z_{C_4} + Z_{L_3}} \tag{29}$$

or

$$E_L = E_{AB} \frac{j\omega L_3}{j\omega L_3 - j\left(\frac{1}{\omega C_2} + \frac{1}{\omega_{C_4}}\right)} \tag{30}$$

As L_3 is an RF choke, it has a large reactance as compared to sum of the reactances of C_2 and C_4. Thus, reactances due to C_2 and C_4 can be neglected and the equation reduces to

$$E_L = E_{AB} \tag{31}$$

This shows that the voltage across the RFC equals the applied primary voltage. The mutually coupled, double tuned circuit has high primary and secondary Q and a low mutual inductance. Thus, while calculating the primary current, we can without any difficulty neglect the impedance transferred from secondary to primary and the primary resistance.

Then I_p is given by

$$I_p = \frac{E_{AB}}{j\omega L} \tag{32}$$

The transformer operation theory states that a voltage is generated in series in secondary winding, when current flows in the primary and voltage so generated is given by

$$E_s = \pm j\, \omega\, M\, I_p \tag{33}$$

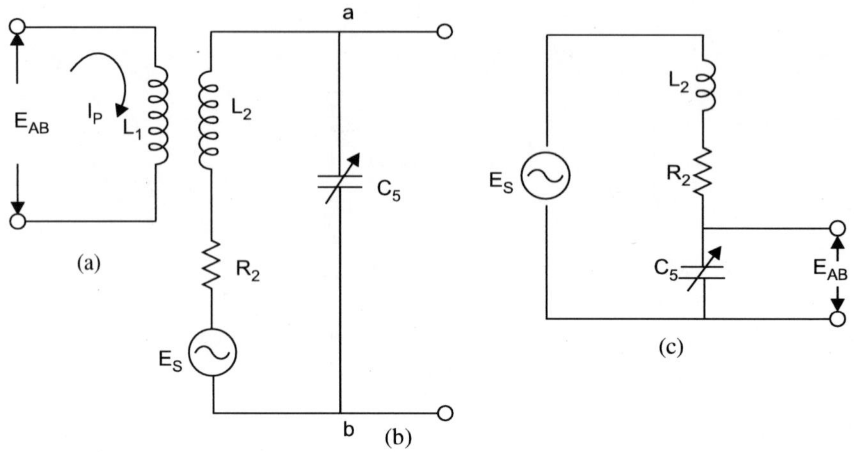

Fig. 2.39: Discriminator Secondary Circuit

Negative sign of voltage depends upon the direction of winding.

Usually, this sign is negative. An equivalent secondary circuit is shown in Fig. 2.39.

We have,
$$E_s = -j\omega M I_p = -j\omega M \frac{E_{AB}}{j\omega L_1}$$

$$E_s = -\frac{M}{L_1} E_{AB} \tag{34}$$

The voltage across the secondary winding can now be calculated from Fig. 2.39c.

$$E_{ab} = E_s \frac{Z_{C_5}}{Z_{C_5} + Z_{L_2} + R_2}$$

$$E_{AB} = \frac{jX_{C_5}(-E_{AB}M/L_1)}{R_2 + j(X_{L_2} - X_{C_5})}$$

$$E_{AB} = \frac{jM}{L_1}\frac{E_{AB}X_{C_5}}{R_2 + jX_2} \tag{35}$$

(Since substituting X_2 for $X_{L_2} - X_{C_5}$)

X_2 can have any value and is dependent on frequency.

The voltages applied to two diodes D_1 and D_2 will now be calculated. These diodes are denoted as E_{ak} and E_{bk}.

$$E_{ak} = E_{ap} + E_L = \frac{1}{2}E_{ab} + E_{AB} \tag{36}$$

$$E_{bk} = E_{bp} + E_L = -E_{ap} + E_L$$

$$E_{bk} = -\frac{1}{2}E_{ab} + E_{AB} \tag{37}$$

It can be seen that the voltage applied to each diode is the sum of the primary voltage and corresponding half secondary voltage. The dc voltage conditions cannot be evaluated as d.c. diode drop is not known. However, it is known that each will be proportional to the peak value of the RF voltage applied to the respective diode. Therefore,

$$E_{a'b'} = E_{a'k} - E_{b'k}$$

or

$$E_{a'b'} \; \alpha \; E_{ak} - E_{bk} \tag{38}$$

At input frequency f_{in} equal to f_c, X_2 becomes zero, so that eqn. (35) reduces to

$$E_{ab} = \frac{jM}{L_1}\frac{E_{AB}X_{C_5}}{R_2}$$

$$E_{ab} = \frac{M}{L_1}\frac{E_{AB}X_{C_5}}{R_2}\angle 90° \tag{39}$$

Equation (39) proves that the secondary voltage E_{ab} leads the applied primary voltage E_{AB} by 90°C.

Therefore, $-E_{ab}$ will lag E_{AB} by 90°. It now becomes possible to add the diode input voltages vectorially as shown in Fig. 2.40.

At this stage, i.e. when $f_{in} = f_c$ the two voltages E_{ak} and E_{bk} will be equal so that the discriminator output will be zero. Thus, at the carrier frequency or in other words an unmodulated carrier frequency produces a zero voltage at the discriminator output. But it will be seen that at any other frequency there will be an output.

If f_{in} is greater than f_c then equation (35), X_{L_2} becomes greater than X_{C_5} so that X_2 becomes positive. Therefore, equation (35) takes the form

$$E_{ab} = \frac{jM}{L_1} \times \frac{E_{AB}X_{C_5}}{R_2 + jX_2}$$

$$E_{ab} = \frac{ME_{AB}X_{C_5}}{L_1 |Z_2|}\frac{\angle 90°}{\angle 0°} \tag{40}$$

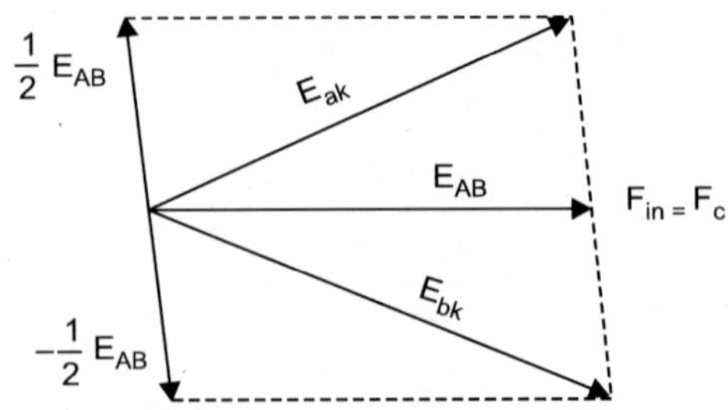

Fig. 2.40

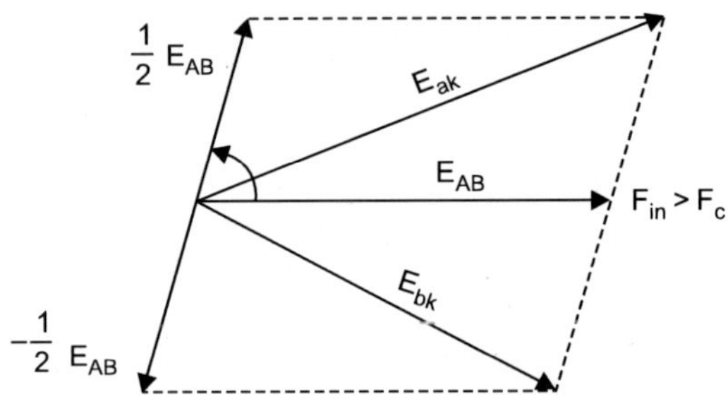

Fig. 2.41

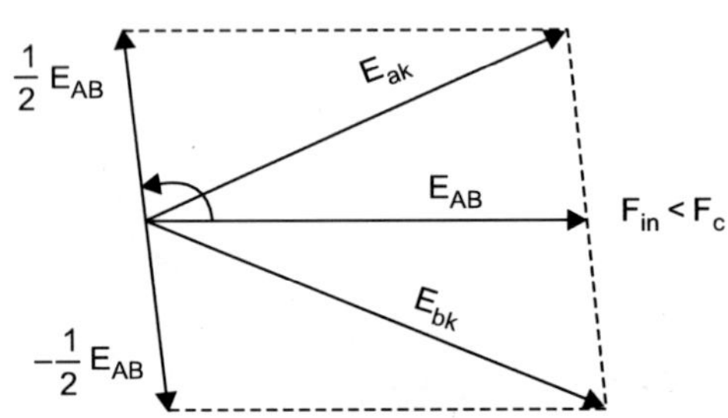

Fig. 2.42

Equation (40) shows that E_{ab} leads E_{AB} by less than 90° and therefore $-E_{ab}$ will lag E_{AB} by more than 90°. Figure 2.41 shows this case. Here, E_{ak} is greater than E_{bk} showing that discriminator output will be positive when f_{in} is greater than f_c. If input frequency is less than f_c, X_2 becomes negative and the angle of impedance Z_2 will also become negative. This causes E_{ab} to lead E_{AB} by more than 90°. The final output of discriminator is found to be negative in this case.

A phase discriminator has only tuned circuits to be tuned to a single frequency, so that its alignment is much easier as compared to balanced slope detector. The linearity is better because the circuit uses less of frequency response and more of primary secondary phase relations which are more linear. A phase discriminator, however, provides no amplitude limiting which of course is a disadvantage with it.

2.34 RATIO DETECTOR

Ratio detector is another type of FM demodulator. Besides demodulating the FM signal, it also provides amplitude limiting. A phase discriminator needs prior limiting, while in ratio detector, prior limiting is avoided.

Figure 2.43 shows circuit diagram of ratio detector.

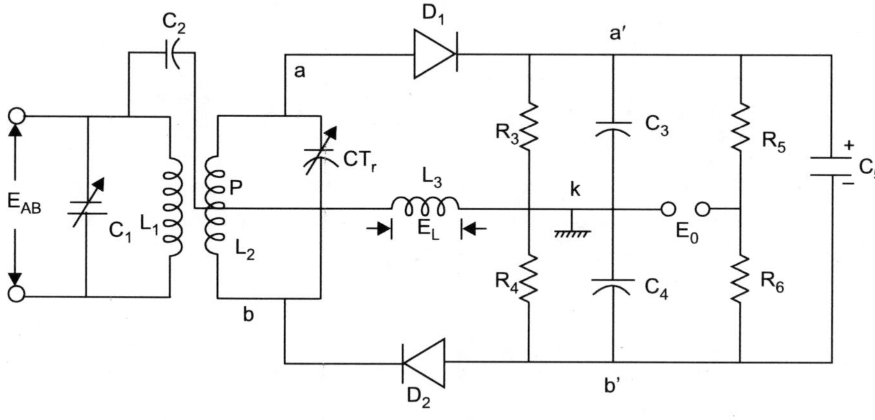

Fig. 2.43: Circuit Diagram of Radio Detector

In Figs. 2.40, 2.41, 2.42 it will be seen that the sum $E_{ak} + E_{bk}$ remains constant, while the difference varies in accordance with the frequency change. Practically the situation is different as there is always some variation in sum voltage also. This does not cause much distortion in a ratio detector but of course little distortion cannot be ruled out. It means that any variations in the sum voltage are spurious and need suppression. For this a discriminator has to be designed whose output remains independent of the input signal amplitude.

To keep the sum voltage constant, some major changes are made in the phase discriminator described before A modified circuit called ratio detector is shown in Fig. 2.43. Three important changes are made to obtain the circuit from a phase discriminator circuit.

(1) A large electrolytic capacitor C_5 is put across the output.

(2) Diode D_2 is reversed.

(3) The points for taking outputs are changed.

With the reversal of diode D_2, point k now becomes positive with respect to point b′ so that the voltage $E_{a'b'}$ now becomes a sum voltage instead of difference voltage of the output of the two diodes. A large capacitor C_5 is put across points a′ and b′ to keep this sum voltage constant. With addition of C_5, the output voltage can no longer be taken from across points a′ and b′, as this voltage will now remain constant. The output voltage now is taken from across the points k and k′. Point k is grounded as it is more convenient of the two for grounding. Assuming $R_5 = R_6$, E_0 can be evaluated as follows:

$$E_0 = E_{b'k'} - E_{b'k}$$

$$= \frac{E_{a'b'}}{2} - E_{b'k}$$

$$= \frac{E_{a'b'} + E_{b'k'}}{2} - E_{b'k}$$

$$E_0 = \frac{E_{a'k} - E_{b'k}}{2}$$

This equation shows that output voltage of ratio detector is proportional to the difference between the individual output voltages of the two diodes, as was the case with phase discriminator. Thus, a ratio detector behaves identically to a phase discriminator for input frequency changes.

Ratio-Detector Provides Amplitude Limiting:
Capacitor C_5 helps to achieve amplitudes limiting in a ratio detector. If the input voltage E_{AB} remains constant, for quite sometime, capacitor C_5 charges upto the potential existing between the points a′ and b′. Since this is dc voltage if E_{AB} is constant, no current flows in the capacitor to change it or flows outside to discharge it. It means that the input impedance of C_5 is infinite.

The total load impedance for the two diode is therefore the sum of resistances R_3 and R_4 as R_5 and R_6 are large. If E_{AB} tries to increase, C_5 tends to oppose any rise in E_0. This operation not only depends on the large time constant, but there is another reason for it. As the inpout voltage rises, excess diode current flows in the capacitor C_5, charging it. The voltge $E_{a'b'}$ remains constant initially, because it is not possible from the voltage across the capacitor to charge instantaneously. Thus, the situation now is that although the diode load current has increased, yet the voltage across the load remains unaltered, meaning that the load impedance has decreased. Due to this, the secondary of the ratio detector gets more heavily damped, thus lowering the Q and so the gain of the amplifier driving the ratio detector. This reduction in gain suitably concentrates the initial increase in the input voltage.

Consider now the situation when the input voltage falls. The diode current falls, but load voltage still remains constant, initially because of capacitor C_5. The effect is that of an increased diode load impedance. This results in reduction in damping and increase in the driver amplifier gain, thus counteracting the falls in the input voltage.

2.35 DEMODULATION OF SSB

Product Demodulator:
SSB demodulation process is quite different from that of ordinary AM demodulation and the circuitry used for SSB is also quite different. The basic demodulator circuit

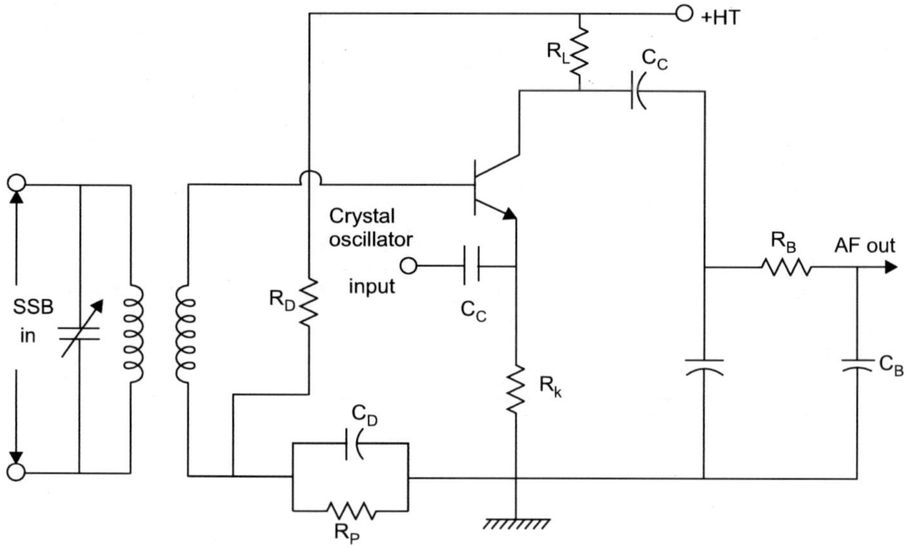

Fig. 2.44: Circuit of Product Demodulator

used for SSB is sort of mixer and is called product demodulator. Instead a balanced modulator can also be used but it is used only in transreceivers where a balanced modulator serves two purposes. A circuit diagram of product detector or product demodulator is shown in Fig. 2.44.

A product demodulator mixes the down converted SSB-signal with crystal oscillator frequency to obtain AF output. This circuit can also be used for demodulating other forms of AM signal. The SSB signal is fed to the base of bipolar transistor through a fixed frequency IF transformer and the crystal oscillator is connected to the emitter through a coupling capacitor C_c. The frequency of crystal oscillator is made equal to the nominal carrier frequency or is derived from the pilot frequency. The IF for a double conversion type SSB receiver fed to the product demodulator is usually about 200 kHz. If USB is fed to detector the IF will be 200.3 to 203.0 kHz for A3J. A crystal oscillator frequency will be 200 kHz. The mixing of these two frequencies results in a number of frequency components which includes the difference frequency also ranging from 0.3 to 3.0 kHz.

This is the wanted AF signal and is selected using a bandpass filter, which rejects all other frequencies. This filter is made up of two capacitors C_B and one resistor R_B. Thus, the wanted AF-signal is made available by the use of product demodulator. In case LSB is used the sidebands stretches from 200 to 202.7 kHz and the oscillator frequency is 203 kHz.

Detection of SSB Using Diode Balanced Modulator:

This circuit is particularly advantageous for a transreceiver as it performs two functions, thus reducing the weight and power consumption of the system. At transmitter side it acts as a modulator and at the receiver side as a detector and can

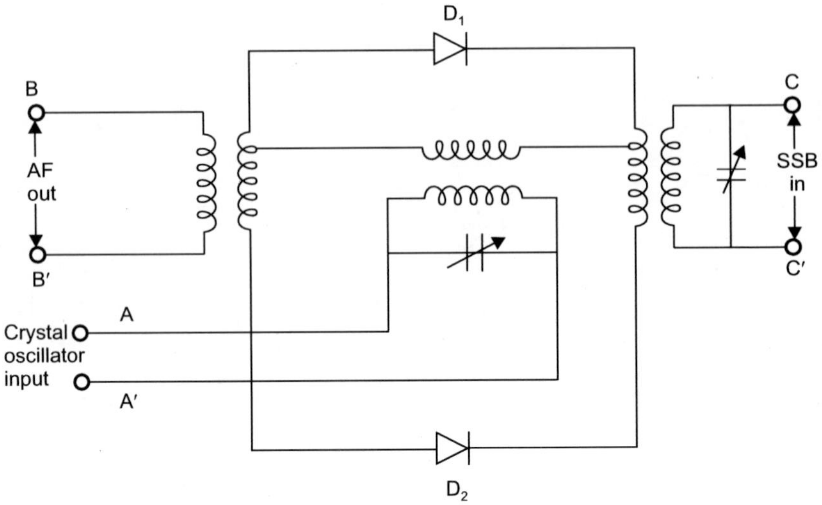

Fig. 2.45

be switched on to either side when required. A diode balanced modulator is shown in Fig. 2.45.

The local oscillator frequency is either 200 or 203 kHz depending upon the sideband to be demodulated. The output of local-crystal oscillator is fed to terminal A A′. In case of modulator the carrier suppressed signal is taken out from terminals CC′, but while using it as a demodulator, we feed in the SSB-signal at these points. This circuit now behaves as a non-linear resistance resulting in sum and difference frequencies at the AF transformer primary. The transformer stops the RF frequencies and allows only AF signals to terminals BB′ thus avoiding the use of a separate low-pass filter. The circuit easily drives out the modulating information on the SSB-signal and works quite similarly to product detector.

2.36 VSB (VESTIGAL SIDEBAND) TRANSMISSION

SSB signals are relatively difficult to generate. The bandwidth of composite video signal is limited to 4 MHz, using maximum modulation frequency of 4 MHz, amplitude modulation of carrier results in two sidebands each of 4 MHZ width so that the total bandwidth requirement for channel amounts to 9 MHz. The bandwidth is large. Bandwidth required may be reduced by utilizing the fact that the entire signal information is contained in each of the two sidebands of the modulated carrier. Thus, we may use only one sideband and the carrier and suppress the other sideband, thereby reducing the bandwidth requirement to half, i.e. MHz. But TV signals extend down to zero frequency and removal of one sideband from the rest becomes practical impossibility. Hence, in practice, only a part of one sideband is suppressed.

More the information that must be sent in given time, the larger is the bandwidth required. The simplify video demodulation in the receiver, the carrier is, in practice, sent undiminished. Because the phase response of filters, near the edges of the flat pass band, would have a harmful effect on the received video signals in TV receiver, a portion of unwanted (lower) sidebands must also be transmitted. The result is VSB-transmission or C3F as shown in Figs. 2.46 and 2.47.

By sending the first 1.25 MHz of the lower sideband (the first 0.75 MHz of it undiminished) if it possible to make sure that the lowest frequencies in the wanted upper sideband are not distorted in phase by the VSB filter. Because only first 1.25

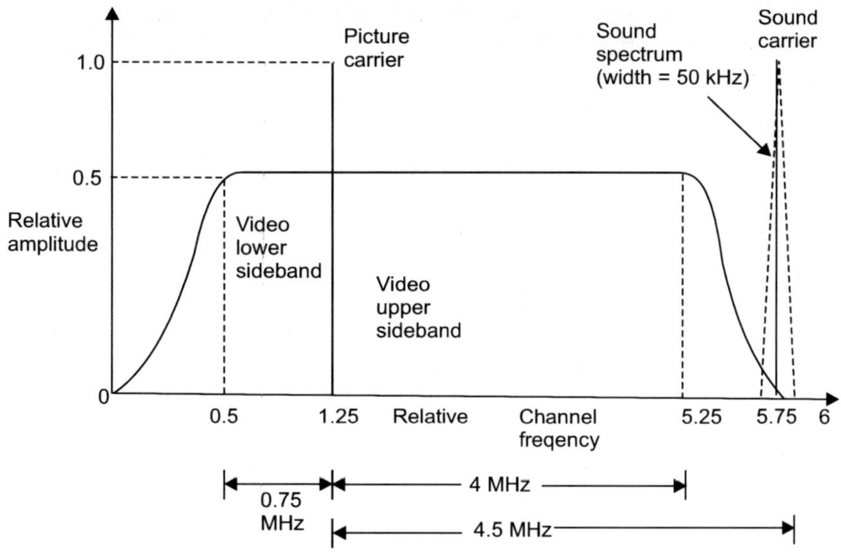

Fig. 2.46: Spectrum of Transmitted Signal

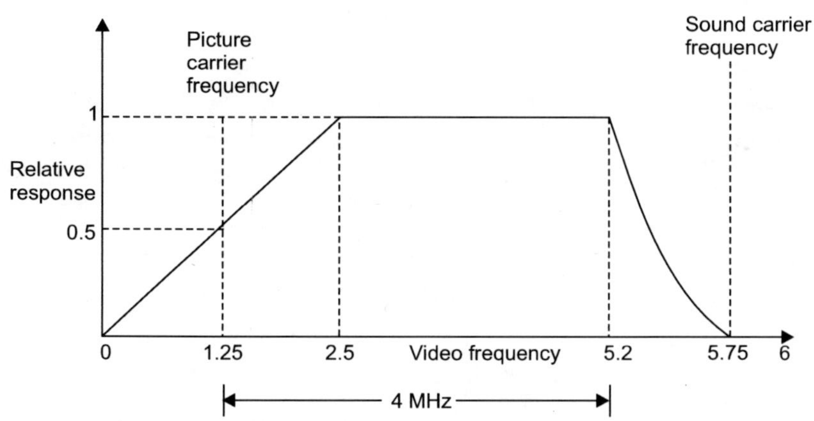

Fig. 2.47: Corresponding Receiver Amplifier Frequency Response

MHz of the lower sideband is transmitted, 3 MHz of spectrum is saved for every TV-channel. Since the total bandwidth requirement of a television channel is now 6 MHz instead of 9 MHz, clearly a great saving has been made, and more channels consequently can be accommodated.

All these signals occupy frequencies near the video transmissions simply because sound is required with the pictures and it would not be very practical to have a completely separate receiver for the sound, operating at some frequency remote from the video transmitted frequencies.

Attenuation is purposely provided for the video frequencies from 0 to 1.25 MHz. The reason is extra power is transmitted at these frequencies. Accordingly, these frequencies would be unduly emphasized in the video output of the receiver if they were not attenuated appropriately.

Numericals

Example 2.8: The voltage of certain amplitude modulated wave is expressed as:

$$v = 10\,(1 + 3\cos(2\pi \times 1000 \times t)) \times [\cos 2\pi \times 10^6 \times t]$$

Calculate (a) m_a = ? (b) f_c and (c) f_m.

Solution:

$$v = 10\,[1 + 3\cos(2\pi \times 1000 \times t))] \times [\cos 2\pi \times f_c \times t] \quad (*)$$

Standard form

$$V = E_c[1 + m_a \cos(2\pi \times f_m \times t))] \times [\cos 2\pi \times f_c \times t] \quad (*)(*)$$

Comparing (*) and (**) we have

(a) $m_a = 3$

(b) $f_m = 1000\text{ Hz} = 1\text{ kHz}$ and

(c) $f_c = 10^6\text{ Hz} = 1\text{ MHz}$

Example 2.9: A bandwidth of 20 MHz is to be considered for the transmission of AM signals. If the highest audio frequencies used to modulate the carriers are not to exceed 3 kHz, how many stations could broadcast within this band simultaneously without interfering with one another?

Solution: Given Total bandwidth = 20 MHz

$$f_{a\,max} = 3\text{ kHz}$$

Find : Number of AM stations.

The maximum bandwidth of each AM station is determined by the maximum frequency of the modulating signal.

$$\text{Station B.W.} = 2 f_{a\,max}$$

$$= 2 \times 3 \times 10^3 = 6 \times 10^3 = 6\text{ kHz}$$

Thus, the number of stations that can broadcast simultaneously without interfering with one another is:

$$\frac{20\times10^6}{6\times10^3} = 3.333\times10^3$$

Number of stations = 3333

Example 2.10: The per cent modulation of an AM wave changes from 40% to 60%. Originally the power content at the carrier frequency was 900 W. Determine the power content at carrier frequency and within each of the sidebands after the per cent modulation has risen to 60%.

Solution: Given $m_1 = 0.40$

$m_2 = 0.60$

$P_{C_{0.40}} = 900$ W

Find : $P_{C_{0.60}}$, $P_{USB_{0.60}}$, $P_{LSB_{0.40}}$

The power content of the carrier of an AM signal remains the same regardless of per cent modulation.

Thus, $$P_{C_{0.60}} = P_{C_{0.40}} = 900 \text{ watts}$$

The power content of each of the sidebands is equal to $m^2 \dfrac{P_C}{4}$

$$\therefore \quad P_{USB_{0.60}} = P_{LSB_{0.60}} = \frac{m^2 P_C}{4}$$

$$= \frac{(0.60)^2 \times (900)}{4}$$

$$\therefore \; P_{USB_{0.60}} = P_{LSB_{0.60}} = 81 \text{ watts}$$

Example 2.11: An SSB transmission contains 10 kW power. This transmission is to be replaced by a standard amplitude – modulated signal with the same power content. Determine the power content of the carrier and each of the sidebands when the per cent modulation is 80%.

Solution: $P_{SSB} = 10$ kW

$m = 0.80$

Find P_C, P_{LSB}, P_{USB}

Since the total power content of the new AM signal is to be the same as the total power content of the SSB signal.

$$P_T = P_{SSB} = 10 \text{ kW}$$

Solving for power contained at the carrier frequency.

$$P_T = P_C + P_{LSB} + P_{USB}$$

$$P_T = P_C + \frac{m^2 P_C}{4} + \frac{m^2 P_C}{4}$$

$$10{,}000 = P_C + \frac{(0.8)^2 P_C}{4} + \frac{(0.8)^2 P_C}{4}$$

$$= P_C + \frac{0.64 P_C}{2} = 1.32 \ P_C$$

$$\frac{10{,}000}{1.32} = P_C$$

$\therefore$ $$P_C = 7575.76 \text{ watts}$$

The power content of the sidebands is equal the difference between the total power and the carrier power

$$P_{SB} = P_T - P_C$$

The power content of the upper and the lower sidebands is equal

$$P_{LSB} + P_{USB} = 10{,}000 - 7575.76$$

$$= 2424.24$$

$$P_{LSB} = P_{USB} = \frac{2424.24}{2}$$

$$= 1212.12 \text{ watts}$$

Thus, $$P_C = 7575.76 \text{ watts}$$

$$P_{LSB} = P_{USB} = 1212.12 \text{ watts}$$

Example 2.12: The aerial current of an AM transmitter is 13 amps when unmodulated and 12 amps when amplitude modulated. Calculate the modulation index.

Solution: Let $I_t = 13$ Amps

$I_c = 12$ Amps

We know that :

$$\left(\frac{I_t}{I_c}\right)^2 = \left(1 + \frac{m_a^2}{2}\right)$$

$$\left(\frac{I_t}{I_c}\right)^2 - 1 = \frac{ma^2}{2}$$

$$\therefore \quad ma^2 = 2\left[\left(\frac{I_t}{I_c}\right)^2 - 1\right]$$

$$\therefore \quad m_a = \sqrt{2\left[\left(\frac{I_t}{I_c}\right)^2 - 1\right]}$$

$$\therefore \quad m_a = \sqrt{2\left[\left(\frac{13}{12}\right)^2 - 1\right]}$$

$$\therefore \quad m_a = 0.5$$

Example 2.13: A 75 MHz carrier having an amplitude of 50 V is modulate by a 3 kHz audio signal having an amplitude of 20 V.

Calculate: (a) Modulation index (m)

(b) What frequencies would show up in a spectrum analysis of modulated wave.

(c) Write trigonometric equations for carrier and the modulating waves.

Solution: (a) From the defining equation of modulation factor.

$$m = \frac{E_m}{E_c}$$

$$\therefore \quad m = \frac{20}{50} = 0.4$$

(b) The frequency content of an AM signal consists of the carrier frequency and the sideband frequencies which result from adding the audio frequency to the carrier and from subtracting the audio frequency from the carrier frequency.

$$f_c = 75 \text{ kHz}$$

$$f_c + f_a = 75 \text{ MHz} + 3 \text{ kHz}$$

$$= 75000 \text{ kHz} + 3 \text{ kHz} = 75003 \text{ kHz}$$

$$f_c - f_a = 75{,}000 \text{ kHz} - 3 \text{ kHz} = 74997 \text{ kHz}$$

Thus, the frequency content of the AM wave is

75.000 MHz

75.003 MHz

74.997 MHz

(c)
$$V_a = E_m \sin 2\pi f_a t$$
$$= 20 \sin 2\pi (3000) t$$
$$V_a = 20 \sin 6000 \pi t$$

and
$$V_c = E_c \sin 2\pi f_c t$$
$$= 50 \sin 2\pi (75 \times 10^6) t$$
$$V_c = 50 \sin 150 \times 10^6 \pi t$$

Example 2.14: The trapezoidal pattern shown in figure results when examining an AM wave. Determine the modulation index. What can be said about the distortion of the AM wave.

Solution: Trapezoidal pattern shown in figure.

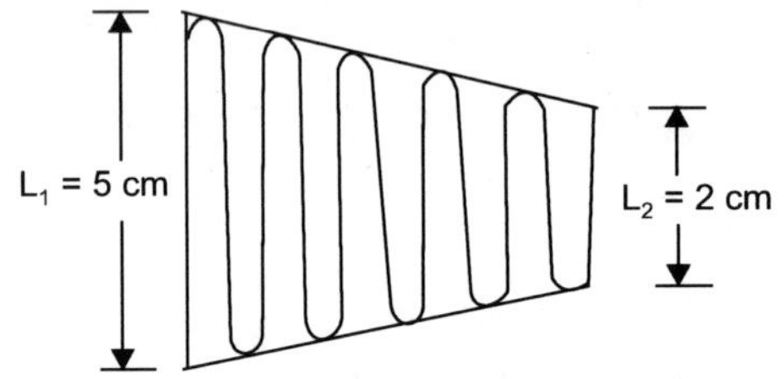

$$m = \frac{L_1 - L_2}{L_1 + L_2} = \frac{5-2}{5+2} = \frac{3}{7}$$

$$\% \ m = \frac{3}{7} \times 100 = 42.9$$

$$m = 42.9 \ \%$$

Regarding distortion: Since the sides of the trapezoid pattern show very little, if any curvature, it can be said that there is very little, if any, distortion of the modulated wave.

Example 2.15: An AM broadcast station operates at its maximum allowed total output of 50 kW and with 95% modulation.

Estimate the power in the intelligence part of the transmitted power.

Solution:

$$P_t = P_c\left[1+\frac{m^2}{2}\right]$$

$$50 \times 10^3 = P_c\left[1+\frac{(0.95)^2}{2}\right]$$

So that

$$P_C = \frac{50\times10^3}{\left[1+\frac{(0.95)^2}{2}\right]}$$

$$= 34.5 \times 10^3 \text{ watts}$$

Hence, the power, in the intelligence part of the output is given by:

$$P_i = (P_t - P_C) = (50 - 34.5)$$

$$P_i = 15.5 \text{ kW}$$

Example 2.16: An AM signal is represented by

$$U(t) = 0.1[1 + 0.1 \cos 25\ 12t + 0.5 \cos 6280\ t]\ .\ \sin (10^7\ t + 45°)$$

State what information can be gathered from above relation.

Sketch the amplitude spectrum of the signal.

Solution: The above relation would yield the following information.

(i) The carrier voltage amplitude is 0.1 V

(ii) Carrier angular frequency is 10^7 rad/sec corresponding to

$$f = 1590 \text{ kHz}$$

(iii) Modulating signal frequencies are 2512 rad/sec and 6280 rad/sec, corresponding to 400 Hz and 1000 Hz.

(iv) For the 400 Hz signal, the depth of modulation would be 0.1 while for the 1000 Hz signal, it would be 0.5.

(v) The sidebands would be 1590 kHz ± 400 Hz and 1590 kHz ± 1000 Hz.

(vi) The sideband amplitude would be $\left(0.1\times\frac{0.1}{2}\right)$ V or 0.005 V in case of 1590 kHz ± 400 Hz and $\left(0.1\times\frac{5}{2}\right)$ V or 0.025 V in the case of 1590 kHz ± 1000 Hz.

(vii) The amplitude spectrum can be drawn in the light of the above data as shown in figure given below:

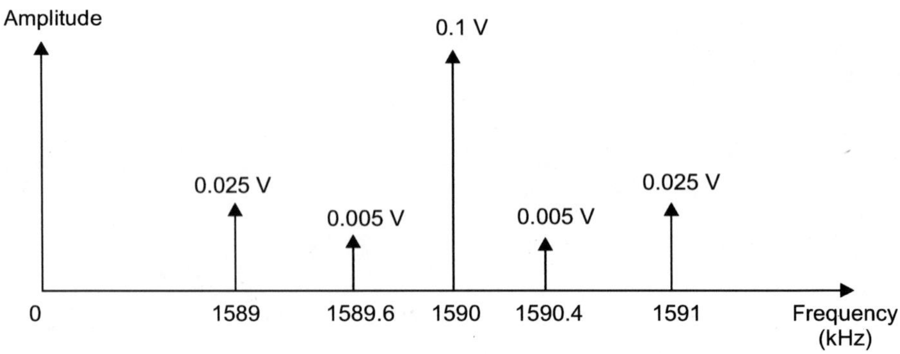

Example 2.17: Determine the power in the sidebands as a percentage of the total power of modulated signal in the case of a carrier amplitude modulated by two sinusoidal signals of different frequencies, with individual modulation depths of 0.3 and 0.4.

Solution: Assume carrier power to be P_c

The total sideband power would be

$$m^2 \overline{x^2(t)} P_C$$

where $m \overline{x(t)} = 0.3\cos\omega_1 t + 0.4\cos\omega_2 t$, here ω_1 and ω_2 are two modulating frequencies

So that $m^2 \overline{x^2(t)} = (0.3 \cos \omega_1 t + 0.4\cos\omega_2 t)^2$, averaged over a suitable period

$$\therefore \quad m^2 \overline{x^2(t)} = \left(\frac{1}{2}\right)[(0.3)^2 + (0.4)^2] = 0.1250$$

Also $\quad 2 P_{sb} = 0.1250 P_c$

and $\quad P_t = P_c + 2 P_{sb}$

$\therefore \quad P_t = P_c + 0.1250 P_c$

$P_t = 1.1250 P_c$

$$\text{Total sideband power} = \frac{0.1250 P_c}{1.125 P_c} \text{ or } 11.1\%$$

Example 2.18: An AM signal is represented by

$$V(t) = [10 + 4 \sin 1000 \pi t] + [\cos (2\pi \times 10^6)t] \text{ V}$$

where t is expressed in second.

Determine frequency of modulating signal, the frequency of carrier, the index of modulation and fractional power in each sidebond.

Solution: Referring to the standard relation for an AM signal.

We have: Angular frequency of the modulating signal = 1000 π rad/sec. corresponding to 500 Hz. Angular frequency of carrier = $2\pi \times 10^6$ rad/sec. corresponding to 10^6 Hz or 1 MHz

$$\text{Also modulation index } m = \frac{4}{10} = 0.4$$

$$\text{Power in carrier} = P_C = (10)^2 = 100 \text{ units}$$

$$\text{Sideband amplitude} = \frac{1}{2} \times 4 = 2 \text{ V}$$

$$\text{Sideband power} = 2^2 = 4 \text{ units}$$

$$\text{Total sideband power} = 2 \times 4 = 8 \text{ units}$$

$$\text{Total signal power} = (100 + 8)\ 108 \text{ units}$$

Hence, fractional power in each

$$\text{Sideband} = \frac{4}{108} = 0.037$$

Example 2.19: A class c amplifier, with an output of 5 kW, operates an efficiency of 80% from a 500 V anode supply.

If the amplifier is to be anode modulated to a depth of 70%, find

(i) the required power output of the modulating signal amplified and

(ii) The load impedance presented to this amplifier.

Solution:

$$P_c = 5 \text{ kW}$$

$$\eta_1 = 80\% = 0.8$$

$$P_{d.c.} = \left(\frac{5}{0.8}\right) = 6.25 \text{ kW} = 6250 \text{ watts}$$

$$V_{d.c.} = 5000 \text{ V},$$

$$I_{d.c.} = \left(\frac{6250}{5000}\right) = 1.25 \text{ A}$$

$$R_{d.c.} = \left(\frac{5000}{1.25}\right) = 4000\ \Omega$$

$$m = 0.7$$

$$P_{SB} = \left(\frac{1}{2}\right) m^2 P_c$$

$$= \left(\frac{1}{2}\right) (0.7)^2 \times 5000 = 1225 \text{ W}$$

$$P_{m\ out} = \left(\frac{1225}{\eta}\right) = \left(\frac{1225}{0.80}\right) = 1531 \text{ watts}$$

$$R_{a.c.} = R_{d.c.} = 4000\ \Omega$$

Example 2.20: A transistor class C amplifier has maximum permissible collector dissipation of 20 watts and collector efficiency of 75%. It is to be collector modulated to a depth of 90%. Calculate (i) Maximum carrier power and (ii) sideband power that are generated.

Solution: Given data
(i) Maximum permissible collector dissipation with no modulation is given by

$$P_{C_{1\,max}} = 20 \text{ watts}$$

Collector efficiency = $\eta_C = 0.75$

Power from collector supply source is given by

$$P_{CC} = \frac{P_C}{\eta_C} = \frac{P_C}{0.75}$$

Maximum permissible collector dissipation with no modulation is given by

$$P_{C_{1\,max}} = (P_{cc} - P_{c\ max})$$

$$= \left(\frac{P_{c\,max}}{0.75} - P_{c\,max}\right)$$

$$\therefore \quad P_{C_{1\,max}} = \left(\frac{1}{0.75} - \frac{1}{1}\right) P_{C_{max}}$$

$$\therefore \quad 20 \text{ watts} = P_{C_{max}} \left[\frac{0.25}{0.75}\right]$$

$$\therefore \quad P_{C_{max}} = 20 \times 3 = 60 \text{ watts}$$

Maximum carrier power = 60 watts

(ii)
$$P_{SB_{max}} = P_{LSB_{max}} = P_{USB_{max}} = P_{C_{max}} \times \frac{m^2}{2} \times \frac{1}{2}$$

$$\therefore \quad P_{SB_{max}} = 60 \times \frac{(0.9)^2}{2} \times \frac{1}{2}$$

$$= 60 \times 0.45 \times \frac{1}{2}$$

$$\therefore \quad P_{SB_{max}} = 2.4 \text{ watts} = 1.2 \text{ watts}$$

Example 2.21: If a 6 MHz band were being considered for use with same standards that apply to the 88–108 MHz band, how many FM stations could be accommodated?

Solution: B.W. = 6 MHz

Find : Number of stations

Each station requires a total bandwidth of 400 kHz; 150 kHz for the signal and a 25 kHz guard band above and below with only alternative channels used

$$\text{Number of stations} = \frac{6 \times 10^6}{400 \times 10^3} = 15$$

Number of stations = 15

Example 2.22: A carrier of 90 MHz with 5 V peak amplitude. A modulating signal of 5 V peak amplitude and frequency of 5 kHz. The frequency deviation constant is 1 kHz/volt. Determine the bandwidth of the frequency modulated signal.

Solution: Given data

$$\delta = 1 \text{ kHz/V},$$

$$f_m = 5 \text{ kHz}$$

We know $\quad m_f = \frac{1}{5} = 0.2$

We have, $\quad (BW)_{FM} = 2(m_f + 1)\, f_m = 2(0.2 + 1) \times 5 \text{ kHz}$

$$(BW)_{FM} = 12 \text{ kHz}$$

Example 2.23: A 1 MHz carrier of 10 V amplitude, when frequency modulated by a 400 Hz, 1 volt modulating signal, undergoes 1 kHz deviation. If the modulating signal frequency changed to 1 kHz with 2 V amplitude, determine the bandwidth of the signal.

Solution: Here $\quad f_c = 1 \text{ MHz}$

$$A_c = 1 \text{ V}$$

$$f_m = 400 \text{ Hz}$$

$$A_m = 1 \text{ V}$$

$$f_m = 1 \text{ kHz}$$

With modulating signal amplitude becoming 2 V, the maximum frequency deviation also change accordingly, increasing from 1 kHz to 2 kHz.
With modulating frequency changing to 1 kHz the modulation index would become 2 so the signal band would be

$$2(2 + 1) = 6 \text{ kHz.}$$

Example 2.24: A varactor diode, with 15 V bias, is used for frequency modulation. With no modulating signal oscillator frequency is 10 MHz.
Compute the maximum frequency deviation when the modulating signal is + 0.15 sin 5000 t V.

Solution:

Now $\quad f_0 = 10 \text{ MHz} = 10^7 \text{ Hz}$

and $\quad x(t) = (0.15 \sin 5000\, t) \text{ V}$

Diode capacitance with 15 V bias, would be

$$C_0 = K(15)^{-1/2}$$

Oscillator angular frequency $= \omega_0 = (LC_0)^{1/2}$

Instantaneous angular frequency

$$= \omega_t(t) = (LC_t)^{-1/2}$$

and
$$C_1 = k[15 + 0.15 \sin 5000\, t]^{1/2}$$
$$= C_0\, [1 + 0.01 \sin 5000\, t]^{-1/2}$$
$$= C_0\, [1 - 0.005 \sin 5000\, t]$$

so
$$w_i(t) = (LC_0)^{-1/2}\, [1 - 0.005 \sin 5000\, t]$$

Hence, value of maximum frequency deviation would be:

$$\Delta f = f_0\, [0.0024]$$
$$= 10 \times [0.0025] = 25 \text{ kHz}$$
$$\Delta f = 25 \text{ kHz}$$

Example 2.25: An AM generator, using Armstrong method employs a crystal oscillator at 500 kHz. The maximum phase deviation is 0.25. The modulating signal frequency is 1 kHz. The PM signal frequency so generated is to be converted into a 100 MHz signal with approximately 75 kHz deviation, suggest a suitable scheme of frequency multiplication and conversion to obtain this signal.

Derive the values of all frequencies for heterodyning from the 500 kHz crystal oscillator.

Solution: Modulation index for

$$\text{PM} = 0.25$$

Modulating signal frequency $= 1000$ Hz

$$\text{Frequency deviation} = (0.25 \times 1000) = 250 \text{ Hz}$$

$$\text{Required deviation} = 75 \text{ kHz} = 75{,}000 \text{ Hz}$$

$$\text{Required frequency multiplication} = \frac{75{,}000}{250} = 300$$

$$\text{Carrier frequency} = 100 \text{ MHz}$$

$$\text{Initial frequency} = 500 \text{ kHz}$$

$$\text{Frequency multiplication} = \frac{100 \times 10^6}{500 \times 10^3}$$

$$= 200$$

Since frequency multiplication are different to meet the two requirements, frequency conversion has to be resorted to the multiplication factors can be:

$$300 = 2 \times 2 \times 3 \times 5 \times 5$$

Multiplying by 2, 3, and 5 kHz and then superheterodyning 500 kHz × 2 × 3 × 5

$$\text{Frequency conversion} = 2 \times 50 = 100 \text{ MHz}$$

Here the frequency convertor output frequency has to be

$$\left(\frac{100}{2} \times 5\right) = 15000 \text{ kH}$$

$$= 15 \text{ MHz}$$

So the required frequency convertor heterodyning frequency would be

$$(15 \pm 10) \text{ MHz} = 25 \text{ MHz or } 5 \text{ MHz.}$$

Example 2.26: An FM signal is produced through the generation of a narrow-band PM signal. The NBM signal is obtained as a combination of a DSB – SC signal and a 90° phase shifted carrier. The depth of modulation in this combined signal, corresponds to m = 0.2. The modulating signal frequency is 100 Hz. It is desired to generate an FM signal with 25 kHz deviation. Suggest appropriate frequency multiplier chain.

Solution: Depth of amplitude modulation

= m = 0.2 it is also m of NBPM also,

$$\text{Maximum phase modulation} = 0.2 \text{ rad}$$

$$\text{Maximum frequency deviation} = (0.2 \times 100) \text{ Hz}$$

$$\text{Required deviation} = 25 \text{ kHz}$$

$$\text{Required frequency multiplication factor} = \left(\frac{25000}{20}\right) = 1250$$

Example: 2.27: The reactance tube modulator shown in figure uses as remote cutoff tube whose transconductance g_m varies from 2500 μs to 3500 μs. Determine range of capacitance it presents.

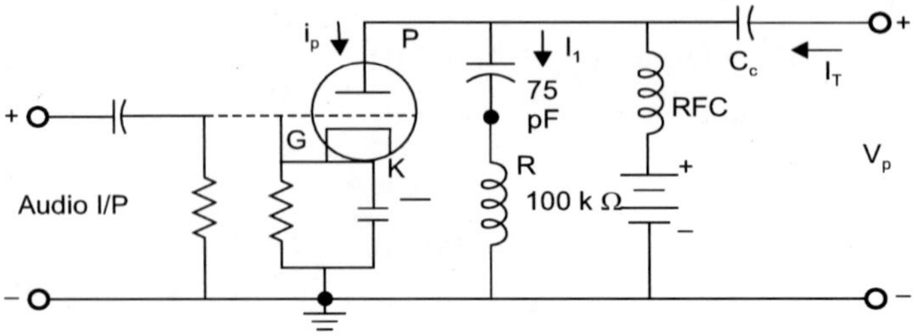

Solution: given

$$g_{m1} = 25000 \ \mu s,$$
$$C = 75 \ pF$$
$$g_{m2} = 3500 \ \mu s$$
$$R = 100 \ k\Omega$$

Find: C_{eq1}, C_{eq2}

Equivalence capacitance of reactance tube modulator is

$$C_{eq_1} = g_{m1} RC$$

Using lower value of g_m,

$$C_{eq_1} = (2500 \times 10^{-6})(100 \times 10^3) \times (75 \times 10^{-12}) = 18.75 \times 10^{-9}$$

$$\therefore C_{eq_1} = 0.01875 \ \mu F$$

High value equivalent capacitance is

$$C_{eq_2} = g_{m2} RC$$
$$= (3500 \times 10^{-6})(100 \times 10^3)(75 \times 10^{-12})$$
$$= 26.25 \times 10^{-9}$$

Thus, highest value reached by C_{eq_2} is

$$\therefore C_{eq_2} = 0.02626 \ \mu F$$

Example 2.28: A diode, used as an envelope detector, has a forward resistance of 1 Ω.

If detector load is a parallel combination of a 1000 Ω resistor and a 1000 pF capacitor find the maximum permissible depth of modulation so as to avoid diagonal clipping, with modulating signal frequency values of (i) 100 Hz (ii) 1 kHz and (iii) 10 kHz.

Solution: (i)

$$f_m = 100 \text{ Hz}$$
$$R = 1000\ \Omega$$
$$C = 10000 \text{ pF}$$

Hence
$$2\pi f_m RC = 2 \times 3.14 \times 100 \times 1000 \times 1000 \times 10^{-12}$$
$$= 6.28 \times 10^{-3}$$

Hence,
$$m_{max} = [1 + (6.28 \times 10^{-3})^2]^{-1/2}$$
$$= 0.994$$

(ii)
$$f_m = 1 \text{ kHz}$$

Here
$$2\pi f_m RC = 6.28 \times 10^{-2}$$

So that
$$m_{max} = [1 + (6.28 \times 10^{-2})^2]^{-2}$$
$$= 0.997$$

(iii)
$$f_m = 10 \text{ kHz}$$

Hence,
$$2\pi f_m RC = 0.628$$

So that
$$m_{max} = [1 + (0.628)^2]^{-1/2}$$
$$= 0.847$$

Example 2.29: A non-linear device, with an output-input relation of $i = [10 + 2V_i + 0.2\ V_i^2]$ mA, is supplied with a carrier of 1 V amplitude and sinusodal signal of 0.5 V amplitude in series.

Estimate the depth of modulation of the AM signal at the output, considering only the frequency components of the AM signal at the output.

Solution: The input signal is given by

$$V_i\,(t) = \cos \omega_c t + 0.5 \cos \omega_m t$$

with the usual notations.

The output current is given by

$$i = 10 + (\cos\omega_c t + 0.5 \cos \omega_m t) +$$
$$0.2\,(\cos \omega_c t + 0.5 \cos \omega_m t)^2 \text{ mA}$$

The carrier frequency term is 2 cos $\omega_c{}^t$, while the $(f_c \pm f_m)$ frequency term is $0.2 \times 2 \times 0.5 \cos \omega_c t \cos \omega_m t$ or $0.2 \cos \omega_m t$.

The remaining terms would not contribute to the AM signal at the carrier frequency f_c. So, the AM signal would be $2 \cos \omega_c t + 0.2 \cos \omega_c t \cos \omega_m t$ corresponding to m = 0.1 or 10%.

Example 2.30: The load on certain diode detector is a 5 kΩ resistor shunted by a 10000 pF capacitor.

Find the maximum depth of modulation that the detector can handle without causing any diagonal clipping when modulation signal frequency is (i) 400 Hz; (ii) 1 kHz; (iii) 10 kHz.

Solution:

$$R = 5\ k\Omega = 5{,}000\ \Omega$$

$$C = 10000\ pF = 10^{-8}\ F$$

$$RC = (500 \times 10^{-8}) = 5 \times 10^{-5}\ sec$$

(i)

$$f_m = 400\ Hz$$

$$(2\pi f_m RC)^2 = (2 \times 3.14 \times 400 \times 5 \times 10^{-5})^2 = 0.016$$

so that

$$m_{max} = [1 + 0.016]^{1/2} = 0.992 \text{ or } 99.2\%$$

(ii)

$$(2\pi f_m RC)^2 = (2 \times 3.14 \times 1000 \times 5 \times 10^{-5})^2 = 0.1$$

so that

$$m_{max} = [1 + 0.1]^{-1/2} = 0.953 \text{ or } 95.3\%$$

(iii)

$$f_m = 10\ kHz$$

$$(2\pi f_m RC)^2 = (2 \times 3.14 \times 10^4 \times 5 \times 10^{-5})^2$$

Hence,

$$m_{max} = 0.301 \text{ or } 30.1\%$$

Example 2.31: Signal to noise of a DSB transmitter is 40 dB with modulation index 0.5. Calculate the new signal to noise ratio if the transmission is changed to SSB with same modulation index.

Solution: Let

$$\left(\frac{S}{N}\right)_{D.S.B.} = 40\ dB,$$

$$m = 0.5$$

$$\left(\frac{S}{N}\right)_{S.S.B.} = ?$$

We know that

$$\frac{\left(\frac{S}{N}\right)_{S.S.B.}}{\left(\frac{S}{N}\right)_{D.S.B.}} = \left(1+\frac{2}{m^2}\right)$$

$$\therefore \qquad \left(\frac{S}{N}\right)_{S.S.B.} = \left(1+\frac{2}{m^2}\right)\left(\frac{S}{N}\right)_{D.S.B.}$$

$$\therefore \qquad \left(\frac{S}{N}\right)_{S.S.B.} = \left(1+\frac{2}{(0.5)2}\right)\times 40 \text{ dB}$$

$$\therefore \qquad \left(\frac{S}{N}\right)_{S.S.B.} = 360 \text{ dB}$$

Example 2.32: In a certain SSB generator of the phase-shift type, the carrier input to one of the modulators fall to 98% of correct amplitude.

Determine its effect on the SSB output.

Solution: Assume reduced carrier input is to be applied to the modulator B.

So, output of modulator A $= \cos \omega_c t \cdot \sin \omega_m t$

Output of modulator B $= 0.98 \sin \omega_c t \cdot \sin \omega_m t$

Difference between outputs $= \cos \omega_c t \cdot \cos \omega_m t - 0.98 \sin \omega_c t \cdot \sin \omega_m t$

$= (0.99 + 0.01) \cos \omega_c t \cdot \cos \omega_m t - (0.99 - 0.01) \sin \omega_c t \cdot \sin \omega_m t$

$= 0.99 \cos (\omega_c + \omega_m)t + 0.01 \cos (\omega_c + \omega_m)t$

Example 2.33: In a series plate modulated amplifier, plate supply voltage is 300 volts and the DC plate current under unmodulated condition is 20 amp. The sinusoidal modulating voltage which appears in the plate circuit of modulated amplifier has amplitude of 150 volts. The unmodulated output carrier power is 4.5 kW. Calculate (i) modulation index (ii) carrier power under modulated condition (iii) plate circuit efficiency and (iv) plate dissipation under unmodulated and modulated conditions.

Solution: (i) Modulation index

$$m_a = \frac{V_{mm}}{V_{bb}} = \frac{150}{300} = 0.5$$

(ii) Carrier power under modulated condition is given by

$$P = P_C\left[1+\frac{m^2}{2}\right]$$

$$= 4500\left[1+\frac{(0.5)^2}{2}\right]$$

$\therefore$ $$P = 4500\ [1 + 0.12]$$

$\therefore$ $$P = 5062 \text{ watts}$$

(iii) D.C. power from plate supply source is given by

$$P_{bb} = V_{bb} \times I_{bb} = 300 \times 20 = 6000 \text{ watts}$$

A.C. output carrier power = 4500 watts

Hence, plate circuit efficiency

$$\eta_p = \frac{4500}{6000} = \frac{45}{60} = 0.75$$

$$\%\ \eta_p = 0.75 \times 100 = 75\%$$

(iv) Plate dissipation without modulation is given by,

$$P_{po} = (P_{bb} - P_c)$$

$$= (6000 - 4500)$$

$$= 1500 \text{ watts}$$

Plate dissipation under modulated condition is given by

$$P_p = P_{po}\left[\frac{1+m_a^2}{2}\right]$$

$$= 1500\left[1+\frac{(0.5)^2}{2}\right]$$

$\therefore$ $$P_p = 1688 \text{ watts.}$$

QUESTIONS

2.1. The expression for an AM wave is given by

$$A_c\ [1 + m \times \sin \omega_m t)]\ \cos \omega_c t$$

(With no restriction on the magnitude of m and $\times \sin \omega_m t$ Write down the expression for the modulated wave from envelope.

2.2. For the modulating signal shown in figure, sketch:

(i) an AM signal with m = 0.5

(ii) an AM signal with m = 1.0 and

(iii) a DSB signal.

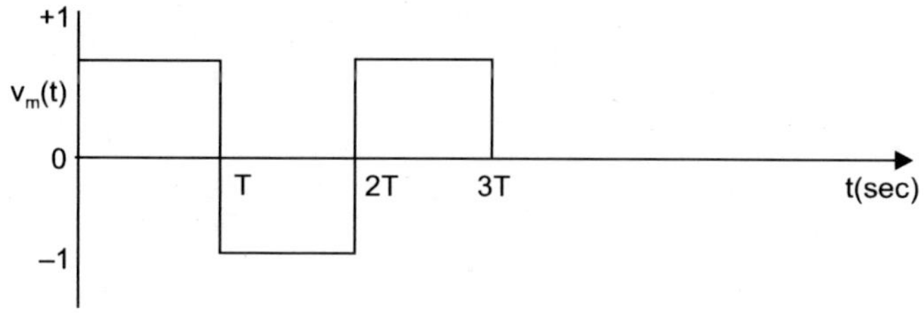

2.3 A combination of telegraph and speech waveforms in the same sideband of an ISB, is undesirable. Explain why it is so.

2.4 Gives the relative merits of low level modulation and high level modulation.

PROBLEMS

2.1. A 6 kHz signal and a 100 kHz signal of amplitude 1 V each, are applied to a product modulator.

2.2. In telephone systems, the speech frequency range is restricted to 300–3400 Hz.

2.3. A 50 MHz carrier is frequency modulated by a 5 kHz modulating signal, with a maximum frequency deviation of 30 kHz.

2.4. There is a two-tone modulating signal with frequencies f_1 = 1 kHz and f_2 = 10 kHz with the individual modulation indices being m_1 = 0.5 and m_2 = 2.0, so that the individual maximum frequency

$$f_1 \times m_1 = 1 \times 0.5 = 0.5 \text{ kHz and}$$

$$f_2 \times m_2 = 10 \times 2 = 20 \text{ kHz}$$

Determine the spectrum of the modulated signal with this two tone modulating signal as well as the bandwidth and compare these with the corresponding characteristics having individual modulating signals.

CHAPTER 3

AM, FM Transmitters and Receivers

3.1 INTRODUCTION

The message that comes from the information source is electrical in nature, it will be unsuitable for immediate sending. Even then, a lot of work must be done to make such a message suitable. This may be demonstrated in single sideband modulation, where it is necessary to convert the incoming sound signals into electrical variations, to restrict the range of the audio frequencies and then to compress their amplitude range. All this is done before any modulation. In wire telephony no processing may be required, but in long distance communications, a transmitter is required to process and possibly encode the incoming information so as to make it suitable for transmission and subsequent reception.

Eventually, in a transmitter, the information modulates the carrier, i.e. is impressed on a high frequency sine wave. The actual method of modulation varies from one system to another. Thus, modulation may be high level or low level and the system itself may be amplitude modulation, frequency modulation, pulse modulation or any variation or combination of these, depending on the requirements.

The transmitter radiates (transmits) modulated signal and is collected by receiving antenna. What must receiver do? The signal at this point is generally quite weak, powers of the order of picowatts being common, the receiver must first amplify the received signal. Since the signals is quite likely to be accompanied by a lot of other (unwanted) signals, probably at neighbouring frequencies, it must be selected and the others rejected.

Finally, since modulation took place in the transmitter, the reverse process of demodulation must be performed in the receiver, to recover the original modulating voltages.

3.2 CLASSIFICATION OF RADIO TRANSMITTERS

Radio transmitters may be classified according to the following methods:

(a) According to the type of modulation used.

(b) According to the service involved.

(c) According to the frequency range involved.

(a) Classification According to Type of Modulation Used:
According to the method of classification, radio transmitters may be classified as below:

(1) *Amplitude Modulated Transmitters:* In such transmitters the modulating signal amplitude modulates the carrier. Such transmitters are used for radio broadcast on long, medium and short waves, radio telephony on short waves, TV picture broadcast on very short waves or ultra short waves.

(2) *Frequency Modulation Transmitters:* In FM transmitter, the signal voltage frequency modulates the carrier. Such transmitters are used for radio broadcast in VHF & UHF range, television sound broadcast in VHF & UHF ranges radio telephone communication in VHF & UHF range over short distances.

(3) *Pulse Modulation Transmitters:* In Pulse Modulation transmitters the signal voltage alters some characteristics of pulse. These characteristics of pulses altered on modulation may be PWM, PPM, PAM, pulse frequency of pulse code (PCM). Accordingly, the various methods of pulse modulation is done.

(b) Classification According to the Type of Service Involved:
According to the type of service involved, radio transmitters may be classified as below.

(1) *Radio Broadcast Transmitters:* These transmitters are designed for transmitting speeches, talks, music, dramas, etc. for information and recreation of people. The electromagnetic energy is so radiated from the transmitting antenna that largest number of persons may be able to receive the broadcast with the help of their radio receivers. These broadcast transmitters may be either amplitude or frequency modulated. The AM transmitters operate on long waves, medium waves and short waves and radiate carrier power as low as about 1 kilowatt and is as high as 100 kW or more. The FM broadcast transmitters operate on very short waves or on ultra-short waves and radiate carrier power of the order of 100 kW or so.

(2) *Radio Telephone Transmitters:* These transmitters are designed for transmitting telephone signals over long distance by radio means. A radio telephone transmitter uses certain special devices as volume compressors, privacy devices, peak limiters, etc. The transmitting antenna is designed for beaming electromagnetic energy into a narrow beam directed toward the distant receiving antenna. Smaller amount of power is thus required. Radio telephone transmitters may be either of amplitude modulation type or of frequency modulation type. The AM telephone transmitters usually work on short wave, have output carrier power of typically a few kWs and are used for point-to-point communication over long distances. The FM radio telephone transmitters usually work on ultra high frequencies, carry small power usually less than 1 kW and are used for communication over short distances not exceeding about 30 kms or 50 kms.

(3) *Radio Telegraph Transmitters:* A radio telegraph transmitter transmit telegraph signals from one radio station to another. It may use either AM or FM.

When point-to-point radio communication is involved, the transmitting antennas are highly directive so that the electromagnetic energy is beamed into a narrow beam directed towards the receiving antenna at the receiving radio station.

(4) *TV Transmitters:* TV broadcast requires two transmitters one for transmission of picture and other for transmission of sound. Both operate in very high frequency or ultra-high frequency range. The picture transmitter is AM by picture signal occupying a band of about 5.5 MHz. VSB technique is used, so the total bandwidth occupied by one TV channel is about 7 MHz. The sound carrier is frequency modulated.

(5) *Radar Transmitters:* Radar may be of two types Pulse radar and CW radar. Pulse radar transmitter uses pulse modulation of carrier. It uses high output power typically 100 kW peak and operates at microwave frequencies typically 3000 MHz. The CW radar transmitter may use FM of carrier voltage or may utilize Doppler effect.

(c) Classification According to Frequency (Carrier) Range Involved:
According to this method of classification, transmitters may be classified as:

(1) *Long Wave Transmitters:* These transmitters operate on long wave, i.e. on frequencies below 300 kHz. Such long wave radio transmitters are used for broadcast in temperate countries, where atmospheric disturbances on long waves are not severe. Since long wave radio signals travels along the surface of earth and are rapidly attenuated, for reasonably high signal strength at the distant receiving aerial, the carrier power radiated from the transmitting aerial must be very large typically 100 kW or more.

(2) *Medium Wave Transmitters:* These transmitters operate on frequencies in the range of 550 to 1650 kHz and are usually used for broadcast. Hence, the band of frequency extending from 550 to 1650 kHz is commonly referred to as Broadcast band. The carrier power may vary from as low as 5 kW to as high as 500 to 1000 kW.

(3) *Short Wave Transmitters:* These transmitters operate on frequencies in short wave range of 3 to 30 MHz. In practice, frequencies beyond 24 MHz are not used. Ionospheric propagation of electromagnetic waves takes place at such short waves.

The attenuation of radio waves travelling from the transmitting aerial to distant receiving aerial through the ionosphere is small. Hence, carrier required to be radiated from the transmitting aerial small.

For national broadcast, the carrier power used may vary from about 1 to 10 kW. For overseas broadcast, certain amount of bearing of power is required to be done.

(4) *VHF & UHF Transmitters:* These transmitters operate either in VHF range or in the UHF range and are used for FM broadcast, TV broadcast, FM–radio telephony, etc.

(5) *Microwave Transmitters:* These transmitters operate on microwave frequencies, i.e. frequencies beyond about 1000 MHz and are used in radar, TV

relay, microwave link between two adjoining countries or between mainland and adjoining island, etc.

3.3 CONSTITUENT STAGES OF AMPLITUDE MODULATED

Radio transmitters:

Figure 3.1a shows block diagram of AM radio transmitter using modulation at high power level of the carrier.

Explanation:

(a) *Master Oscillator:* It generates oscillations of desired frequency with high constancy of frequency. The generated frequency is required to remain constant within close limit in spite of variations in the supply voltage, ambient temperature.

(b) *Buffer or Isolating Amplifier:* If the master oscillator directly drives a harmonic generator or class C power amplifier which may draw current, then power is drawn from master oscillator. This results in loading of master oscillator which in turn causes variation of effective resistance of the tank circuit of the oscillator and hence results in frequency variation.

(c) *Harmonic Generator:* Usually, master oscillator generates voltage at a frequency which is a sub-multiple of carrier frequency, i.e. the frequency of radiated power. Basically, these harmonic generators are class-C tuned amplifier in which the output RF voltage is first distorted through class C operation and then the tuned circuit in the output circuit of the amplifier selects the desired harmonic frequency.

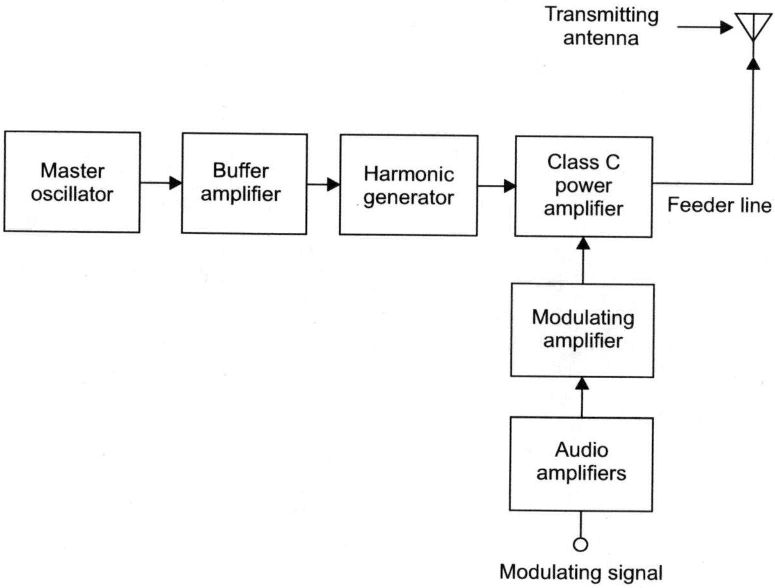

Fig. 3.1(a): Block Diagram of AM Radio Transmitter Using Modulation at High Carrier Power Level

(d) *Class C Amplifier:* RF voltage generated by the master oscillator has usually very small power, of the order of few watts. The power level is required to be raised to final high value in a chain of class C amplifier having high output circuit efficiency of the order of 70%. In general, first few stages of class C amplifier act as harmonic generators as well.

(e) *Modulated Amplifier:* This is class-C tuned amplifier usually of push-pull type and is modulated by audio modulating voltage from modulating amplifier. High efficiency series plate modulation is most popularly used in high power radio broadcast and radio telephone transmitters. Grid biased modulation and suppressor grid modulation are sometimes used particularly for modulation at low levels. In small transistorized radio transmitter, collector modulator or base modulator or both may be used.

(f) *Modulating Amplifier:* This is usually a class-B push pull amplifier and feeds audio power into the modulation amplifier in the plate circuit, control grid circuit or suppressor grid circuit depending upon the method of modulation used. Class-B operation is generally used because of high plate circuit efficiency. However, class-A modulating amplifier are also sometimes used particularly in low power transmitters.

The transmitter shown in Fig. 3.1a uses the so called "high power level modulation" system in which the carrier voltage is modulated at highest power level. An alternatives system called the "low power level modulation" system consists in modulating the carrier at low power level and carrier power is subsequently raised to the desired level in class B tuned power amplifiers as shown in Fig. 3.1b.

Such tuned power amplifiers cannot use class-C operation because class-C operation cannot reproduce the variation in the amplitude of the carrier voltage since plate current in class-C amplifier flows in the form of pulses. Such low

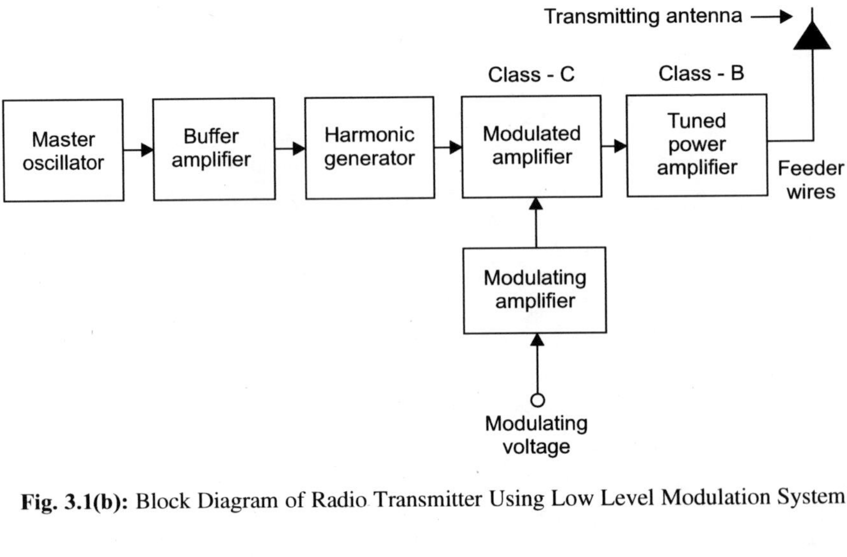

Fig. 3.1(b): Block Diagram of Radio Transmitter Using Low Level Modulation System

power level modulation system was extensively used in the early stages of development of radio communication but has now been almost completely replaced by high power level modulation system.

3.4 FREQUENCY SHIFT KEYING (FSK)

This consists in using one carrier frequency for the mark interval and another carrier frequency for the space interval. Figure 3.2 shows the block diagram of a typical FSK-transmitter.

Here DC telegraph signal is applied to a varactor diode changing its effective capacitance.

The varactor diode is placed in shunt with the tuned circuit of a 200 kHz oscillator. The frequency of this oscillator then suddenly change as the d.c. telegraph signal changes. Let during space interval the oscillator frequency be 200 kHz and let it increase to 200 kHz + Δ f Hz during mark interval. The 200 kHz oscillator output and a crystal oscillator output heterodyne in a mixer to produce a sum or difference frequency. Let the crystal oscillator frequency be 1 MHz. Then the output sum frequency becomes 1.2 MHz. This output of the mixer is fed to chain of harmonic generators to raise the frequency to the desired value and a chain of class-C amplifiers to raise the carrier power to the desired value. The output carrier is then fed to the transmitting antenna. In the mixer, the frequency shift Δ f doesn't alter. However, this frequency shift gets multiplied in harmonic generator and finally assumes the desired value. If the harmonic generator consists of two doublers and a tripler, then the carrier frequency increases to $1.2 \times 2 \times 2 \times 3 =$ 14.4 MHz and the frequency shift increases to Δ f $\times$ 12 Hz. If Δ f is 100 Hz, then

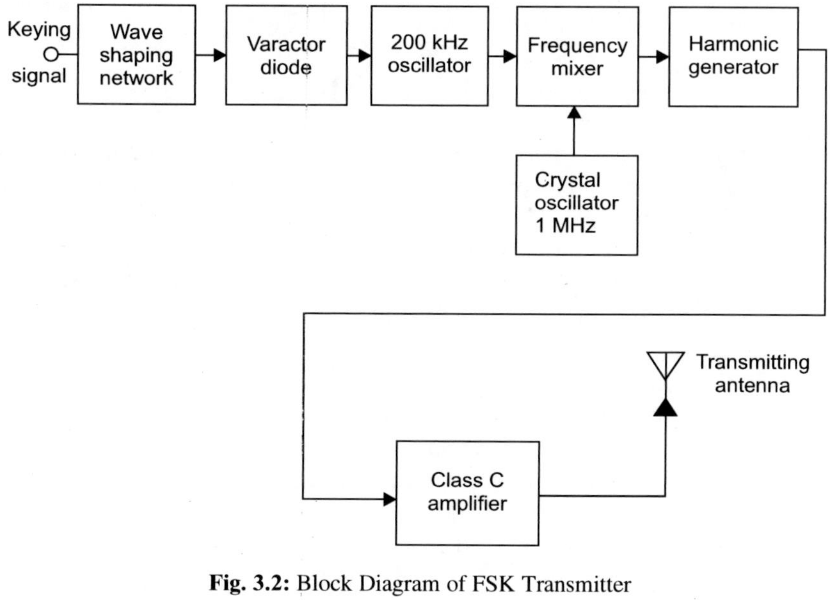

Fig. 3.2: Block Diagram of FSK Transmitter

in the output the frequency shift becomes 100 × 12 = 1200 Hz. In practice, the frequency shift of the radiated wave lies in the range of 500 to 2000 Hz, typical value being 1100 Hz.

Frequency shift keying is almost always used in all radio telegraph systems using automatic printing equipment. FSK being basically a frequency modulation system, possesses inherent discrimination against noise if the noise is weaker than the signal. This helps in reducing errors in automatic printing equipment.

3.5 FM TRANSMITTER USING REACTANCE MODULATOR

Figure 3.3 shows block diagram of FM transmitter using reactance modulator. It is simple 96 MHz FM transmitter using reactance modulator.

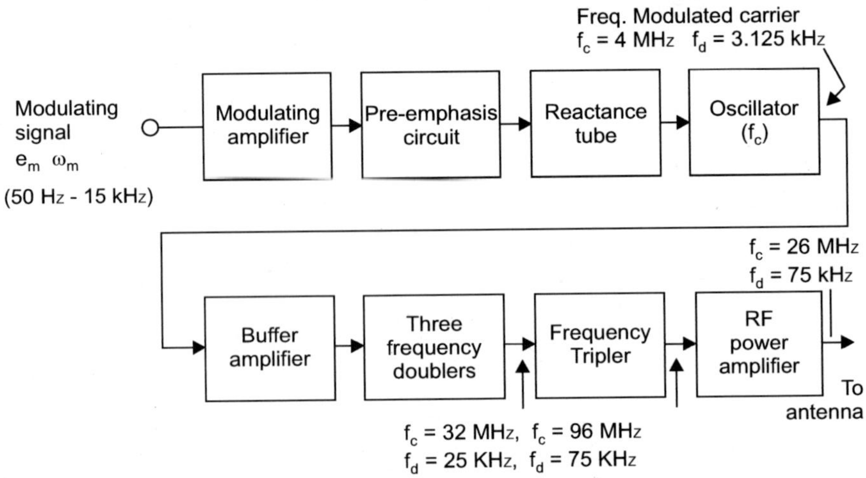

Fig. 3.3: Block Diagram of FM Transmitter

In this transmitter the reactance tube frequency modulates the master oscillator of carrier frequency say 4 MHz, producing frequency deviation of say 3.125 kHz.

The master oscillator is followed by a buffer amplifier to avoid loading the master oscillator. Buffer amplifier is followed by a chain of frequency multipliers to raise the carrier frequency to the required frequency level. Since the frequency multipliers follow the modulated master oscillator, the carrier frequency as well as the frequency deviation get raised by the same factor say n. Thus, the requirement of frequency deviation in the reactance tube modulator is reduced by a factor n. Here, three frequency doublers and one frequency tripler. Thus, n = 24. At the output of master oscillator f_c = 4 MHz and frequency deviation f_d = 3.125 kHz.

At the output of frequency multipliers, the resulting carrier frequency f_c = 4 × 24 = 96 MHz while the resulting frequency deviation = 3.125 × 24 = 75 kHz.

Frequency multipliers are followed by a chain of RF power amplifiers to raise the carrier power to the desired level. The final output has then fed to the transmitting antenna.

3.6 RECEIVERS

Receiver does the function to pick up the desired radio frequency signal (modulated) and recovers from it the original modulating signal.

3.6.1 Basic Functions of Radio Receiver

(1) Intercepting the incident electromagnetic waves to produce the desired RF signal.
(2) Amplifying this RF signal.
(3) Demodulating this RF signal to get back the original modulating voltage.
(4) Amplifying and then feeding this modulation frequency voltage to loudspeaker to recreate the original programme.

3.6.2 Classification of Radio Receivers

Radio receivers are generally classified according to the type of traffic they are designed to handle. Accordingly they may be classified as:

(1) *AM Broadcast Receivers:* These are meant for listening to broadcast of speech or music radiated from amplitude modulation broadcast transmitter operating on long wave, medium wave or short wave bands.

(2) *FM Broadcast Receivers:* These are used for receiving broadcast programmes for FM broadcast transmitters operating in VHF or in UHF bands.

(3) *TV Receivers:* These receivers are used for receiving TV broadcast in VHF or in UHF bands.

(4) *Code-Receivers:* These are, in general, simple superheterodyne receivers with the addition of IF beating oscillator to produce audio beat note with IF signal. Other code receivers are meant for receiving code signals, i.e. radio telegraph signals only and consists of an oscillating detector with amplifier stages.

(5) *Radar Receivers:* These are receivers used for receiving–Radar (radio detection and ranging) signals.

(6) *Communication receivers:* These are superheterodyne receivers used for reception for code (telegraph) and short-wave telephone signals and include circuit refinements such as IF beating oscillator for code reception, noise limiters or noise suppressors, band spread for fine tuning, crystal filter for high and adjustable selectivity, sensitivity, sensitivity control volume expander, interchannel noise suppressor, tuning indicator, etc.

These communication receivers are therefore, more costly and complicated and their operation involves some technical knowledge and skill possessed by radio operators.

3.6.3 Salient Features of Broadcast Receivers

Broadcast receivers, both of AM & FM types are meant for use in home mostly for entertainment purpose. Following are the features of broadcast receivers.

(1) *Simplicity of Operation:* These receivers are required to be handled by listeners who have little technical knowledge and hence simplicity of operation is

essential. Simplest broadcast receiver, therefore has only three controls, namely (a) bandswitch for selecting band of frequencies, (b) tuning control for tuning to the desired station, and (c) volume control for adjusting the volume of reproduced sound programme.

(2) *Good Fidelity:* Since these broadcast receivers are primarily designed for entertainment purpose, these should have good electrical fidelity, i.e. a reasonably large and uniform frequency response over almost the entire audio frequency band. In AM broadcast the maximum audio modulating frequency is 5 kHz. Hence, a receiver with good fidelity must reproduce all these frequency terms. In FM broadcast, maximum modulation frequency allowed is 15 KHz. Hence, FM broadcast receivers are designed to reproduce faithfully the entire audio frequency band upto 15 kHz. FM broadcast thus permits better fidelity.

(3) *Good Selectivity:* By selectivity of radio receivers is meant its ability to discriminate the desired signal from unwanted signal at other frequencies.

(4) *Average Sensitivity:* Broadcast receivers should have reasonably high sensitivity so that it may have good response to the desired signal of medium and low strength but should not have excessively high sensitivity otherwise it will pick up all the undesired electrical disturbances produced in the vicinity.

(5) *Adaptability to Different Types of Aerials:* A broadcast receiver should be designed to operate satisfactorily with any type of aerial.

3.7 BASIC FUNCTIONS OF AM RECEIVER

A radio receiver in its most elementary form performs the following four essential functions:

(1) *Reception:* This consists in receiving or picking up energy from the various electromagnetic waves radiated by radio transmitter. This function is performed by the receiving antenna. When electromagnetic waves strike the antenna, a voltage of the wave frequency is induced in the antenna.

(2) *Selection:* This consists in selecting or responding the desired radio wave with the exclusion of all others. Thus, at any instant of time, a large number of electromagnetic waveforms of different radio stations at different frequencies are intercepted by the antenna and each of these induce voltage in antenna. The selector circuit or tuner in the form of a parallel tuned circuit responds to the desired signal only and rejects all other signals.

(3) *Detection or Demodulation:* The desired signal in the form of a modulated carrier voltage is detected in a detector circuit to recover the original modulating voltage.

(4) *Reproduction:* This consists in feeding the detected signal to a loudspeaker or headphones to reproduce the sound waves giving the original programme.

3.8 VARIOUS TYPES OF RECEIVERS

Practically there are two types of receivers:

(1) TRF (Tuned Radio Frequency) receivers.

(2) Superheterodyne Receiver.

Commercially only the superheterodyne receiver is used mostly, however, the operation of TRF receiver is much simpler comparatively. TRF receiver has a number of shortcomings due to which superheterodyne receiver is always preferred over TRF.

3.8.1 TRF Receiver

Initially, when radio receivers came into operation, i.e. before World War II, most of the receivers were of TRF type. A TRF receiver was introduced as an improvement over the existing crystal, regenerative and superregenerative receiver. A block diagram of TRF receivers is shown in Fig. 3.4a.

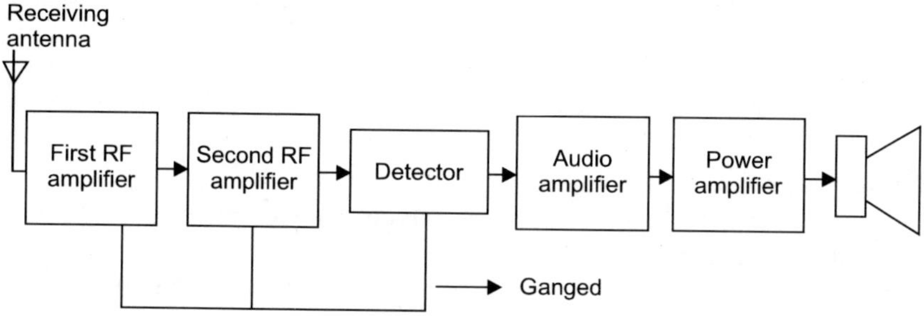

Fig. 3.4(a): Block Diagram of TRF Receiver

A TRF receiver has two to three RF amplifiers all tuned together (ganged). These RF amplifiers are used to select the incoming signal and amplify it. The amplified signal is fed to a demodulator and then to the audio amplifier. The audio amplifier amplifies the signal to the required level which is then fed to a loudspeaker. TRF receivers were simple to design and align at broadcast frequencies (535 to 1640 kHz) but at higher frequencies, it presented some difficulties. The main problem is of high gain at one frequency, i.e. RF frequency, which can easily cause instability. Usually, the RF amplifier have a gain of the order of 10,000. Thus, if only 1/10,000 of the output of last stage somehow finds its path back at the input of the first stage with correct phase, the system starts oscillating at a frequency at which the polarity of the supurious feedback is positive.

At higher operating frequencies, it becomes very difficult to suppress these oscillations. Besides, TRF receiver suffered from other drawbacks, such as a variation in its bandwidth over the tuning range, its inability to achieve sufficient selectivity at high frequencies because of the use of large number of a single tuned circuits. It was not found feasible to use double tuned RF amplifiers in TRF receivers because of the difficulty of making several double tuned amplifiers tune in unison.

Consider a tuned circuit of typical receiver required to have a bandwidth of 10 kHz at 535 kHz i.e. the lower side of medium wave band. The Q of the circuit must be;

$$Q = \frac{f}{\Delta f} = 53.5$$

The other end of medium wave broadcast band requires a

$$Q = \frac{1640}{10} = 164$$

This variation of Q is not possible to achieve in practice. Various losses at higher frequencies prevent the Q to increase beyond about 120 resulting in BW of

$$\frac{1640}{120} = 13.7 \text{ kHz}$$

This causes the receiver to pick up adjacent stations to the one, tuned on the receiver. At about 36.5 MHz, i.e. the upper end of the short-wave broadcast band, the required Q will be $\frac{36500}{10} = 3650$.

Such a high value of Q is impossible to achieve with the ordinary tuned circuits.

3.8.2 Superheterodyne Receiver

Practically all radio receivers are superheterodynes. While this receiver system is more complicated than the TRF, the ability to operate satisfactorily on any frequency with a constant value of selectivity and good sensitivity makes it highly desirable. Fig. 3.4b and 3.4c shows block diagram of superheterodyne receiver.

Functions of Different Stages:

Antenna: It intercepts the electromagnetic waves. Voltages induced in the antenna are communicated to the receiver input circuit by means of a feeder wire or led in wire. A parallel tuned circuit at the input of the receiver responds only to voltage at the desired carrier frequency and rejects voltages at all other frequencies. The voltages so picked up is fed to the input of RF amplifier stage.

RF Amplifier: This stage is generally a tuned voltage amplifier tuned to desired carrier frequency. The chief functions of RF amplifier stage are:

1. To amplify the input signal voltage to a suitable high level before feeding it to the frequency mixer which contributes large noise. Thus, $\left(\frac{S}{N}\right)$ ratio is improved.
2. To provide discrimination or selectivity against image frequency signal and intermediate frequency signal.

Frequency Converter Stage: This consists of a local oscillator and frequency mixer. To the frequency mixer are fed both the local oscillator voltage as well as signal voltage. The mixer being a non-linear device produces at its output the various intermodulation terms. The difference frequency voltage is picked up by the tuned circuit in the output circuit of the mixer. This difference and frequency is called intermediate frequency, the value of which is constant for a receiver. For all wave receivers, typical value of IF is 465 kHz or 456 kHz. Sometimes two separate transistors are used as local oscillator and frequency mixer but more often only one transistor functions both as local oscillator and frequency mixer.

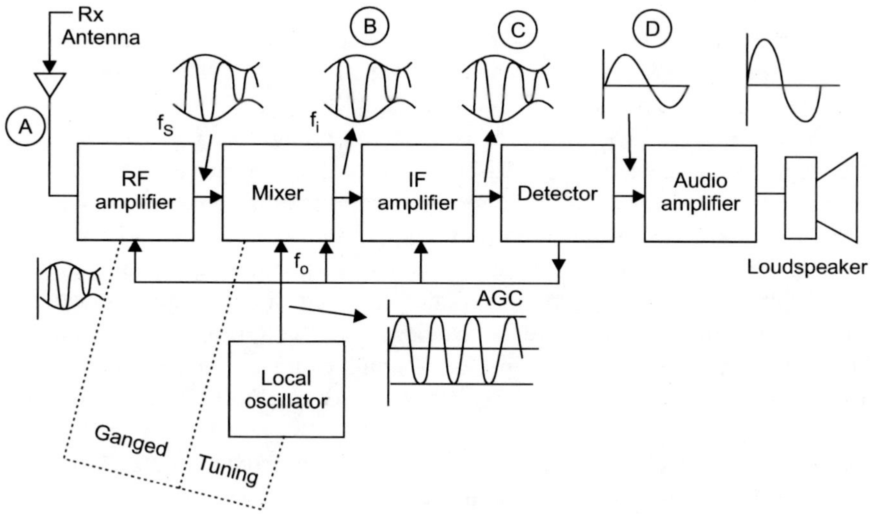

Fig. 3.4(b)

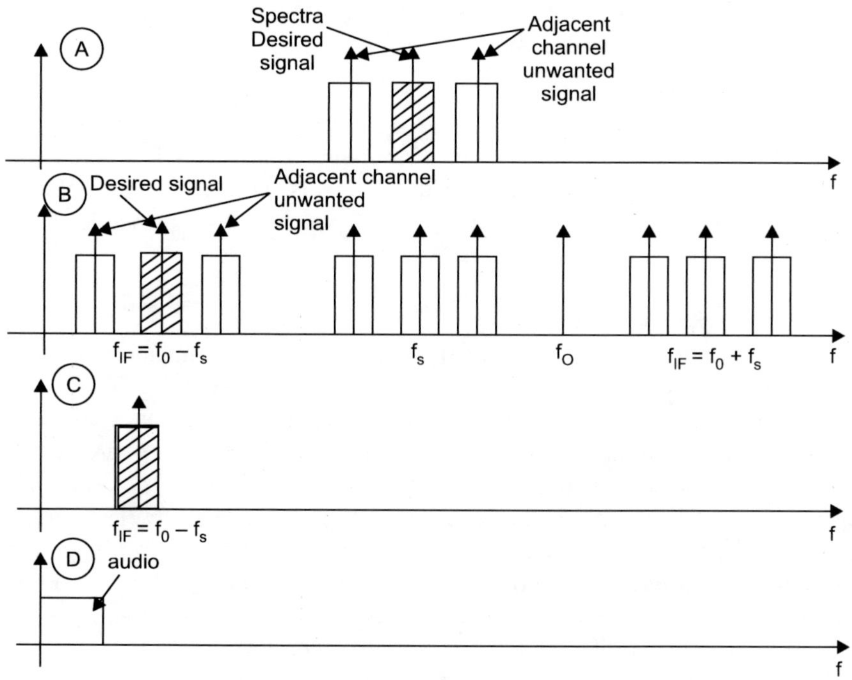

Fig. 3.4(c)

Such a transistor is then referred to as frequency converter transistor. Thus, with the help of frequency converter stage RF signal of any carrier frequency are converted into similarly modulated fixed frequency IF signal.

IF Amplifier Stage: It consists of two or more stages of fixed frequency tuned voltage amplifier having a 3-dB bandwidth of 10 kHz for AM broadcast. This IF amplifier provides most of the receiver amplification and selectivity.

Detector: Output of last IF amplifier stage is fed to this detector which is generally linear diode detector. Output of this detector is the original modulation frequency voltage. For satisfactory operation of this stage, i.e. for linear detection, it is necessary that the carrier voltage fed to it be at least 1 volt. Hence, the preceding amplifier stages must be designed to provide enough gain so as to feed a carrier of at least 1 volt to the detector for the weakest signal desired to be received by the receiver.

Audio-Frequency Amplifier: Audio-frequency output from the detector is fed to the AF amplifier which provides additional amplification. Usually, one stage of audio voltage amplifier is used followed by one or more stages of audio power amplifiers.

Loudspeaker: Amplified audio output voltage of AF amplifier is fed to loudspeaker through the impedance matching transformer. The loudspeaker reproduces the original programme.

3.9 AM RECEIVERS

These receivers are used to produce the programme of speech or music radiated by AM broadcast transmitter. Most of the receivers, whether they involve AM, FM or any other modulation technique, use superheterodyne principle. We will discuss each stage of AM receiver in detail.

3.9.1 Ferrite Rod Aerial

This is the form of two sections of coil wounded on high permeability ferrite rod of circular cross-section of length typically about quarter metre. The aerial is housed within the receiver itself and is used for reception of medium wave signals. Figure 3.5 shows the diagram of ferrite rod aerial and its coupling circuit.

The ferrite rod is kept horizontal and is usually mounted at the centre and in several cases, it is capable of being oriented slightly in azimuth. The aerial has maximum pick up when the rod is normal to the direction of propagation of electromagnetic waves and its axis coincides with the magnetic vector. Accordingly, it is ideally suited for receiving medium wave signal travelling along the ground with electric vector vertical and magnetic vector horizontal.

The magnetic vector of the incident electromagnetic waves produces magnetic flux through the ferrite core. Since the permeability of ferrite core is very high, these results in a very heavy concentration of magnetic flux through core. This time varying magnetic flux threads through the two sections of the coil and induces relatively high voltages, one section of the coil is tuned to the signal frequency "f_s"

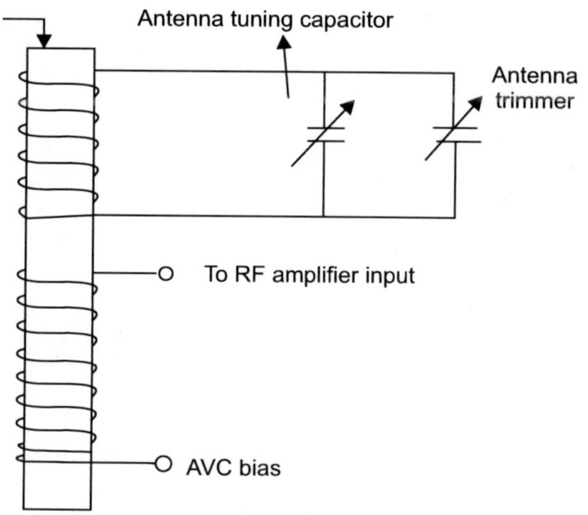

Fig. 3.5: Ferrite Rod Aerial and Its Coupling Circuit

with the help of one section of the ganged tuning condenser kept in the shunt with it. The voltage developed across the other section is fed to input of the RF amplifier.

3.9.2 RF Amplifier

Figure 3.6 shows circuit diagram of one-stage of RF amplifier using NPN transistor in CE-configuration.

It uses a parallel tuned circuit as the load impedance. This parallel tuned circuit is tuned to the signal frequency. The output from the receiving antenna is transformer coupled to the base of this amplifier. The secondary is tuned to the signal frequency with the help of section of ganged tuning capacitor. The variable air capacitor (tuning capacitor) in the input tuned circuit and output tuned circuit are ganged together. In addition, small trimmer capacitor are connected in shunt with these tuning capacitor sections for the purpose of RF alignments. Self-bias is provided with the help of R_1 and R_2 and R_e C_e assembly. A decoupling network consisting of R_b and C_b is placed in the collector supply lead.

Amplified RF signal developed across the collector tuned circuit is coupled through a step-down transformer to the input of the frequency mixer. This step-down transformer provides the impedance match between the high impedance of RF amplifier collector circuit and the low impedance of the base to emitter circuit of the following stage. The collector is connected to a suitable point on the primary of the output transformer so that the load impedance to the collector is optimum.

The RF amplifier is helpful to improve S/N ratio in the output of the receiver. The signal voltage is raised to suitably high level with the help of this RF amplifier before it is fed to the frequency mixer stage which contributes most of the noise generated in the receiver. If the signal is fed directly to the mixer, S/N ratio at the

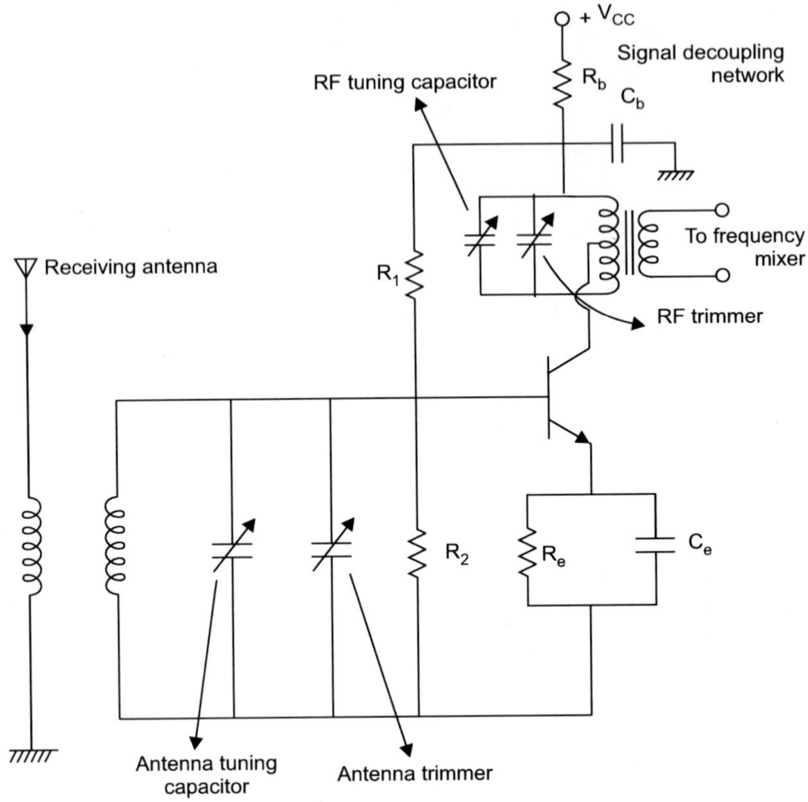

Fig. 3.6: Circuit of RF Amplifier Stage

input of mixer stage is poor and any amount of subsequent amplification does not improve the ratio.

3.9.3 Frequency Mixer and Frequency Converters

In the frequency mixer shown in Fig. 3.7a, the RF signal voltage of frequency "f_s" and local oscillator voltage of frequency f_0 are applied at its input.

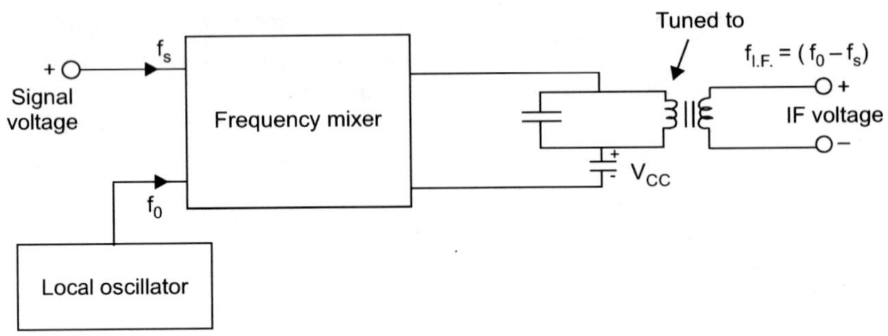

Fig. 3.7(a): Schematic Diagram of Frequency-Mixer

The frequency mixer uses a device which presents non-linear dynamic characteristics. Hence, the two input voltages beat together or heterodyne within the mixer to produce the output current having components of frequency f_s, f_0, $(f_0 + f_s)$, $(f_0 - f_s)$. Out of all these terms, one of interest in the component of difference frequency $(f_0 - f_s)$. This component may be selected by using a tuned circuit in the output circuit tuned to this difference frequency, called intermediate frequency.

Every RF signal irrespective of its frequency is reduced to this standard IF. Accordingly, it is necessary that the local oscillator frequency f_0 be made to vary in such a way as to maintain the difference frequency $(f_0 - f_s)$ always equal to the IF.

Frequency Converter Using Transistor:

A frequency converter uses a transistor which serves both as frequency mixer and local oscillator.

Figure 3.7b shows basic circuit of frequency converter using transistor.

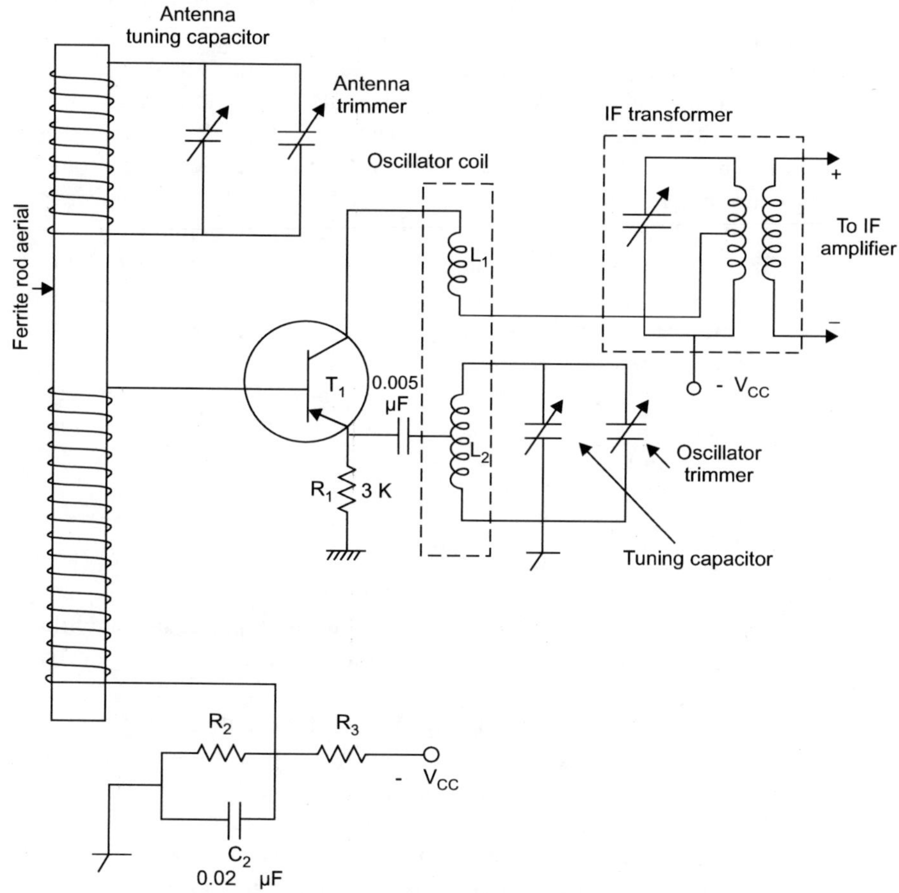

Fig. 3.7(b): Frequency Converter Using Transistor

The RF voltage from receiving antenna or from RF amplifier is fed to the base of transistor T_1, while the local oscillator voltage gets fed to the emitter of the same transistor. Coil L_1 and L_2 constitute the oscillator coils. Coil L_2 is tuned with the help of a tuning capacitor and trimmer and is inductively coupled to coil L_1. Thus, there results a positive feedback from coil L_1 to L_2 resulting in desired oscillations.

Coil L_2 along with the tuning capacitor and trimmer determine the frequency of local oscillator.

Transistor T_1 acts as internal amplifier for the feedback oscillator. The resulting local oscillator voltage causes the collector current to have component of local oscillator frequency f_0. The collector current is also controlled by the RF voltage of frequency f_s applied at the base. Further, the transistor is so biased that the operation is over the non-linear region of its characteristics. Hence, the resulting collector current contains a d.c. component plus components of frequencies f_s, f_0 $(f_0 - f_s)$, $(f_0 + f_s)$. The parallel tuned circuit in collector circuit constituting the load is tuned to the desired difference frequency $(f_0 - f_s)$ and this frequency constitutes IF. This tuned circuit thus develops across it voltage of IF only and rejects all other frequency components. This tuned circuit constitutes the primary of IF transformer whose secondary is left untuned. The IF voltage developed across secondary is fed to IF amplifier. The IF transformer is step-down transformer. This provides current gain and also impedance match between the high impedance of collector circuit and low impedance of the IF amplifier input circuit.

The stabilizing resistor R_1 is not shunted by a bypass capacitor otherwise the oscillator voltage across tuned circuit coil L_2 finds an a.c. short circuit through this bypass capacitor. Resistors R_2 and R_3 forms a potential divider across the supply voltage V_{CC} and provides necessary emitter bias.

Frequency converter circuit is most popularly used in all cheaper type of broadcast receivers since this circuit needs only one transistor. Further in these receivers no RF amplifier is used.

3.9.4 IF Amplifier

This stage of a superheterodyne receiver usually consists of two or more stages of tuned small signal amplifier using CE transistors. The output of the frequency mixer developed across the tuned circuit is coupled to the input of this amplifier while the IF amplifier output developed across the load circuit of last stage drives the detector.

Figure 3.8 shows circuit diagram of single tuned, two stage IF amplifier

The use of single tuned circuit results in an increase in gain and hence better receiver sensitivity. A double tuned circuit, but still the bipolar transistor amplifiers use single tuned circuit for interstage coupling. The only reason for this use of single tubed circuits is that, greater gain can be achieved in this way by use of tapping coils in tuned circuits. This tapping can be used to obtain maximum power transfer and a reduction of the damping of the circuit involved.

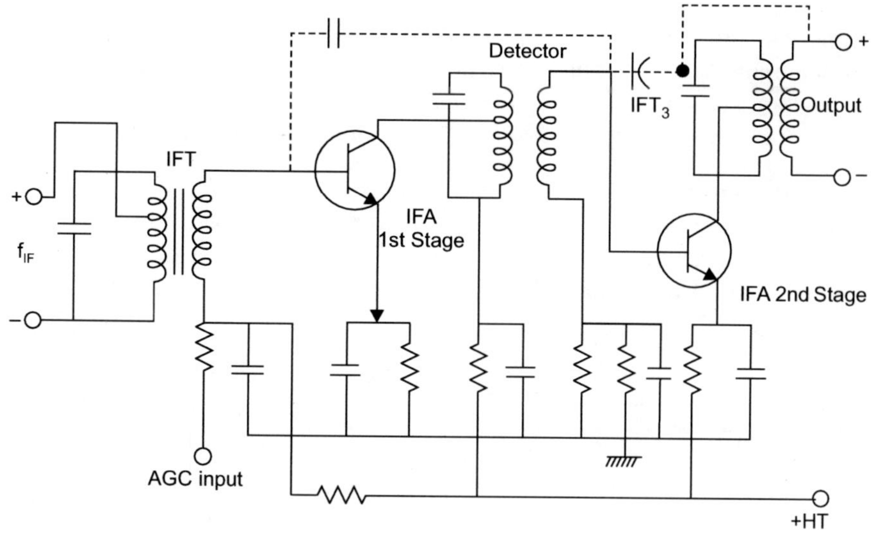

Fig. 3.8: Single Tuned, Two Stage IF Amplifier

It is well known fact that bandwidth of tuned circuit depends on its loaded Q, which is turn depends on the unloaded Q and external damping resistance. Since transistor impedances are low, tapping is employed along with somewhat lower inductances, than would have been used with tube circuits. In case a double tuned transformer is used, both sides of it may have to be tapped, rather than just single side as it is with a single tuned transformer. Thus, a reduced voltage would be applied to each transistor, thereby reducing the gain. Neutralization has to be done, depending on the frequency and type of transistor used.

3.9.5 Automatic Volume Control (AVC) or Automatic Gain Control (AGC)

As a result of synchronous fading or general fading, the amplitude of IF carrier at detector input may vary as much as 30 or 40 dB and result in corresponding variations in general level of reproduced programme at the receiver output. At the carrier minimum, loudspeaker output becomes inaudible and gets merged in noise. At the carrier maximum, the output of loudspeaker becomes intolerably large. The commonest simplest and the most effective means adopted to counteract the general fading is the AVC or AGC and is universally used in all broadcast and radio telephone receivers.

Principle of AVC:

The principle of AVC consists (1) in deriving by rectification of carrier voltage in linear diode detector, a d.c. voltage proportional to the carrier amplitude and (2) in applying this d.c. voltage as a reverse bias at the input of RF amplifier, frequency mixer and IF amplifier. Then if the carrier amplitude increases the AVC bias increases and gains of all tuned stages preceding the detector decreases, resulting in reduction in carrier amplitude at the input of the detector bringing it back to its normal value. Reverse action takes place if the carrier amplitude suddenly decreases. Thus, AVC smoothens out the variations in the carrier amplitude to a very large extent.

Linear Diode detector with π-filter and simple AVC:
For providing better removal of RF components from the modulation frequency output, it is preferred to use π-filter instead of a simple capacitor filter.

Figure 3.9a shows circuit of this linear diode detector with π-filter and simple AVC.

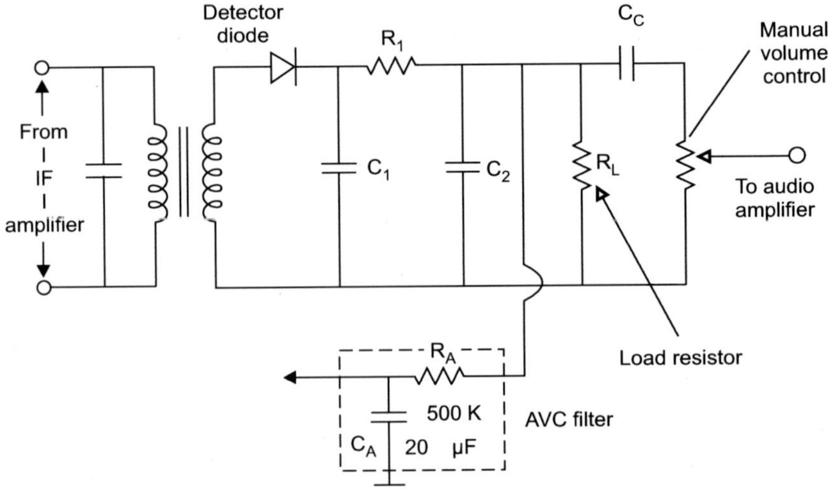

Fig. 3.9(a): Linear Diode Detector with π-Filter and Simple AVC

Manual volume control is generally provided by using a variable resistor at the input of the first audio amplifier.

AVC-characteristics:
Simple AVC system suffer from the drawback that the AVC becomes operative even for weak signals with the result that receiver gain starts falling as soon as detector diode starts developing the output. The AVC characteristic, i.e. the curve giving audio output of receiver plotted against input carrier voltage for simple AVC system given by curve S. An ideal AVC system, on the other hand, should remain inoperative until the input carrier voltage reaches a reasonably large predetermined voltage. Subsequently, the AVC should come into operation to maintain output level constant in spite of variation in input carrier level.

Figure 3.9b shows AVC characteristics.

Delayed AVC:
Figure 3.9c gives simple arrangement for producing delay in AVC operation. With

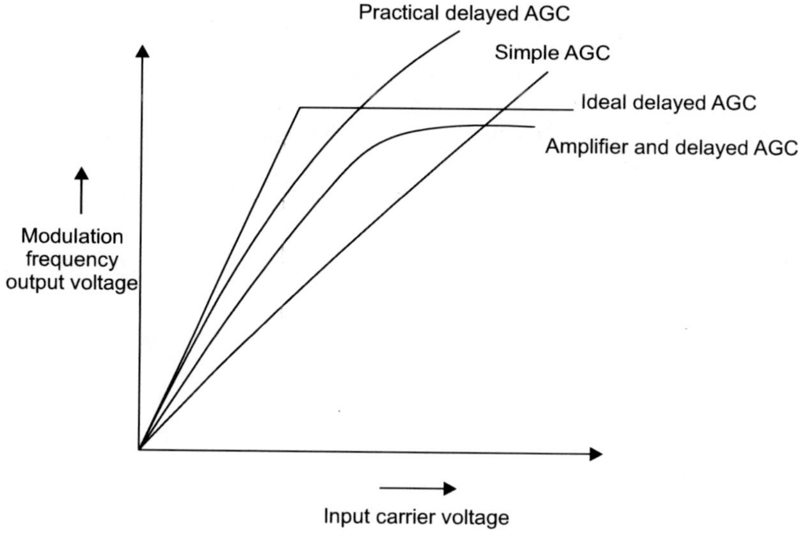

Fig. 3.9(b): AVC Characteristics

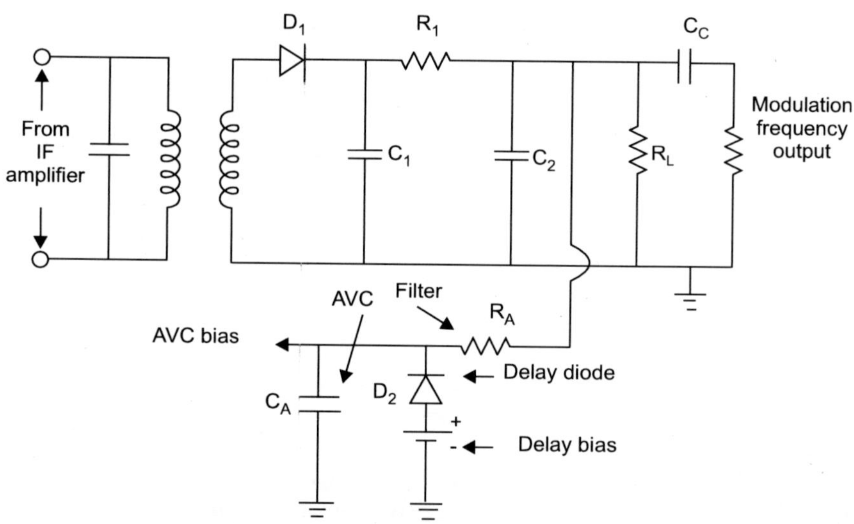

Fig. 3.9(c): Linear Diode Detector with Delayed AVC

zero and small input signals, diode D_2 conducts so that the AVC bias just equals the potential of cathode of this diode. Thus, AVC remains fixed at a low positive value. As the input carrier voltage increases, the AVC bias developed due to rectification of carrier voltage in detector diode D_1 increases. When this rectified bias magnitude exceeds the magnitude of the positive cathode voltage of diode D_2 then diode D_2 ceases to conduct and AVC system functions normally.

3.9.6 Tone Control

Tone control in receiver enables the listener to adjust relative amplitudes of higher frequency tones in the reproduced programme according to the nature of the programme and according to individual's taste. Several circuits for tone control have been designed. The one most commonly used is shown in Fig. 3.10.

A ckt consists of simple series combination of a fixed capacitor and variable resistor placed in shunt with the load resistor R_c in an audio voltage amplifier.

Thus, as the value of R is increased, relative amplitudes of higher frequency terms gets increased.

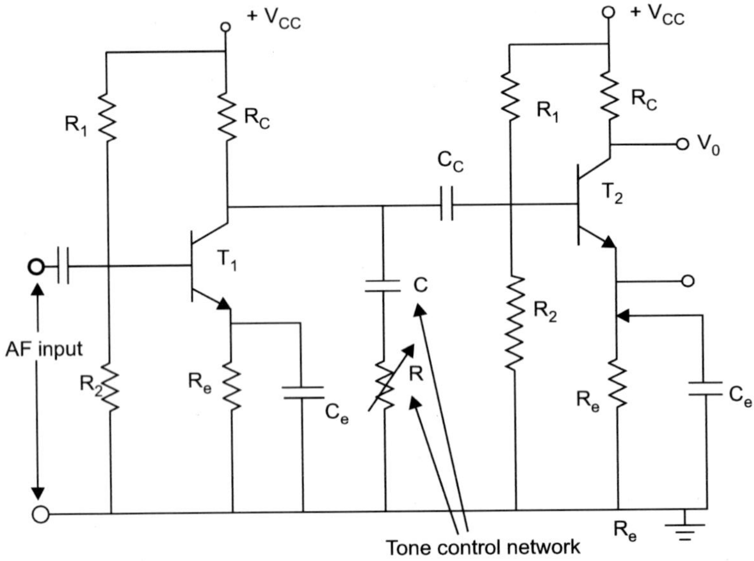

Fig. 3.10: Tone Control Circuit

3.9.7 Band Spread Tuning

In short-wave band, the frequency interval between adjacent stations in relation to the carrier frequency becomes small. The process of tuning of these stations then becomes tedious, very slight movement of tuning capacitor makes a large variation in tuned circuit frequency. It is therefore advantageous to spread out the tuning range so that considerably large movement of tuning dial results in only a small frequency change. Band spread is generally obtained through one of the following means:

1. Using auxiliary variable vernier trimmer capacitors ganged together. In this case, the receiver is first tuned approximately by main tuning control or coarse tuning control and then the receiver is tuned accurately with the help of vernier tuning control or fine tuning control. The method is generally employed in communication receivers.
2. By allocating full tuning band to a small range of frequencies. Thus, instead of normal 3-bands, a receiver may have the following 8 bands.

Band 1-13 metre band
Band 2-16 metre band
Band 3-19 metre band
Band 4-25 metre band
Band 5-31 metre band
Band 6-41 metre band
Band 7-49-90 metre band
Band 8-Medium wave band

Thus, band 2 covers only a few stations centred about 16 metres wavelength. Such a narrow band coverage with full variation of the tuning condenser can be achieved by switching in auxiliary reactance into the tuning circuit as shown in Fig. 3.11 a.

Figure 3.11b also shows bandspread with permeability tuning.

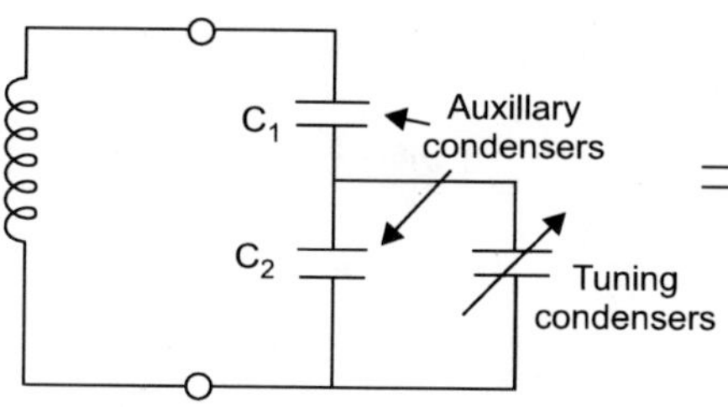

Fig. 3.11(a): Bandspread with Condenser

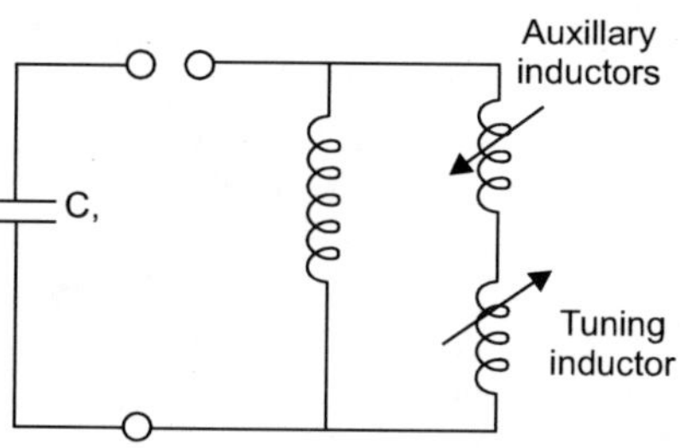

Fig. 3.11(b): Bandspread with Inductor Tuning

This type of bandspread is generally employed in broadcast receivers. Maintenance of frequency calibration in this bandspread tuning is rather difficult because small percentage band change in frequency due to temperature variation or aging may shift the pertinent band entirely outside the tuning range of the receiver.

3.9.8 Noise-Limiter

Noise voltages of impulsive nature are more annoying. A noise limiter simply removes from the output any voltage in excess of a certain predetermined level. This level is adjusted and is usually adjusted to a value slightly greater than the normal audio frequency signal voltage.

Figure 3.12 gives a simple circuit arrangement. D_1 is the noise limiter diode placed in shunt with the audio frequency coupling network between two audio amplifier stages of the receiver. The cathode of this limiter diode is given a positive bias V_0 with respect to ground. This bias determines the level of limiting.

If the positive voltage on limiter diode anode exceeds this bias, then the limiter diode conducts and bypasses the voltage. The bias is usually adjusted to a value only slightly greater than normal AF-signal level at this point in the receiver. Thus, the noise voltage amplitude is limited to a value only slightly above the signal level. Since, each stage of AF amplifier inverts the polarity of voltage, it is necessary to

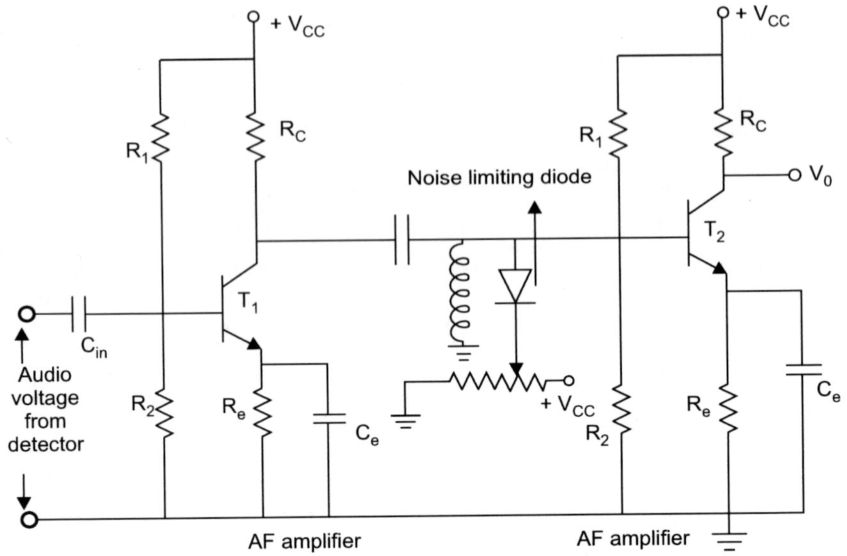

Fig. 3.12: Noise Limiter

ensure that the noise pulse obtained after rectification in detector is positive going at the limiter diode anode.

3.10 COMMUNICATION RECEIVERS

Communication receivers are high quality, short wave, multipurpose AM superheterodyne receivers designed for multipurpose reception over several methods of transmission such as broadcast signals, radio telephony. These receivers includes special features for noise-free uninterrupted reception of these signals and also use circuits to ensure accurate tuning. To provide these features, these receivers, therefore employ following circuits:

(a) Double superheterodyne reception
(b) Adjustable and high selectivity.
(c) Bandspread tuning
(d) Noise limiter or noise suppressors
(e) Beat frequency oscillator
(f) AFC (Automatic Frequency Control)
(g) Tuning indicator
(h) Calibration oscillator.

Figure 3.13 shows block diagram of communication receiver.

Description:

(a) *Input-Stage:* Normally communication receiver has one or sometimes two stages of RF amplification. Regardless of number of input stages, a system of band changing has to be employed if the same receiver has to be used to cover a

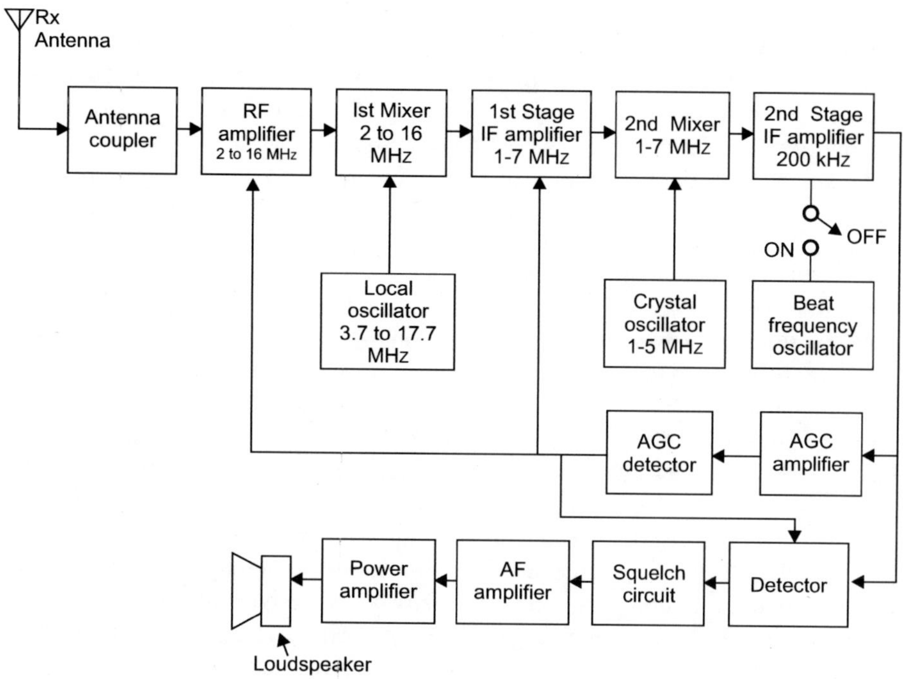

Fig. 3.13: Block Diagram of Communication Receiver

large frequency range, as mostly the situation is. The problem is that the normal variable capacitor cannot be used to cover a frequency ratio of more than 2 : 1, particularly at high frequencies. Thus, for various bands, particular RF stages, mixer and local oscillator coils, will have to be switched or frequency synthesis will have to be used.

In order to obtain maximum efficiency from different antenna system at different frequencies, a matching network is provided between the antenna and first RF stage. For this purpose different sockets for trimmer capacitors, tapped transformers or sometimes complete matching networks are provided. The coupling network, if adjustable is normally not tuned for optimum results in the middle of each band.

(b) *Double Conversion:* When communication receiver has two IF as shown in block diagram it is said to have a double conversion system. Communication receivers required to have very high performance use double conversion. The first IF is high, generally several MHz and second one quite low (of the order of 200 kHz). The working upto first mixer stage or even first IF amplifier stage is similar to the basic superheterodyne receiver. The output of RF amplifier is mixed in first mixer with the local oscillator frequency. The IF generated by this mixing is quite higher than 455 kHz of the basic domestic superheterodyne receiver. The first IF is amplified by high frequency IFA and its output is fed to second mixer. Here second IF is generated by mixing the first IFA signal with second local oscillator frequency, which usually uses a crystal oscillator. The second low IF is amplified by a low frequency IFA, which is next detected as usual.

The use of double conversion has been necessitated because of the considerations in selection of high or low IF. In ordinary receivers, a compromise is made in selection of IF, because there are equally good reasons for both higher and lower IF. By using double conversion method, we try to make use of all good points of both the choices of high and low IF. The high first IF pushes the image frequency further away from the signal frequency and thus permits its better attenuation. The low second IF, on the other hand, gives a good selectivity and good adjacent channel rejection besides the other merits of low IF. It is clear that the higher IF has to come first, otherwise the image frequency will not be sufficiently rejected at the input and will become inextricably mixed with the proper signal. Any number of IF stages after this will not be able to reject the image frequencies.

The overall effect of double conversion is higher image and adjacent frequency rejection, which cannot be achieved by a simple superheterodyne receiver. This method is not used for domestic receivers or any other receiver working in medium frequency band. However, receivers working in short-wave band can use double conversion advantageously, as the band is too overcrowded.

(c) *Blocking:* If a very weak signal is tuned, the AGC generated is proportionately low, thus keeping the gain of RFA high. Consider a situation of a very strong unwanted signal existing near the wanted signal. This signal, if not properly dealt can produce a substantial AGC, reducing the gain of the receiver and thus causing the wanted signal to be inaudible. This situation is unwelcome and causes trouble. A receiver whose AGC system shows very little reaction to the nearby spurious signal is said to have good blocking. To have good blocking a very high IF rejection of adjacent signals is required. Good blocking is more important for SSB receivers or receivers operating in crowded band.

(d) *Squelch Circuit: (Muting or Quieting):* Consider a sensitive receiver switched on and the signal being absent, i.e. no transmission on a given channel. A sensitive receiver will produce very high noise which is very annoying. This is because in the absence of any signal, AGC disappears thus causing the receiver to operate at its maximum sensitivity. The low level of noise at the input of receiver gets amplified with the receive operating on maximum gain in many cases the noise is not of much significance, but in many other cases this noise causes much trouble. Example: in the communications receivers used by police, ambulances, coast radio stations airborne receivers, etc. These receivers are required to tuned and manned at all times, but transmission is sporadic. To solve this problem, a squelch or muting or quieting circuit is used. Squelch circuit enables the receiver output to remain cut off unless the carrier is present. This system helps in increasing the efficiency of the operator.

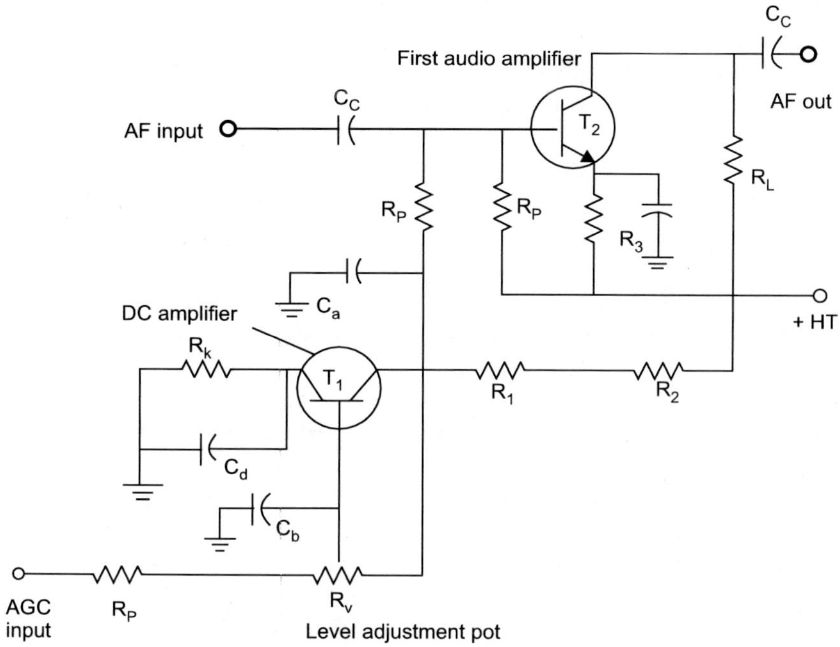

Fig. 3.14: Squelch Circuit

Figure 3.14 shows typical squelch circuit.

A circuit consists of a dc amplifier to which AGC is applied and which operates on the first stage of audio amplifier of the receiver. When AGC is absent or it very low, the dc amplifier T_1 draws current so that the voltage drop across its load resistor R_1 cuts off the audio amplifier T_2 and so no signal or noise is passed. When AGC voltage attains sufficient level so as to cut off T_1 then dc amplifier stops drawing collector current, so that the bias now on T_2 is its self bias only which is made available to it by the by-passed emitter resistor R_3 and by the base potentiometer resistors. The audio amplifier now works as though the squelch circuit was absent from the scene. R_v is a variable dropping resistor whose main job is to ensure that the high tension supplied to the collector and base potentiometer of T_2 is higher than the high tension supplied indirectly to its emitter. Manual adjustment of R_v allows the cut-in bias of T_1 to be varied so that quieting may be applied for a range of AGC values. This facility is important for reception of weak stations, which will not generate sufficient AGC so as to cut off T_1.

3.11 FM RECEIVER

FM receiver also uses the superheterodyne principle and is quite similar to an AM-receiver. However, there are some basic differences between FM and AM receivers such as:

(a) Much higher operating frequencies in FM.

(b) Need for limiting action in FM.
(c) Need for de-emphasis in FM.
(d) Different methods of FM detection.
(e) Different methods of obtaining AGC in FM.

3.11.1 Block Diagram of Superheterodyne FM Receiver

Figure 3.15 shows block diagram of superheterodyne FM receiver:

(1) *RF Amplifier or Signal Frequency Amplifier:* It serves the same function as in AM superheterodyne receiver. Thus, it serves (a) to raise the signal level appreciably before the signal is fed to the mixer and (b) to discriminate against the image signal.

But in FM broadcast, the signal bandwidth is large being 150 kHz, as against 10 kHz in AM broadcast. Hence, RF amplifier must be designed to handle this large bandwidth.

(2) *Frequency Mixer:* It performs the usual function of mixing or heterodyning the signal frequency voltage and the local oscillator voltage to produce the difference frequency voltge which is the IF-voltage. Since FM broadcast takes place either in VHF band or UHF band, single transistor frequency converter is not used. A separate local oscillator is always used and another transistors serves as the frequency mixer.

The IF used in FM receivers is higher than that in AM receivers operating at short waves. Typical value of IF is 12 MHz. This high IF helps in image rejection.

(3) *Local oscillator:* A separate local oscillator is always used. At ultra-high frequencies, it is preferred to keep the local oscillator frequency smaller than the signal frequency by an amount equal to IF.

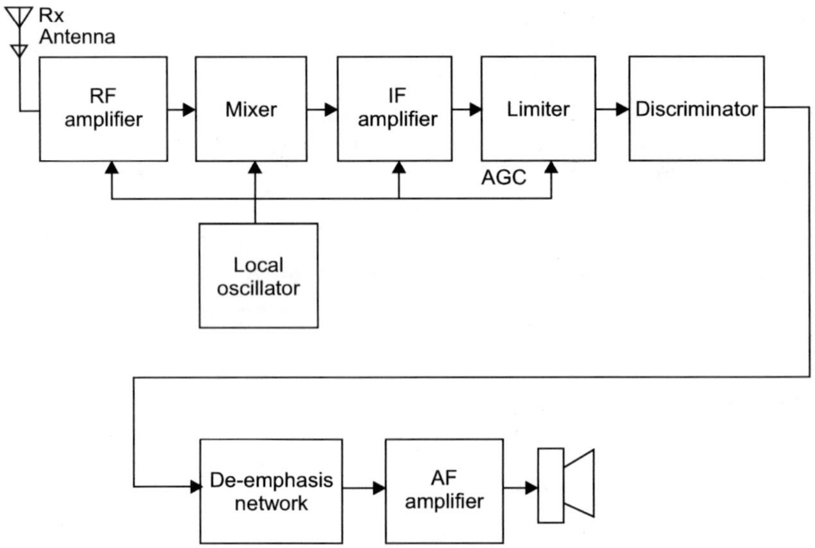

Fig. 3.15: Block Diagram of Superheterodyne FM Receiver

(4) *IF Amplifier:* A multistage amplifier is used to provide large gain. Further this IF amplifier should be designed to have high overall bandwidth of the order of the 150 kHz. Since the overall bandwidth decreases as the number of stages in cascade increases, it is necessary to design individual stages to have correspondingly higher bandwidth than the overall bandwidth desired. Double tuned circuits may be used but it is preferred, particularly at higher freqeuncies in the UHF range, to use stagger tuned single tuned circuit which are found to produce more gain bandwidth product than conventional double tuned circuits.

(5) *Limiter:* The IF amplifier is followed by a limiter which limits the IF voltage to a predetermined level and thus removes all amplitude variations which may be incidently caused due to changes in the transmission path or by man-made static or natural static.

(6) *FM-Detector:* This extracts the original audio modulation frequency voltage from FM carrier voltage. A discriminator is used as FM detector.

(7) *De-Emphasis network:* High frequency signals are boosted at the transmitter which is to be recovered at a receiver. The amount by which the high frequency signal is boosted must be recovered by the same amount at receiver. It is done by de-emphasis network.

(8) *Audio Frequency Amplifier:* The output of FM detector is fed to an audio frequency small signal amplifier and one or more audio frequency large signal amplifiers. The output audio voltage is then fed to loudspeaker. In FM broadcast, the maximum modulating frequency permitted is 15 kHz, and hence the audio frequency amplifier must be designed to accommodate such large bandwidth. Similarly, the loudspeaker must be capable of reproducing all high frequency tones upto 15 kHz. Often two or more loudspeakers are used, each reproducing a limited range of frequencies.

3.11.2 Amplitude Limiter

To avoid distortion, any AM present in FM signal must be removed before it is being fed to the demodulator. For this an amplitude limiter is inserted in the circuit in between the IFA and the demodulator. This is important as most FM demodulators react to amplitude variations as well as frequency changes. Thus, limiter acts as a clipping device which causes the output to remain constant irrespective of the changes in the input signal amplitudes. Most limiters behave in this way provided that input voltage remains within a certain predetermined range. The common types of amplitude limiters use two separate methods, namely the lead-type bias and early saturation.

Operation of Limiter:

Figure 3.16a shows a limiter using FET.

It can be seen that the drain supply voltage has been dropped through a resistor R_P. Also, the bias on the gate is leak type bias supplied by the parallel $R_g - C_g$ combination. Considering the high operating frequency, a capacitor C_N is used for neutralisation.

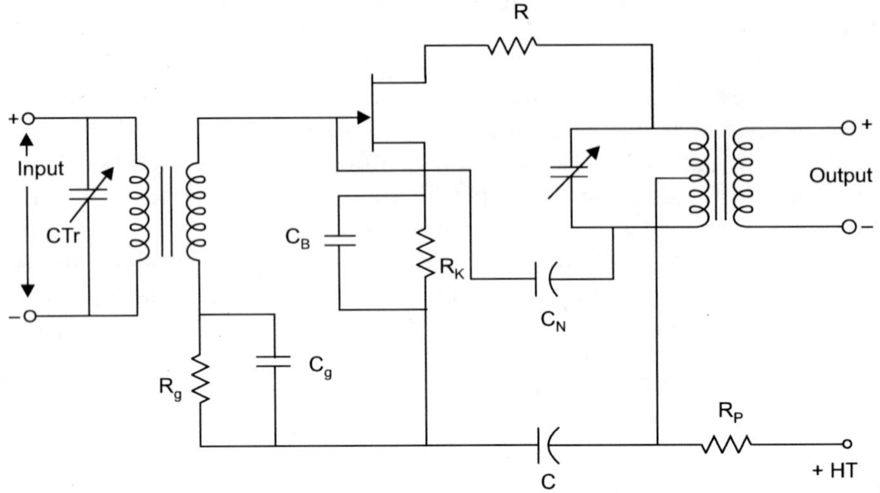

Fig. 3.16(a): A FET Amplitude Limiter

When an increase in signal voltage, current flows in $R_g - C_g$ bias circuit, resulting in a negative voltage across capacitor. The bias on the FET is increased in proportion to the size of the input voltage. As a result the gain of the amplifier gets reduced and the output voltage tends to remain constant. This method provides an insufficient limiting and it occurs for large input voltages. Early saturation method is a much better method. The saturation of the output current is achieved using low drain supply voltage. For this, a drain dropping resistor is put in the circuit (Fig. 3.16 (a)). The supply voltage is kept nearly one half of the normal dc drain voltage. As a result of early saturation, limiting is achieved with conveniently low input voltages. It is quite possible that the gate drain section may become forward biased under saturated conditions resulting in a short circuit between that input and output. To avoid this, a resistance "R" is placed between drain and its tank circuit.

Figure 3.16b and c shows the transfer characteristics for amplitude limiter and typical amplitude limiter response characteristics.

From figure it is clear that limiting action takes place only for a certain range of input voltages. Outside this range, output voltage varies with input. Referring to Fig. 3.16b, we see that as input increases from point a to b, output current increases along with it. Thus, from point a to b, no limiting action takes place. From point b onwards, we see that output remains constant indicating that limiting action has started. Point b at which limiting action starts is called *limiting-threshold*. From points c to d, no rise in output takes place except that the output current flows for some smaller duration of time like a class-C amplifier. As in class-C amplifier, the tank circuit is used here to get a sinusoidal output using the current pulses of Fig. 3.16c.

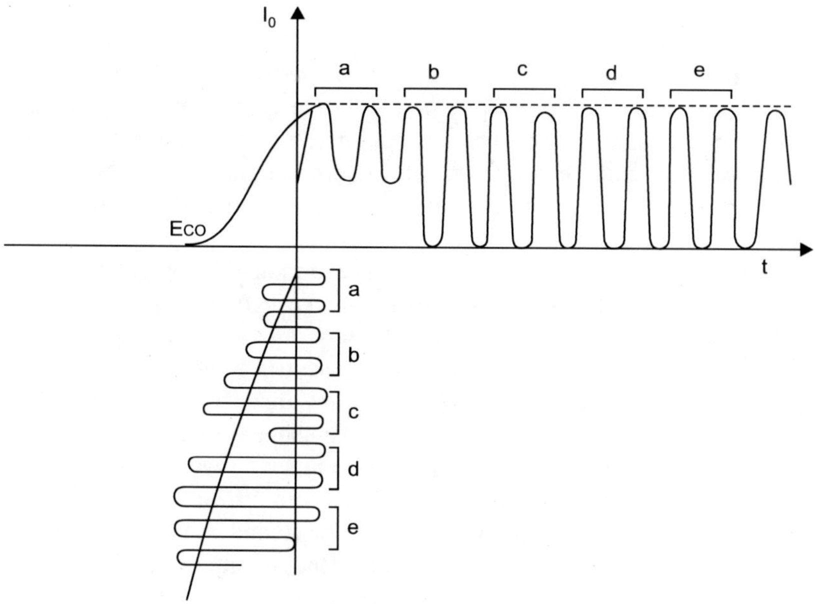

Fig. 3.16(b): Transfer Characteristics of Limiter

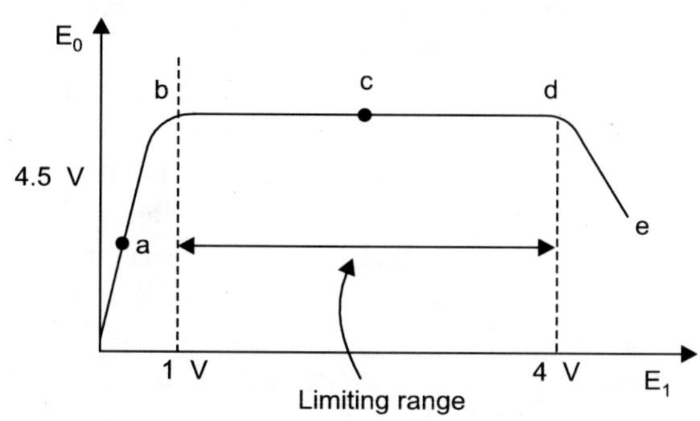

Fig. 3.16(c): Amplitude Limiter Response Characteristics

After point d, the angle of conduction of current becomes so small that the voltage across the tank circuit falls till point e. Thus, the output voltage falls, and no longer remains constant. The range of input voltage in between which limiting takes place (point "b" to 'd") is called limiting range.

3.12 MEASUREMENT OF RECEIVER PERFORMANCE

(Standard tests on Radio Receivers)

Measurement of overall performance of radio receivers is necessary not only to evaluate the relative merits of a particular circuit design but also to indicate the usefulness of a receiver under any special operating conditions.

In case of broadcast receivers, exact location of a particular receiver is never known and a properly designed receiver will give satisfactory performance at all locations. Measurement of the receiver performance would therefore give an idea of a good design. On the other hand, a designer of a communication receiver known approximate working conditions of the receiver and the performance tests would show the extent to which receiver would be suitable under those operating conditions.

In general, a radio receiver has to perform several distinct functions. Firstly, it should be capable of receiving satisfactorily the weakest possible signal and provide sufficiently high output. Secondly, it must receive only the signal desired by the user and reject all other signals that may be present in atmosphere at that time. Thirdly, the signal received must be free from any distortion and should have a wave-shape that resembles closely with the wave-shape of the modulating signal at the transmitter. Fourthly, the signal should be received with the least possible noise. Lastly, the receiver should consume a minimum of the power. This is especially important for sets working on dry cells or secondary batteries. Following tests are usually carried out on receiver to judge the quality of the receiver set.

(1) *Sensitivity:* Sensitivity of a radio reciever is a measure of its ability to receive weak signals. Since a receiver receives very small signals, amplifies them and converts them into original modulating signal, it is the receiver gain that basically determines the sensitivity of a radio receiver.

Sensitivity of AM radio broadcast receiver is defined as the input carrier signal modulated to 30% with a modulating signal of 400 Hz which when applied to an antenna terminals of the receiver produces a standard output. For FM-broadcast receivers, sensitivity may be defined as the input carrier voltage frequency modulated by a 400 Hz signal so as to produce a deviation of 22.5 kHz, required to produce a standard output at the receiver output points when all receiver controls are positioned for maximum output.

For receivers capable of giving an undistorted output of 1 watt or more, the standard output is taken to be 500 MW. For receivers, having an undistorted output power lower than 1 watt, the standard output is taken as 50 MW.

The sensitivity of radio receivers usually lies in the range of a few microvolts and may be expressed in decibels below 1 volt or in terms of the signal power at the receiver input points described in decibels below 1 mw (dbm). Since the receiver gain is not constant over a receiver frequency band, the sensitivity is different at different frequencies, in a given receiver. A typical sensitivity curve is shown in Fig. 3.17a whole the experimental set up for the measurement is shown in Fig. 3.17b.

(2) *Selectivity:* Selectivity of a radio-receiver is its capability to pick up a desired signal while rejecting all other signals. Since, the selection of signal is done by resonant circuits, the receiver selectivity is entirely dependent upon the frequency

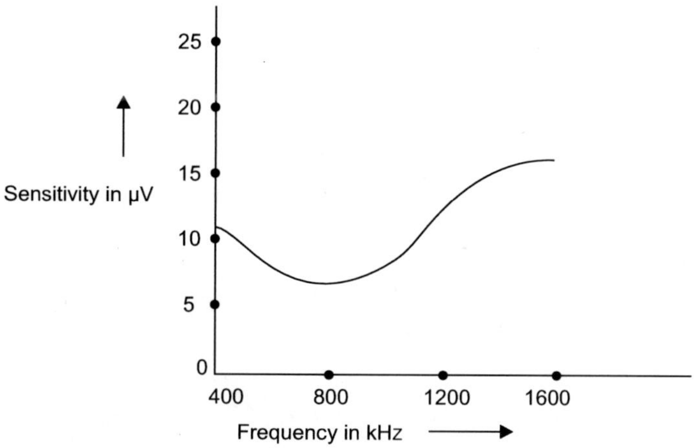

Fig. 3.17(a): Sensitivity Curve for MW Band of a Broadcast Receiver

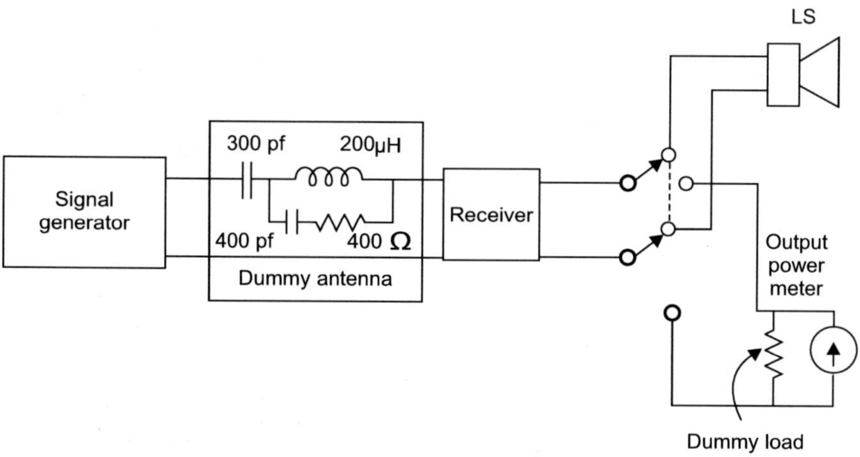

Fig. 3.17(b): Experimental Setup for Sensitivity Measurement

response of these circuits. A resonant circuit with a high magnification factor Q is more selective than one with low Q.

Selectivity of receiver may be expressed in terms of the ratio of the receiver sensitivity when a standard modulated signal off tuned by a specific amount from the frequency to which the receiver is tuned to its sensitivity when the signal is correctly tuned.

For selectivity measurement, the receiver is tuned to a frequency at which the measurement is to be made. Sensitivity measurement is then made using the setup described in Fig. 3.17b. Signal generator frequency is then changed on the upper and lower side of the receiver frequency and signal input is adjusted to give a standard output in each case. Selectivity curve is shown in Fig. 3.17c.

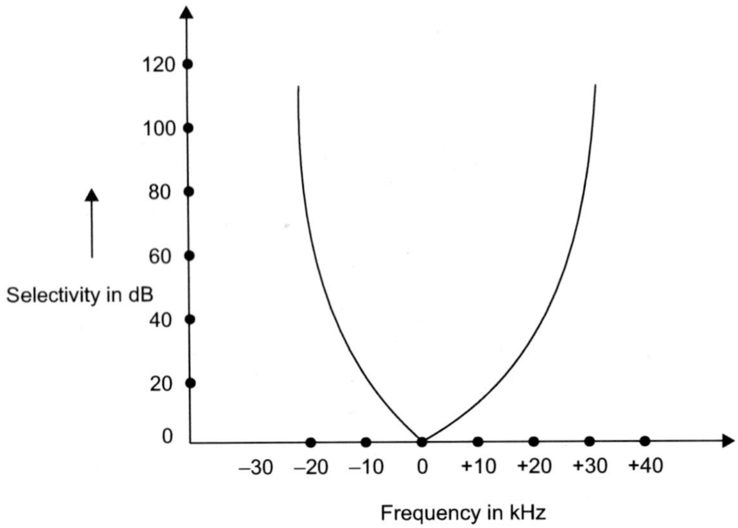

Fig. 3.17(c): Selectivity Curve of a Broadcast Receiver

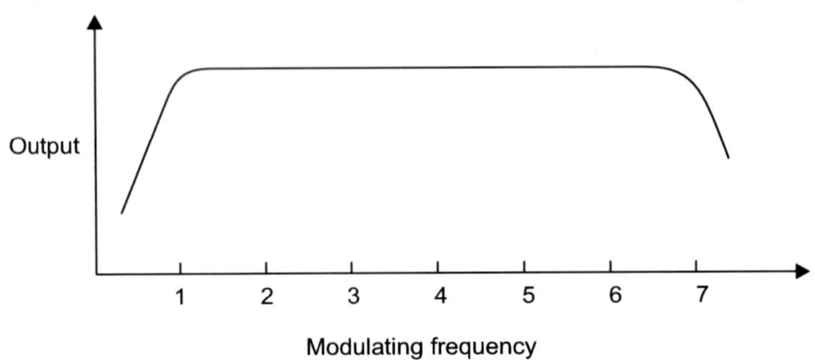

Fig. 3.17(d): Fidelity Curve

(3) *Fidelity:* Fidelity of receiver is the ability of receiver to reproduce correctly different modulating frequency components in an input signal. The experimental setup is same as that used in sensitivity measurement but the carrier signal is modulated by an external audio signal in this case. The receiver is correctly tuned to the signal frequency and standard output is obtained at 400 Hz. Modulating frequency is then varied over the entire audio band and variation in the receiver output is noted down.

Fidelity curve is then plotted for receiver output versus modulating signal frequency. Overall fidelity of a typical receiver is shown in Fig. 3.17d.

(4) *Signal to Noise Ratio (S/N ratio):* S/N ratio may be defined as the ratio of the signal to noise power at the receiver output. Setup is made as per sensitivity

measurement and unmodulated carrier of any frequency is given to receiver to which the latter is also tuned. The receiver output then corresponds to internal noise which modulates the carrier. Noise power measurement is made preferably by the use of a thermocouple instrument. New modulation at 400 Hz is switched on and degree of modulation is adjusted to give the same output power as obtained in the first case. Noise input voltage is then given by $E_n = m\ E_c$ where m is the depth of modulation and E_c is the carrier voltage. E_c is quite large in comparison to E_n. AVC line is disconnected while making this measurement.

S/N voltage ratio at receiver output may then be expressed as

$$\text{S/N voltage ratio} = \frac{m_a\ E_0}{E_n}$$

where, E_0 & m_a are the carrier amplitude and degree of modulation respectively; of the input signal frequency for which S/N ratio is desired.

3.13 NOISE SOURCES

There are number of ways of classifying noise. Classification can be done according to the type, source, effect or relation to the receiver, depending upon the circumstances.

We subdivide noise into broad sections, namely noise resources which are external to the receiver and noise produced within the receiver itself.

External noise cannot be reduced except by changing the location of the receiver or the entire system. Internal noise on the other hand can be easily evaluated and can be reduced to great extent.

3.13.1 External Noise

The various types of noise created outside the receiver come under this category and include atmospheric, man-made, an extraterrestrial noise.

3.13.2 Atmospheric Noise

A very easy way of feeling the effect of atmospheric noise is by just switching in the broadcast receiver for short-wave reception. A wide variety of sounds can be heard other than the original signal, which interfere with it. Most of these spurious sounds are due to the random radio waves which induce voltage in the antenna and the majority of these radio waves come from natural resources of disturbances called static. It originates from thunderstorms discharges between clouds, etc.

This type of noise is random in nature but its intensity varies spasmodically. Field strength of this type of nose is approximately proportional to the frequency, so that this noise interferes more with the reception of radio than with TV. Static from distant sources varies according to the propagating weather conditions. Hence, the level of the static rises at night time.

3.13.3 Extraterrestrial Noise

The sources for extraterrestrial noise are so many that it is not possible to pen all them. This type of noise is being divided into two subgroups.

Cosmic Noise: Under normal condition there is a constant noise radiation from any large body which is at a high temperature. Thus, the large number of distant stars all emit radiation whose effect can be easily felt on earth. The noise received is called thermal (or black body) noise and is quite uniformly distributed over the entire sky.

Solar noise: Under quite conditions, sun emits is a constant noise simply it being a large body at a very high temperature. It thus radiates over a very broad frequency spectrum which includes nearly all the freqeuncies used for communication purposes. However, the sun is a variable star and so it undergo cycles at the peak of which electrical disturbance occurs, such as sunspots corona flares, etc. Even though this additional noise energy comes from a very small part of the sun, but its proportion to the noise of quiet sun is many times more.

Since noise is more predominant in frequency range from about 8 MHz to about 1.43 GHz the latter frequency corresponding to 21 cm hydrogen line. It is strongest in the range from 20 to 120 MHz.

3.13.4 Man-made Noise

Man-made noise arises from electrical devices and so is especially strong in cities, particularly in industrial areas. Man-made noise lies approximately in the region of 1 to 600 MHz. Man-made noise in urbanised areas for outstrips the noise from any natural source.

Main sources of man-made noise are electrical arcs in electric motors, switch-gear, leakage and corona discharge from high voltage transmission lines, fluorescent lights, ignition system in auto-engines, electric call bells, aircraft ignition and various other electric machines. Interference from-man made noise is increasing at a rapid rate and there is no way out to reduce it. It can be minimized at the receiver output by proper shielding.

3.13.5 Internal Noise

Internal noise can be defined as the noise generated by any of the active or passive devices in the receiver. This types of noise is random and so it cannot be treated on individual voltage basis, but can be easily studied on statistical basis. Internal noise is distributed randomly over the entire frequency spectrum. This random noise power is proportional to the bandwidth over which it is measured.

3.13.6 Thermal Noise

Conductor contain a large number of free electrons and ions strongly bound by molecular forces. The ions vibrate randomly about their normal positions. However, this vibration being a function of the temperature. Continuous collisions in between the electron and the vibrating ions takes place. Thus, there is a continuous transfer of energy between the ions and electrons. This is the source of resistance in conductor. The movement of electron constitute a current which is purely random in nature and over a long time averages zero. There is random motion of the electrons which gives rise to noise voltage, called thermal noise. The noise generated in any resistance due to random motion of electrons is called *thermal noise* or white or *Johnson noise.*

The analysis of thermal noise is based on the kinetic theory. It shows that the temperature of a particle is way of expressing its internal kinetic energy. Thus, temperature of a body can be said to equivalent to the statistical rms value of velocity of motion of particles in the body. At 273°C (or zero degree Kelvin) the kinetic energy of the particles to a body becomes zero. Thus, we can relate the noise power generated by a resistor to be proportional to its absolute temperature.

Noise power is also proportional to the bandwidth over which it is measured. From this discussion, we can write.

$$P_n \,\alpha\, T\,B$$

$$P_n = K\,T\,B \qquad (1)$$

where; P_n = Maximum noise power output or resistor.

K = Boltzmann's constant = 1.38×10^{-23} joules/kelvin.

T = Absolute temperature.

B = Bandwidth over which noise is measured.

From equation (1) an equivalent circuit can be drawn as shown in Fig. 3.18.

From Fig. 3.18 the resistance equivalent noise voltage E_n can be calculated. R_l is considered to be absolutely noiseless and it receives the maximum noise power generated by R. According to maximum power transfer theorem R should be equal to R_l. Then;

$$P_n = \frac{E^2}{R_l} = \frac{E^2}{R} = \frac{(E_n/2)^2}{R} = \frac{E_n^2}{4R}$$

$$\therefore E^2 = 4R\,P_n = 4R\,K\,T\,B \qquad \text{(since } P_n = KTB\text{)}$$

$$E_n = \sqrt{4\,K\,T\,R\,B} \qquad (2)$$

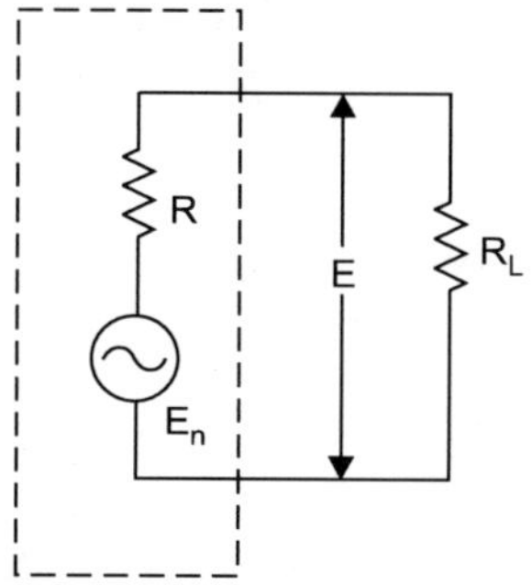

Fig. 3.18: Equivalent Circuit of Resistance Noise Generator

Example 3.1: An RF amplifier is having an input resistor of 8 KΩ and works in frequency range of 12 to 15.5 MHz. Calculate the rms noise voltage at the input of this amplifier at an ambient temperature of 17°C.

Solution:

$$E_n = \sqrt{4KTBR}$$

$$= \sqrt{4 \times 1.38 \times 10^{-23} \times (17 + 273) \times (15.5 - 12) \times 10^6 \times 8 \times 10^3}$$

$$= \sqrt{4 \times 1.38 \times 290 \times 2.5 \times 1 \times 10^{-14}}$$

$$E_n = 21.17 \ \mu V$$

The rms noise voltage due to kinetic energy of particles is so small that it cannot be measured by ordinary voltmeter. For measurement of this voltage an electronic a.c. voltmeter has to be used.

3.13.7 Shot Noise

This is another type of noise found in communication receivers. Shot noise is due to amplifying devices, i.e. due to active devices only.

The fluctuating electrons in vacuum tube constitutes a randomly varying noise current superimposed on the direct current of the output electrode of the amplifying device. When amplified this noise sounds as a shower of lead shots falling on metal sheet giving rise to its name as shot noise. Shot noise behaves exactly as thermal noise, although its origin is different. There are number of variables involved in the generation of shot noise in either tubes or transistors and so approximate equations are used to calculate it. Shot noise current is difficult to add to the thermal noise in calculations and so for all devices with the exception of a diode, shot noise formulae have to be simplified.

For a diode the formula is

$$i_n = \sqrt{2ei_p B}$$

where, i_n = RMS shot noise current
e = charge on the electron = 1.6×10^{-19} coulombs.
B = bandwidth of the system
i_p = direct diode current.

3.13.8 Flicker Noise

Thermal and shot noise are considered to be the most important sources of noise in electronic circuitry. In the low frequency range (below about 5 kHz) it is noticed that the random fluctuation of the anode current is much larger than that expected due to shot noise alone.

Schottky explained this effect as due to the random changes in emissivity of the cathode, resulting in flicker noise. The change in emission is thought to be due to the appearance or departure of foreign atoms at the cathode surface. The life-time of these atoms is comparatively longer and so this type of noise is more pronounced at low frequencies. This noise behaves as $\frac{1}{f}$. Flicker noise is also proportional to junction temperature of a transistor.

3.14 NOISE CALCULATION

3.14.1 Addition of Noise from Several Sources

Consider two separate thermal noise sources in series, with the noise voltage as $E_{n1} = \sqrt{4KTR_1B}$ and $E_{n2} = \sqrt{4KTBR_2}$. The total voltage or the equivalent effect of these two sources can be expressed as

$$E_{nt} = \sqrt{E_{nt}^2 + E_{n2}^2} = \sqrt{4KTBR_1 + 4KTBR_2}$$

$$= \sqrt{4KTB(R_1 + R_2)}$$

$$E_{nt} = \sqrt{4KTBR_t} \qquad (3)$$

where $R_t = R_1 + R_2 + \ldots + R_n$

Equation (3) is valid for any number of sources. Thus, to find total noise voltage generated by a number of sources in series be easily expressed in terms of the total equivalent noise resistance.

Example 3.2: Calculate the noise voltage at the input of a RF amplifier using an active device with 280 Ω equivalent noise resistance and a 310 Ω input resistor. The frequency range of operation of the amplifier is from 10 to 14 MHz. Take temperature as 27°C.

Solution:

$$E_{nt} = \sqrt{4KTBR_t}$$

As two resistances come in series

$$R_t = 280 + 310 = 590\ \Omega$$

$$T = 273 + 27 = 300°\ K$$

$$B = 14 - 10 = 4\ MHz$$

$$\therefore \quad E_{nt} = \sqrt{4KTBR_t}$$

$$= \sqrt{4 \times 1.31 \times 10^{-23} \times 300 \times 4 \times 10^{6} \times 590}$$

$$= \sqrt{4 \times 3 \times 1.38 \times 4 \times 59 \times 10^{-14}}$$

$$E_{nt} = 6.25\ \mu V.$$

Equivalent circuit of two stages of cascade amplifier:

As we have been able to successfully add noise due to two series resistors, we can similarly add noise due to several amplifiers in cascade by little modified version of the formula used before. The situation is shown in Fig. 3.19.

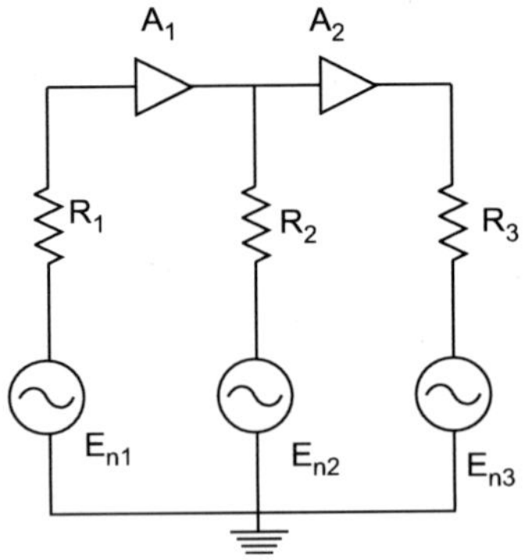

Fig. 3.19: Equivalent Circuit of Two Stages of Cascade Amplifier

The circuit shows that each amplifying stage has input and output resistances. The object now is to find the equivalent effect of all the resistances from noise point of view. One way of solving the problem is to find the noise generated by input resistance of first stage, multiply it with gain of the second stage, add this voltage to that generated at the input of second stages and then multiply by the gain of second stage to find noise output voltage after two stages. Similar method can be used for any number of stages, but this only results in the output noise voltage while we are interested in the input noise voltage of the equivalent input noise resistor for all the stages. Thus, we can replace all the noise generated in the receiver by just one resistance at the input of the receiver.

Let this resistance be termed "R_{eq}" located at the input. R_{eq} is good parameter for comparing the performance of various receivers and permits a quick calculation of the lowest input signal which this receiver may amplify without drawing it with noise.

Let

A_1 = gain of the first stage of amplifier

A_2 = gain of second stage of amplifier.

R_1 and R_2 = two input noise resistances.

R_3 = output resistance.

Noise voltage due to R_3 is E_{n3}.

$$E_{n3} = \sqrt{4KTBR_3}$$

This noise voltage E_{n3} can be transferred at the input point of second stage by dividing E_{n3} by A_2.

$$\frac{E_{n3}}{A_2} = E_{n3} = \frac{\sqrt{4KTBR_3}}{A_2} = \sqrt{4KTBR_3}$$

Here R_3 is resistances which is placed at the input point of second stage would generate a noise voltage equivalent to that if R_3 was present at the input of second stage

$$\therefore\ R_3' = R_3 / A_2^2 \tag{4}$$

Equation (4) illustrate that when a noise resistance is transferred to the input from output stage of an amplifier it has to be divided by square of the gain of that stage. The noise resistance actually present at the output of the first stage (or the input of the second stage) is R_2. Thus, the equivalent noise resistance at the output of first stage is given by

$$R_{eq} = R_2 + R_3 = R_2 + \frac{R_3}{A_2^2}$$

Similarly, we can now find R_2' as

$$R_2 = \frac{R_{eq}'}{A_1^2} = \frac{R_2 + {R_3}/{A_2^2}}{A_1^2} = \frac{R_2}{A_1^2} + \frac{R_3}{A_1^2 A_2^2}$$

The noise resistance actually present at the input of first stage is R_1. Therefore, total resistance at the input of the first stage R_{eq} is given by

$$R_{eq} = R_1 + R_2 = R_1 + \frac{R_2}{A_1^2} + \frac{R_3}{A_1^2 A_2^2}$$

This argument can be extended to any number of amplifier stages in cascade.

Example 3.3: The first stage of a two stage RF amplifier has a voltage gain of 15, a 600 Ω input resistance, a 2000 Ω equivalent noise resistance and a 30 kΩ output resistance. For second stage, similar values are 20, 100 kΩ, 8 kΩ and 1.4 mΩ respectively. Calculate the overall equivalent input noise resistance.

Solution:

$$R_1 = 600 + 2000 = 2600\ \Omega$$

$$R_2 = \frac{30\times100}{30+100}+8 = 23.079+8$$

$$R_2 = 31.08 \text{ k}\Omega,\ R_3 = 1.4 \text{ m}\Omega$$

$$R_{eq} = 2600+\frac{31080}{(15)^2}+\frac{1.4\times10^6}{(15)^2\times(20)^2}$$

$$R_{eq} = 2600 + 138.13 + 15.56 = 2753.69\ \Omega$$

This example shows that the resistor located at the input of the first stage is the greatest contributor of noise voltage. Thus, it is not necessary to consider the noise effect of more than two stages as it becomes insignificant.

3.15 NOISE FIGURE

3.15.1 Signal to Noise Ratio (S/N Ratio)

In cases when it is difficult to obtain equivalent noise resistance or in some particular cases we use a parameter called S/N ratio for evaluation or comparison performances various circuits.

Signal to noise ratio often written as S/N ratio is defined as the ratio of signal power to noise power at the same point. Mathematically:

$$\frac{S}{N} = \frac{P_s}{P_n} = \frac{E_s^2/R}{E_n^2/R} = \left(\frac{E_s}{E_n}\right)^2$$

S/N ratio should be maximum.

4.15.2 Noise Figure

During the process of transmission and recovery of information, it is necessary to use amplifier. At the transmitter end the signals are amplified to a suitable level before transmission and at the receiver side amplifiers are used to strengthen the signal. The output of amplifier is always an imperfect replica of the input. There are two main reasons for this imperfectness.

First reason is due to frequency response characteristics, example linearity of the amplifier which results in distortion of the output signal as compared to the input. Second imperfectness is due to the noise added to the output signal by the amplifier itself. This noise is mainly due to thermal and shot noise. The noise can be reduced to quite a low value by proper design of receiver.

Thus, to evaluate the performance of a receiver design we need such a parameter which takes into consideration the overall damaging effect of the noise in the entire receiver system or amplifier. A most widely used parameter called "noise figure" describes completely the above-mentioned points.

Noise figure (F) is defined as the ratio of the signal-to-signal power supplied to the input terminals of a receiver or amplifier to S/N ratio power supplied to the output or load resistor.

Mathematically:

$$F = \frac{(S/N)_{input}}{(S/N)_{output}}$$

When the amplifier introduces no noise, noise figure becomes unity.

Noise figure involves the total signal and noise powers at the input & output. On the other hand, noise figure can be defined at a single frequency which represents the ratios of powers at a given frequency. The noise figure usually measured practically is called integrated noise figure which corresponds to a band of frequency. In general, it is the single frequency noise figure which can be easily computed from other parameters. Noise figure may be expressed as an actual ratio or in decibels, as may be convenient. The noise figure of a practical receiver can be designed to as low as few decibels upto frequencies of the order of 1 GHz by suitable choice of the device.

3.15.3 Calculations of Noise Figure

The amplifier or a receiver can be analyzed as a single unit for noise figure purposes. Consider a four terminal network of Fig. 3.20.

Let the input impedance of the four terminal network be R_i and its output impedance be R_0. This four terminal network is fed from a single source (in case of receiver, this generator is an antenna) with internal impedance R_a.

The calculation procedure can be broken into a number of steps as

(a) To determine signal input power (P_{si})

(b) To determine the noise input power (P_{ni})

(c) To calculate the input S/N ratio from the ratio of P_{si} and P_{ni}.

(d) To determine signal output power (P_{s0})

(e) Write P_{no} for the noise power output to determine afterwards.

(f) Calculating the output $\frac{S_0}{N_0}$ from the ratio of P_{So} and P_{no}.

(g) To determine the generalized form of noise figure from steps 3 and 6.

(h) Calculating P_{no} from R_{eq} and then substituting it into the general equation for noise figure. "f" to obtain the actual formula, or first determine P_{no} by measurement to substitute it to obtain the formula for "f".

Figure 3.20 shows block diagram for noise figure calculation.

From Fig. 3.20 it can be shown that input signal voltage E_{si} can be expressed as

$$E_{si} = \frac{E_s R_i}{R_a + R_i}$$

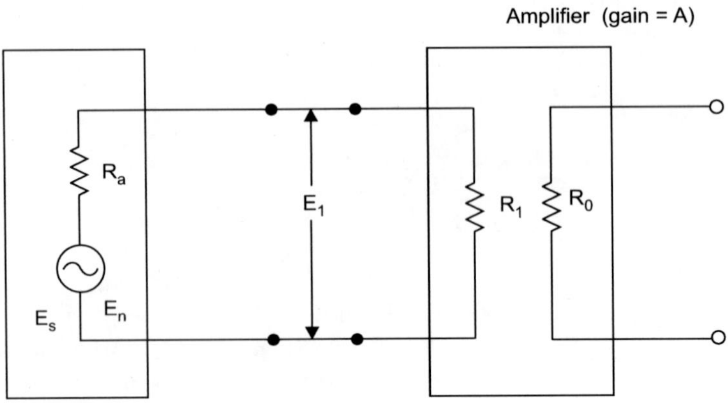

Fig. 3.20: Block Diagram for Noise Figure Calculation

Also
$$P_{si} = \frac{(E_{si})^2}{R_i} = \left(\frac{E_s R_i}{R_a + R_i}\right)^2 \cdot \frac{1}{R_i}$$

$$P_{si} = \frac{E_s^2 R_i}{(R_a + R_i)^2}$$

Similarly, noise voltage can be expressed as

$$E_{ni}^2 = 4KTB \cdot \frac{R_a R_i}{R_a + R_i}$$

then;
$$P_{ni} = \frac{(E_{ni})^2}{R_i} = 4KTB \cdot \frac{R_a \cdot R_i}{R_a + R_i} \frac{1}{R_i}$$

$$P_{ni} = \frac{4KTBR_a}{R_a + R_i}$$

Thus, S/N ratio at the input will be

$$\frac{S_i}{N_i} = \frac{P_{si}}{P_{ni}} = \frac{E_s^2 R_i}{(R_a + R_i)^2} \div \frac{4KTBR_a}{(R_a + R_i)}$$

$$\frac{S_i}{N_i} = \frac{E_s^2 R_i}{4KTBR_a (R_a + R_i)}$$

The signal power at output will be

$$P_{so} = \frac{E_{so}^2}{R_o} = \frac{(AE_{si})^2}{R_o}$$

$$= \left(\frac{AE_sR_i}{R_a + R_i}\right)^2 \cdot \frac{1}{R_o}$$

$$P_{so} = \frac{A^2E_s^2R_i^2}{(R_a + R_i)R_o}$$

It is difficult at this stage to calculate the output noise power, for time being let us assume it to be just P_{no}.

Then, the output S/N ratio results as

$$\frac{S_o}{N_o} = \frac{P_{so}}{P_{no}} = \frac{A^2E_s^2R_i^2}{(R_a + R_i)^2R_oP_{no}}$$

Noise figure is finally obtained as:

$$F = \frac{\frac{S_i}{N_i}}{\frac{S_o}{N_o}} = \frac{E_s^2R_i}{4KTBR_a(R_a + R_i)} + \frac{A^2E_s^2R_i^2}{(R_a + R_i)^2R_oP_{no}}$$

$$F = \frac{P_{no} \cdot R_o(R_a + R_i)}{4KTBR_a\ R_iA^2} \tag{5}$$

In this equation, P_{no} is such a parameter which has to be substituted here from the knowledge of equivalent noise resistance or by practically measuring it. Consider the first method, i.e. calculating P_{no} by equivalent noise resistance method. The equivalent noise resistance of an amplifier or receiver is the sum of the input terminating noise resistances of the first stage, together with the noise resistances of previous stages transferred to the input. All these resistances mentioned above can be represented by an equivalent lumped resistance R_i which is responsible for the entire noise generation in the received. These noise resistances have to be added to the parallel combination of R_a and R_i. In order to correlate noise figure and equivalent noise resistance we define another term R'_{eq}, which is noise resistance that does not include R_i and is given by

$$R'_{eq} = R_{eq} - R_i$$

The total equivalent noise resistance of the receiver will be

$$R = R'_{eq} + \frac{R_aR_i}{R_a + R_i}$$

The equivalent noise resistance presumed to be generated at the input of the receiver will be

$$E_{ni} = \sqrt{4KTBR}$$

As the receiver is now considered to be completely noiseless, the noise output power can be expressed as

$$P_{no} = \frac{E_{no}^2}{R_0} = \frac{(AE_{ni})^2}{R_o}$$

$$P_{no} = \frac{A^2 4KTBR}{R_o} \tag{6}$$

The final result of noise figure is obtained by substituting equation (6) in equation (5)

$$F = \frac{(R_a + R_i)R_o}{4KTBA^2R_aR_i} \cdot P_{no}$$

$$= \frac{(R_a + R_i)R_o}{4KTBA^2R_aR_i} \times \frac{A^2 4KTBR}{R_o}$$

$$= R \cdot \left(\frac{R_a + R_i}{R_a\ R_i}\right)$$

$$= \left(R'_{eq} + \frac{R_aR_i}{R_a + R_i}\right)\frac{R_a + R_i}{R_a\ R_i}$$

$$F = 1 + \frac{R'_{eq} \cdot (R_a + R_i)}{R_a\ R_i} \tag{7}$$

Equation (7) indicates that for noise figure to be minimum for a given antenna resistance R_a the ratio $(R_a + R_i)/R_i$ must also be a minimum. For this to be true, we make R_i much larger as compared to R_a so that the ratio $(R_a + R_i)/R_i$ near unity. This is a mismatched condition and results in noise figure formula.

$$F = 1 + \frac{R'_{eq}}{R_a}$$

This relation is valid for highly mismatched conditions and for matched conditions this formula is not true.

Noise Figure Measurement

We know the equation for noise current generated by vacuum diode as an $i_n = \sqrt{2 V_{ip}\, \delta f}$. Plate noise current is adjusted by means of arrangement formed by d.c. source and potentiometer connected to cathode. Depending upon the position of potentiometer, plate current changes because emission of electrons from cathode depends upon amount of supply voltage.

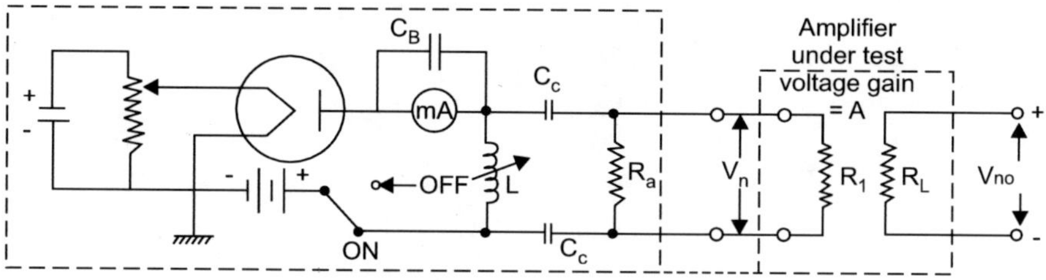

Fig. 3.21: Diode Noise Generator

Output capacitance of diode and associated circuit of diode noise generator is resonated at operating frequency by means of variable inductance. Equivalent noise resistance of diode is R_a and terminal resistance is R_t. Thus, equivalent noise voltage supplied to receiver input by noise generator is given by

$$V_{n_i} = I_{n_i} \cdot Z_{n_i} = I_{n_i} \times \frac{R_a R_t}{R_a + R_t}$$

$$= \frac{R_a\ R_t}{R_a + R_t} \times \sqrt{2 V_{i_p}\, \delta f}$$

Keep noise generator circuit OFF, measure output noise voltage. Now switch on noise generator circuit and adjust filament voltage by means of filament potentiometer so that diode plate current begins to flow. Adjust filament voltage in such a way that output noise voltage developed across R_L should be greater than twice noise voltage developed across R_L when filament supply was OFF

We know that noise figure is given by

$$\text{Noise figure} = F = \frac{R_L (R_a + R_t) \times P_{no}}{A^2 4KT \times \delta f\ R_a\ R_t} \tag{8}$$

We have

$$P_{no} = \frac{(V_{no})^2}{R_L} = \frac{(AVn_i)^2}{R_L}$$

$$\therefore \qquad P_{no} = \frac{A^2 \left(\frac{R_a R_t}{R_a + R_t} \right)^2 2 V_{i_p}\, \delta f}{R_L} \tag{9}$$

Putting the value of P_{no} from equations (9) and (8), we have

$$F = \frac{R_L(R_a + R_t)}{A^2\, 4KT\delta f\, R_a R_t} \frac{A^2\left(\dfrac{R_a R_t}{R_a + R_t}\right)^2 2V_{i_p}\,\delta f}{R_L}$$

$$F = \frac{2V_{i_p}\left(\dfrac{R_a R_t}{R_a + R_t}\right)}{4KT}$$

For $R_t >> R_a$, we have:

$$F = \frac{2V_{ip}(R_a)}{4KT} = \frac{2\times(1.6\times10^{-19})\times i_p \times R_a}{4\times290\times1.38\times10^{-23}}$$

$$F = (R_a\, i_p)\,(2 \times 10)$$

$$F = 20\, R_a\, i_p$$

From this equation it is seen that i_p is measured in milliamperes and R_a is measured in ohms.

3.16 NOISE TEMPERATURE

Thermodynamic consideration have shown that the power spectrum of thermal noise is constant over a very wide frequency range. Noise having such a spectrum, contains all frequency components in equal proportion and is designated white noise by analogy to white light. The available thermal noise power is P_n = KTB watts which depends only on temperature and bandwidth. A thermal resistor, therefore delivers a maximum of KT watts per unit bandwidth, regardless of the value of R.

Clearly, temperature is the fundamental parameter of thermal noise. But there are other white noise sources which are non-thermal in the sense that the noise power is not related to a physical temperature, still we can speak of the noise temperature T_n of any white noise source, thermal or non-thermal by defining

$$T_n = \frac{P_n}{KB}$$

where, P_n is maximum noise power the sources can deliver in bandwidth B.

For thermal sources T_n is a physical temperature for non-thermal source it is a measure of the available noise power. This means that a non-thermal white noise source having equivalent internal resistance R_n and noise temperature T_n can be replaced by a thermal resistor $R = R_n$ at physical temperature $T = T_n$ without altering the thermal properties.

It can be shown that the equivalent noise temperature T_n can be expressed in terms of noise figure as $T_n = T_0 (f - 1)$.

where, T_0 = 17°C or 290 K. Hence, once noise figure is known equivalent noise temperature may be calculated.

Numericals

Example 3.4: In a certain system, the S/N power ratio is 10 and the bandwidth is 10,000 Hz. Find the maximum permissible information rate. What will be the effect on the system, if S/N falls to a value of 5?

Solution: Here the maximum permissible information rate is given by

$$
\begin{aligned}
R_M &= B \log_2 (1 + S/N) \\
&= 10000 \times \log_2 (1 + 10) \text{ bits/sec.)} \\
&= 10000 \times \log_2 (11) \\
&= 10000 \times \frac{\log_{10} 11}{\log_{10} 2} = 10000 \times 3.45943 \\
&= 34594.3 \text{ bits/sec}
\end{aligned}
$$

If S/N becomes 5, then the channel capacity or maximum rate R_M becomes 10000 $\log_2$ (6) bits/sec.

$$= 25849.6 \text{ bits/sec.}$$

Thus, the rate of signalling would exceed the channel capacity, causing excessive errors to appear.

Example 3.5: What will be the system bandwidth for 30 dB signal to noise ratio, if it is desired to transmit digital data at a rate of 30000 bits/sec?

Solution: The channel capacity, expressed by Hartley Shannon law, is given by

$$C = B \log_e [1 + S/N]$$

where B is the channel bandwidth and S/N is the signal to noise ratio. This relationship applies to discrete as well as continuous channels. The law is important since it correlates best reliable information transmission with the channel parameters and since it specifies optimum exchange of bandwidth and signal noise ratio for a given information rate.

Assuming a channel with bandwidth B = 30 Hz and (S/N) = 1

$$
\begin{aligned}
C &= 30 \times 10^3 \log_e (1 + 1) = 30 \times 10^3 \log_e 2 \\
&= 30 \times 10^3 \times 0.693147
\end{aligned}
$$

$= 20794.4$ bits/sec.

So for standard voice telephone circuit, if the power is increased, by factor of 1000 (corresponding to a S/N ratio of 1000 or 30 dB), the bandwidth can be reduced to

$$B = 3 \text{ kHz}.$$

Example 3.6: Find the minimum S/N power ratio in a system with bandwidth of 3 Hz and an information transmitting rate of 5 bits/second.

Solution: The rate of information transfer (R_M) is given by

$$(R_M/B) \leq (S + N)/N \leq [\log_2 (S + N) - \log_2 (N)] \text{ bits.}$$

So
$$5 B \leq B \log_2 (1 + S/N)$$

$$1 + S/N \geq 32$$

$$\therefore (S/N) \geq 14.9 \text{ dB}$$

Example 3.7: The modulated carrier power in an AM transmitter is 40 kW. If carrier is modulated by three modulating sources with $m_1 = 0.5$, $m_2 = 0.7$ and $m_3 = 0.4$, estimate the total modulated output power.

Solution:
$$m = \sqrt{m_1^2 + m_2^2 + m_3^2}$$

$$= \sqrt{(0.5)^2 + (0.7)^2 + (0.4)^2}$$

$$\therefore \quad m = 0.487$$

so that
$$P_t = P_c + P_c\left(\frac{m^2}{2}\right) = P_c\left[1 + \frac{m^2}{2}\right]$$

$$= 50\left[1 + \frac{(0.9487)^2}{2}\right]$$

$$\therefore \quad P_t = 72.5 \text{ kW}$$

Example 3.8: Estimate the relative total power of the carrier and sidebands in a 10 kW transmitter with $m_f = 0.5$.

Solution: From Bessel's function table for $m_f = 0.5$ carrier amplitude $= E_c = 0.98$.

Also the only significant sideband J_1 with a relative amplitude of 0.12.
So the carrier power is $(0.98)^2 \times (10) = 9.604$ W

Also power of each sideband = $(0.12)^2$ (10) kW
= 144 kW

Hence, total power is (9604 + 144 + 144)
= 10 kW

Example 3.9: Find image rejection in dB in the case of an AM broadcast receiver tuned to 500 kHz and Q value of 50 and I.F. of 465 kHz.

Solution: The relative response is given as

$$A_2 = \left[\frac{1}{\sqrt{1+y^2Q^2}}\right]$$

where $$yQ = \left[\frac{f_0}{f_s} - \frac{f_s}{f_0}\right]Q$$

$$= \left[\frac{1430}{500} - \frac{500}{1430}\right] \times (50) = 126$$

so that $$A_r = \frac{1}{\sqrt{[(112+(126)^2)+1]}} = -42 \text{ dB}$$

Example 3.10: A ratio receiver is tuned to 555 kHz, with its oscillator frequency being 1010 kHz. Determine:

(i) Image frequency
(ii) The rejection ratio, if the load Q of R.F. section is 40 and
(iii) Rejection ratio at 20 MHz.

Solution: Now F_{si} = 555 + $2f_i$

$f_i = (f_o - f_s)$

= (1010 – 555)

= 1465 kHz and

$$r = \left[\left(\frac{1465}{555}\right) - \left(\frac{555}{1465}\right)\right]$$

∴ r = [(2.64 – 0.38)] = 2.26

Now rejection ratio is given by

$$\alpha = [1 + r^2 Q^2]^{1/2}$$

$$= [1 + (2.26)^2 \times (40)^2]^{1/2}$$

∴ α_{dB} = 20 log (90.4) = 39.1 dB

This ratio value would be adequate for a domestic receiver.

with $f_s = 20\,\text{MHz}$

we have $f_{si} = [20 + 0.910] = 20.91\ \text{MHz}$

so that
$$r = \left[\frac{20.91}{20} - \frac{20}{20.91}\right]$$
$$= [1.0455 - 0.9565]$$
$$r = 0.089$$

Hence,
$$\alpha = [1 + (0.089)^2 \times (40)^2]^{1/2}$$
$$\alpha = 37$$
$$20 \log 37 = 11.35\ \text{dB}$$

This rejection ratio is not sufficient at 20 MHz. Hence, an R.F. amplifier definitely required in the H.F. band.

Example 3.11: Find the image frequency and the local oscillator frequency in the case of an AM broadcast receiver with the desired signal frequency of 550 kHz and an I.F. of 455 kHz.

Solution:
$$f_i = f_s + 2\text{I F.}$$
$$= [550 + (2 \times 455)] = 1450\ \text{kHz}$$
$$f_{lo} = f_s + \text{I.F.}$$
$$= (550 + 455) = 1005\ \text{kHz}$$

Example 3.12: Estimate the image frequency rejection in an AM superheterodyne receiver, if the input tank circuit has a Q of 125 and the receiver is tuned to 1200 kHz.

Solution:
$$f_i = f_s + 2\ \text{I.F.} = 1200 + (2 \times 455)$$
$$= 2110\,\text{kHz}$$

so
$$\rho = \left[\left(\frac{f_i}{f_s}\right) - \left(\frac{f_s}{f_i}\right)\right]$$
$$= \left[\left(\frac{2100}{1200}\right) - \left(\frac{1200}{2100}\right)\right]$$
$$= 1.99$$
$$\alpha = \sqrt{1 + Q^2\rho^2}$$

$$= \sqrt{1+(125)^2 \times (1.19)^2}$$

$$\alpha = 148.8$$

Example 3.13: Calculate the r.m.s. noise voltage generated in a bandwidth of 10 kHz by a resistor of 1000 Ω, maintained at 17°C. Find the available noise power over this bandwidth.

Solution: The mean squared voltage is given by

$$V_n^2 = 4\,KTRB$$

$$F_2 = 1.38 \times 10^{-23}\ J/°K$$

$$T = 17°C$$

$$= (273 + 17)°\ K = 290\ K$$

$$R = 1000\ \Omega \text{ and } B = 10^4\ Hz$$

$$V_n^2 = 4 \times 1.38 \times 10^{-23} \times 290 \times 1000 \times 10^4$$

$$\text{So, r.m.s. noise voltage} = 0.40\ \mu V$$

$$\text{Available noise power} = KTB$$

$$= 1.38 \times 10^{-25} \times 290 \times 10^4\ W$$

$$= 4 \times 10^{-17} \text{ watts.}$$

Example 3.14: The parallel tuned circuit, at the input of a radio receiver has Q factor of 80 and is tuned to 120 MHz by means of capacitance of 25 PF.

If the receiver channel bandwidth is restricted to 10 kHz, find effective noise voltage of the tuned circuit at room temperature.

Solution: Here $R_d = \dfrac{Q}{\omega_o c} = \dfrac{30}{2\pi \times 120 \times 10^6 \times 25 \times 10^{-12}}$

$$= 1.59\ k\Omega$$

and $\Delta f = 10^4\ Hz$

so that $(V_n)^2 = 4\ R_D\ kT\ \Delta f$

$$= 4 \times 1.59 \times 10^3 \times 4 \times 10^{-21} \times 10^{14}$$

$$= 0.254 \times 10^{12}$$

$$V_n = 0.5\ \mu V$$

Example 3.15: Determine overall noise figure referred to the input in the case of an amplifier with a noise figure of 9 dB and an available power gain of 15 dB followed by a mixer stage with a noise figure of 20 dB.

Solution: Now $F = \left[F_1 + \left(\dfrac{F_2 - 1}{G_1}\right)\right]$

where, $F_1 = 9$ dB, corresponding to 7.94:1 power ratio

$F_2 = 20$ dB corresponding to a 31.62 : 1 power ratio.

So $$F = \left[7.94 + \frac{(100-1)}{31.62}\right] = 11.07$$

This is overall noise factor.
Hence, overall noise figure is:

$$F\,dB = 10 \log (11.07) = 11.44 \text{ dB.}$$

Example 3.16: Determine the open circuit r.m.s. voltage measured of a typical 10 kΩ thermal resistance at a temperature of 290°K, if the associated voltmeter has a bandwidth of 100 kHz.

Solution: $KT_o = 1.37 \times 10^{-23} \times 290$

$= 4 \times 10^{-21}$ w/Hz

$v_n^2 = 4 \times 10^4 \times 1.38 \times 10^{-23} \times 290$

$= 16 \times 10^{-17}$ V²/Hz

So $v_n = \sqrt{16 \times 10^{-17} \times 10^5}$

∴ $v_n = 4\ \mu V$

Example 3.17: Estimate the value of the minimum input signal required for a 20 dB signal to noise ratio in a receiver with a pass band of 10 kHz and an input circuit having a resistance of 10 kΩ at room temperature (290°K).

Solution: The mean-squared noise voltage is given by;

$$E_n^2 = 4\,KTRB$$

$$= 4 \times 1.372 \times 10^{-23} \times 290 \times 10^4 \times 10^4$$

$$= 1.59 \times 10^{-12} \text{ volts.}$$

Hence, $E_n = 1.26\ \mu V$

For a 20 dB (S/N) ratio, the voltage ratio would be 10 : 1, so that the input signal must be not less than

$$= 12.6\ \mu V$$

Example 3.18: In a two stage amplifier, the first stage has a voltage gain of 10, a 600 Ω input resistor, 1600 Ω equivalent noise resistance and 27 kΩ output resistance. The second stage has the corresponding parameter value of 25, 81 kΩ, 10 kΩ and 1 MΩ. Determine (i) the equivalent input noise resistance of two stage amplifier and (ii) the noise figure of the amplifier driven by a generator having internal resistance of 50 Ω.

Solution: (a) The noise resistance, actually present at the input of the first stage would be

$$F_1 = 600 + 1600$$
$$= 2200\ \Omega$$

The second stage has a total resistance of

$$R_2 = \left[\frac{27 \times 81}{27 + 81}\right] + 10$$
$$= 30.2\ \text{k}\Omega$$

Also $\quad R_3 = 1 \times 10^6\ \Omega$

So, the equivalent noise resistance of the whole cascaded amplifier at the input of first stage would be:

$$R_{eq'} = \left[2200 + \left(\frac{30200}{102}\right) + \left\{\left(\frac{1000000}{(10)^2 (25)^2}\right)\right\}\right]$$
$$= 2518\ \Omega$$

Since $\quad R_{eq} = \left[R_1 + \left(\frac{R_2}{A_1^{\,2}}\right) + \left(\frac{R_3}{A_1^{\,2} A_2^{\,2}}\right)\right]$

(ii) Here $\quad R_{eq'} = (R_{eq} - R_t)$

$$= (2518 - 600) = 1918\ \Omega$$

The noise figure is given by

$$F = \left[1 + \frac{R_{eq}}{R_a}\right]$$

$$F = \left[1 + \left(\frac{1918}{50}\right)\right]$$

$$F = 39.4 \text{ corresponding to } 15.84 \text{ dB}$$

Here $\quad R_a = 50\ \Omega$

Example 3.19: The noise output of a resistor is amplified by a noiseless amplifier having a gain of 60 dB and bandwidth of 20 kHz. A meter reads to the output of amplifier one millivolts.

(a) If bandwidth of amplifier is now reduced to 5 kHz, its gain remaining constant, what does the meter read?

(b) If the resistor is operated at 80°C, what is its resistance?

Solution: Given

$$G = 60,$$
$$B = 20 \text{ kHz}$$
$$E_{no_1} = 1 \text{ mV}$$

Part (a) $E_{no_1}^2 = 4R_n KT_o GB$

$$E_{no_2}^2 = 4R_n KT_o GB - \Delta B$$

$$\frac{E_{no_2}^2}{E_{no_1}^2} = \frac{4R_n KT_o G(B-\Delta B)}{4R_n KT_o GB}$$

$$\frac{E_{no_2}^2}{E_{no_1}^2} = \frac{G(B-\Delta B)}{GB}$$

$$\frac{E_{no_2}^2}{(1\times10^{-3})^2} = \frac{(20-5)\times10^3\times60}{(20)\times10^3\times60}$$

$$E_{no_2}^2 = \frac{15}{20}\times10^{-6}$$

$$E_{no_2}^2 = \frac{3}{4}\times10^{-6}$$

$$E_{no_2} = \sqrt{(0.75)\times10^{-6}}$$

$$E_{no_2} = 0.866 \text{ mV}$$

$$E_{no_1}^2 = (GE_{ni})^2 = G^2\times4\text{ KTBR}$$

Part (b)

$$(1\times10^{-3}) = (60)^2\times4\times1.38\times10^{-23}\times290\times80/17\times20\times10^3\times R$$

$$\therefore \quad R = \frac{(1\times10^{-3})^2}{(60)^2\times4\times1.38\times10^{-23}\times290\times80/17\times20\times10^3}$$

$$\therefore \quad R = 0.1845 \text{ M}\Omega$$

QUESTIONS

3.1. Enumerate the main advantages and disadvantages of high-level modulation as compared with low level modulation in a radio transmitter.

3.2. How is the value of the highest Q of the tank circuit, affected by the modulating frequency, the frequency deviation and the carrier frequency in the FM transmitter.

3.3. Explain the basic superheterodyne principle in an AM receiver, with the aid of a block diagram.

3.4. Explain the terms
   (i) Selectivity
   (ii) Sensitivity
   (iii) Fidelity
   (iv) Noise figure of radio receiver.

3.5. Explain the role of a noise limiter in an AM receiver.

3.6. Give reasons why a radio detector does not require an amplitude limiter.

PROBLEMS

3.1. The output voltage of a transmitter operating into a 600 Ω resistive load is given by

$$E_o = 1000\ [1 + 0.3 \sin 6280\ t] \sin 12.58 \times 10^7\ t$$

Find:

   (i) carrier and modulating frequencies.
   (ii) the carrier power output and
   (iii) the mean power output

[Ans. : 20 MHz, 1 kHz, 833 W, 870.5 W]

3.2. An FM system operates with a deviation of 30 kHz and highest modulating frequency is 3 kHz.

Neglecting the effect of pre-emphasis and assuming random noise and the use of a limiter in the FM receiver, estimate the signal to noise ratio improvement as compared with an AM system using the same carrier power. [Ans. : 24.77 dB]

3.3. A superheterodyne receiver, with an R.F. amplifier stage and IF value of 455 kHz, is tuned to 15 MHz. If the receiver is required to have an image rejection value of 41.58 dB, determine the Q value which would be the same for the R.F. as well as the mixer tuned circuits.

3.4. Find the image frequency rejection in dB's of a double superheterodyne with a first I.F. of 2 MHz and a second I.F. of 200 kHz, and RF amplifier with its tuned circuit Q value of 75 (same as that of a mixer), tuned to a 30 MHz signal.

3.5. A diode detector has a dynamic characteristics given by:

$$i = 10^{-5}\ v + 5.5 \times 10^{-6}\ v^2$$

For small positive inputs to the diode.

Determine the different frequency components and their voltage amplitudes across a 60 kΩ load, with an AM wave of 2.5 V amplitude, peak carrier is modulated to a depth of 80% and applied to the diode.

CHAPTER 4

Pulse Modulation

4.1 INTRODUCTION

Previous chapters have discussed the modulation of the amplitude, the frequency or the phase of sinusoidal carrier signal to transmit a low frequency analog information signal, such as voice. This chapter will discuss several systems which modulate some characteristics of a continuous train of discrete pulses as a carrier instead of sine wave.

Pulse modulation is an analog system even though discrete valued samples of continuous analog signal are used in modulation process. These discrete samples can take on any value within the range of the continuous signal. This is one of the ways in which pulse transmission differs from digital transmission, in which a discrete value can only be done of a finite set of numbers.

4.2 PULSE TRANSMISSION

A perfect rectangular pulse is shown in Fig. 4.1a. It has a definite time duration of width τ and a constant amplitude during that time period. A Fourier analysis of this pulse shows that it has a continuous frequency spectrum. If such a pulse is passed through a low-pass filter, then all frequencies above filter cut off frequency will be eliminated and the pulse at the output will be rounded off as in Fig. (4.1b). Since pulse generating equipment is far from perfect this is the type of pulse that will be produced in practice. It will still have constant amplitude during the pulse, except for the lead and tail edge. If the rise and fall time of the edges are very much shorter than the pulse width τ, we can treat the pulse as a perfect one and ignore the rise and fall time.

Figure 4.1 shows pulse transmission (a) rectangular pulse (b) a rectangular pulse which has not been severely distorted by bandwidth limitations, the intercepts for normal, marking and spacing bias are shown. (c) a rectangular pulse distorted by transmission over a facility with too low a bandwidth, the threshold intercepts for a normal bias are shown. (d) a reconstructed pulse with normal bias delayed but unchanged in length form (b). (e) a reconstructed pulse lengthened by marking

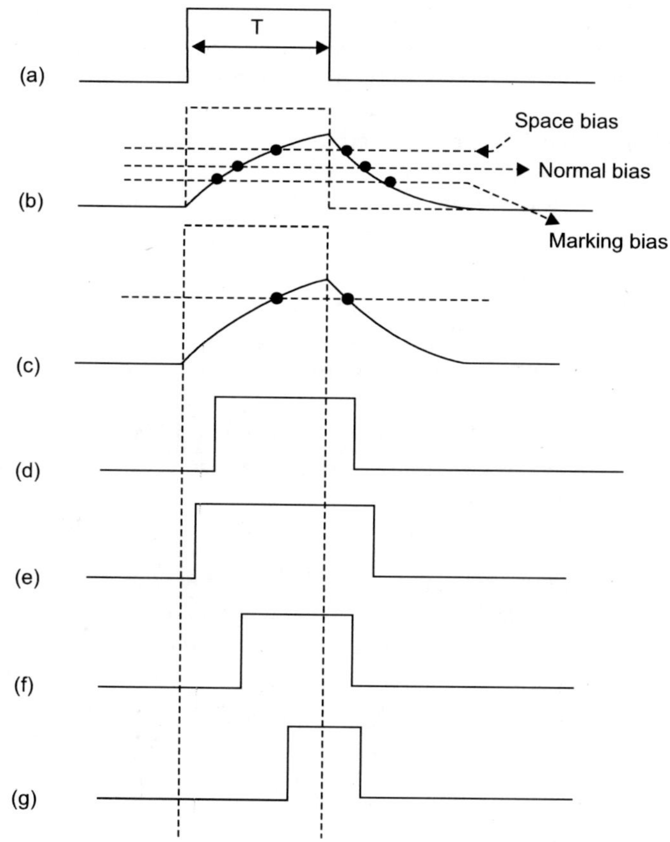

Fig. 4.1: Pulse Transmission Distortion

bias. (f) a reconstructed pulse shortened by spacing bias. (g) a reconstructed pulse resulting because of reduced pulse amplitude with normal bias in (c).

If the transmission channel has a bandwidth that is too narrow or that introduces excessive phase distortion the pulse will be smeared and reduced in amplitude, as shown in Fig. 4.1c. If the distortion is too severe, it will be impossible to detect the pulse at the receiving end. Noise introduced during transmission may produce an unwanted pulse or cancel a wanted pulse resulting in transmission error, as well as increasing the pulse distortion.

To reduce the effects of distortion it is necessary to regenerate pulse signals at repeater stations or at the receiver. If the repeater or receiver is improperly adjusted, a further form of pulse distortion occurs. This is illustrated in Fig. (4.1b) (d), (e) and (f) and is termed bias distortion. The pulse of Fig. (4.1a) is transmitted and becomes distorted because of the bandwidth and phase characteristics of the cable. Receiver bias is the minimum level of pulse amplitude to which the receiver will respond. If the receivers is adjusted for normal bias Fig. (4.1b), then for equally spaced (square) pulses, the pulse area above the bias level will be equal to the area below for symmetrical pulses and the regenerated pulse shown in Fig. 4.1d will be delayed, but not lengthened or shortened. If the bias is set too low, then the pulse area above bias will be larger than that below and the period of the regenerated pulse will be both lengthened and delayed as shown in Fig. 4.1.

This is referred to as marking bias distortion. If the bias is set too high then the pulse is both delayed and shortened as shown in Fig. (4.1f). This is called spacing bias distortion. A similar type of distortion is produced when the received pulse is severely reduced in amplitude, but the bias is kept at the normal level. The result is that the pulse area above bias is smaller than that below, causing the same type of delay and shortening as spacing bias distortion shown in Fig. 4.1g.

The effects of distortion can be minimized by proper receiver adjustment. This is done by sending over the transmission line a standard square wave signal at the proper transmission speed. The sampling speed and bias of the receiver are then adjusted until the receiver produces an undistorted signal from this transmission.

4.3 INTER SYMBOL CROSSTALK (INTER SYMBOL INTERFERENCE)

Unwanted signals such as noise and crosstalk can be added to the signal during transmission which may mask the pulse so that it cannot be retrieved. Figure 4.1b shows the typical shape of a pulse which has been passed through a bandwidth limited channel. When a pulse is passed through a low pass filter or limited channel, its corners are rounded off its amplitude is reduced and it is smeared by unequal time delays into the adjacent time slot. Figure 4.2b shows the result of increasingly lower filter cut off frequency on the pulse shape. Note that the pulse 'tail' endures long after the nominal pulse period and can interfere with a succeeding pulse. This

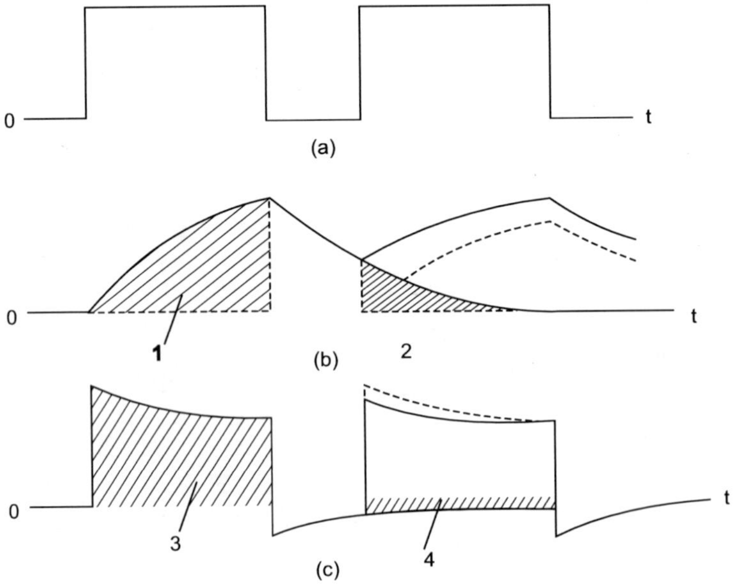

Fig. 4.2: Pulse Transmission Distortion

effect is known as intersymbol crosstalk or intersymbol interference. The succeeding pulse has energy from the first pulse added to it. In a pulse amplitude modulated system, this would cause serious distortion.

In time division multiplexed pulse amplitude modulated system, this form of crosstalk can couple signal energy from one channel into another which maybe of sufficient amplitude to be intelligible, hence the name 'crosstalk'.

Figure 4.2 shows Intersymbol crosstalk (a) Two adjacent pulses (b) upper frequency limitation causes energy from area 2 of the first pulse to be added to the second pulse increasing its amplitude. (c) lower frequency limitation causes energy from area 4 of the first pulse to be subtracted from the second pulse decreasing its amplitude.

4.4 PULSE MODULATION

AM & angle modulation may be called continuous wave (CW) modulation since some parameters of continuous high frequency carrier wave is varied in accordance with the modulating signal. If instead of continuous wave, a train of pulses is employed and some characteristics of these pulses is varied in accordance with the instantaneous value of modulating signal, the modulation is called Pulse-Modulation.

Usually, the pulses are quite short compared to the time between them, so a pulse modulated wave is "off" most of the time. The time interval between pulses can be filled with sample values from other messages, thereby permitting the transmission of many messages on the communication system.

The different characteristics of pulses are

(1) Amplitude, (2) Duration and (3) Position.

Any of these characteristics can be varied in proportion to the instantaneous value of the message and accordingly different types of pulse modulation systems are possible such as

(a) Pulse amplitude Modulation (PAM)

(b) Pulse Duration Modulation OR Pulse Width Modulation (PWM)

(c) Pulse Position Modulation (PPM)

These different types of pulse modulation systems can be categorized as analog pulse modulation systems.

There is another class of pulse modulation called digital or coded pulse modulation. One such example is pulse coded modulation (PCM).

Pulse wave contains appreciable d.c. and low frequency components. Hence, efficient transmission therefore needs a second operation, namely CW modulation to provide complete frequency translation. Thus, pulse modulation is a message processing technique rather than modulation in the usual sense.

4.5 SAMPLING THEOREM

Pulse modulation may be used to transmit analog information such as continuous speech or data. It is a system in which continuous waveforms are sampled at regular intervals.

Let m(t) be a signal which is bandlimited such that its highest frequency spectral component is f_m. Let the values of m (t) be determined at regular intervals separated by time $T_s \le \frac{1}{2} f_m$, i.e., the signal is periodically sampled every T_s seconds. Then these samples m (nT_s), where m is an integer, uniquely determine the signal and the signal may be reconstructed from these samples with no distortion.

The time T_s is called the sampling time. It should be noted that the theorem requires the sampling rate to be rapid enough so that at least two samples are taken during the course of the period corresponding to the highest frequency spectral component. It can be proved mathematically that the minimum allowable sampling rate is $2f_m$. This minimum sampling rate is known as Nyquist rate. For standard audio frequency range of 300 to 3400 Hz. A sampling rate of 8000 samples per second is used.

A signal with highest frequency "ω" must be sampled at minimum rate of 2 W sample per second with the samples in the form of impulses having area proportional to the amplitude of the sampled signal at the instant of sampling. If these impulses are passed through an ideal low-pass filter having cut off frequency W, original signal can be reconstructed.

Figure 4.3 shows a typical signal and its sampled output.

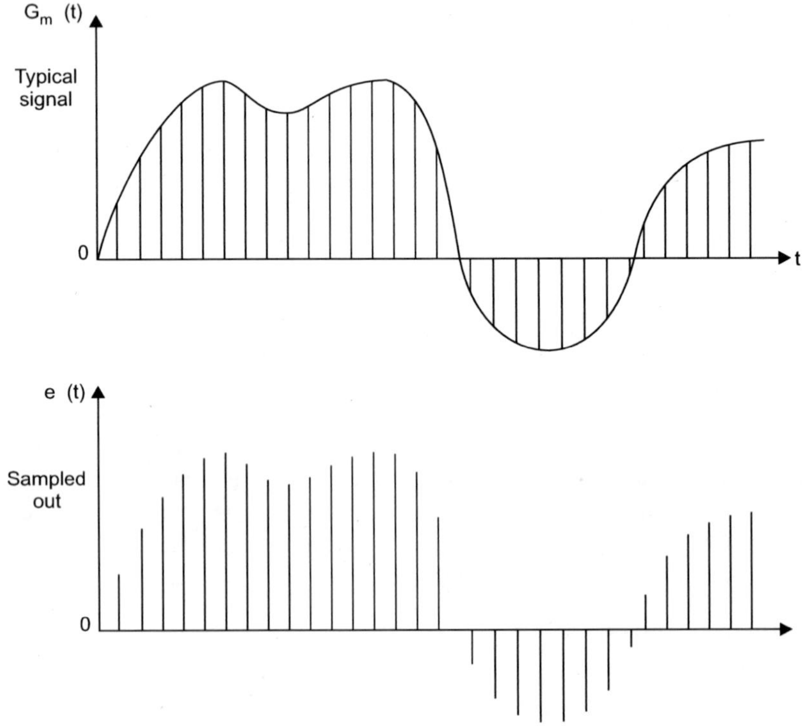

Fig. 4.3: Typical Signal and its Sampled Output

4.6 TYPES OF PULSE MODULATION SYSTEM

Information regarding the signal is transmitted only at the sampling time along with any synchronizing pulses that may be required. At the receiver end the original waveforms may be reconstructed from the information regarding the samples, if these are taken frequently enough.

4.6.1 Pulse Amplitude Modulation (PAM)

It is a system of pulse modulation in which the signal is sampled at regular intervals and each sample is equal to the amplitude of the signal at the instant sampling. A continuous signal e(t) can be sampled by a train of pulses represented by the function p(t) in a sampler or multiplier. The sampled signal $e_f(t)$ may be directly transmitted along a pair of wires. But it cannot be easy transmitted directly by electromagnetic waves in free space since the frequency spectrum of such sampled signal is concentrated at lower frequencies, which requires impracticably large antennas.

Hence, the spectrum is translated to a higher frequency range by amplitude modulation technique.

Figure 4.4a shows PAM and Fig. 4.4b shows PAM transmitter.

The frequency spectrum of PAM signals $e_s(t)$ is a repetition of the spectrum e(t) but with a decaying amplitudes. The rate of decay depends upon the width of the pulses. As the pulses are made wider, the spectrum decays faster. This suggests a method of recovering back the original signal from the sampled signal by passing

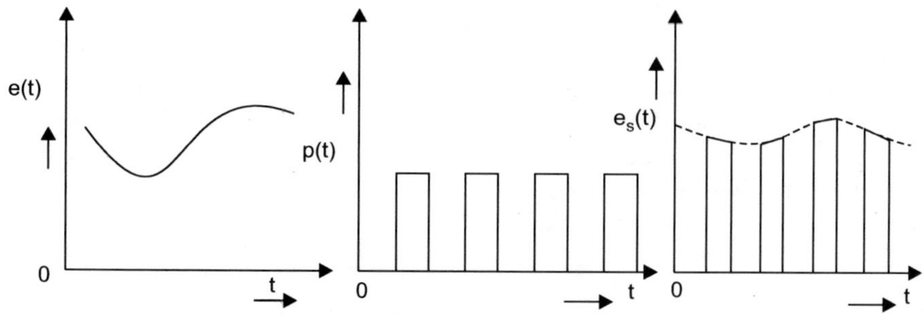

Fig. 4.4(a): PAM

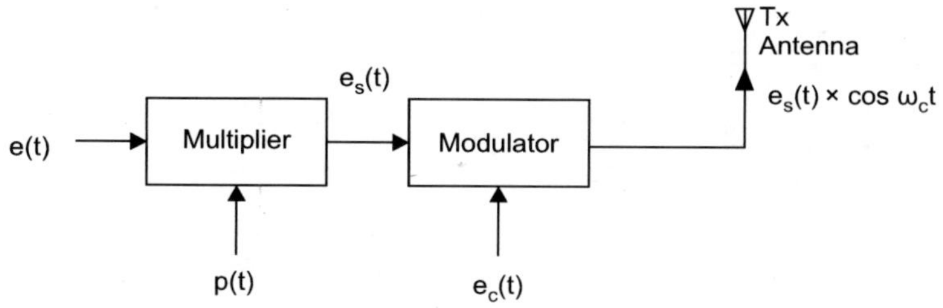

Fig. 4.4(b): PAM/AM Transmitter

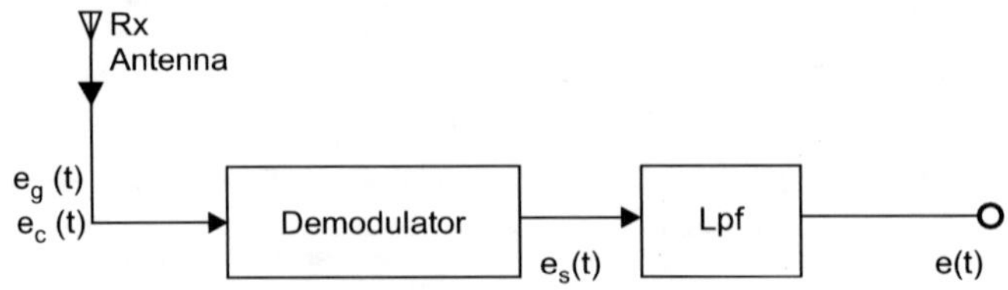

Fig. 4.4(c): Detection of PAM/AM

it through a low-pass filter (LPF) of bandwidth 2 Wm. At receiving end, the signal is demodulated to retranslate the spectrum of its original position. The output of the demodulator is the pulse train of varying amplitudes or the signals $e_s(t)$. The signal e(t) is recovered from $e_s(t)$ by filtering through a low-pass filter as shown in Fig. (4.4c).

4.6.2 Pulse Width Modulation (PWM)

In PAM the amplitude of the pulses is varied in accordance with the instantaneous value of modulating signal and time characteristics of pulses, namely width or duration of pulse and position of pulses remains constant. If any of the time characteristics is varied keeping the amplitude of pulses fixed, pulse time modulation (PTM) results. PTM can be of two types (1) pulse width modulation (PWM) or pulse duration modulation (PDM) (2) pulse position modulation (PPM).

In PWM systems, the amplitude and time of commencement of each pulse is fixed, but the width of each pulse is made proportional to the instantaneous values of the amplitudes of the signal. The time slot allowed for the modulated pulse is small compared to the time between pulses, the variation in impulse duration is much less than sampling time (T). In practice, the leading edge of each pulse is usually fixed and only the trailing edge is modulated as shown in Fig. 4.5a where PAM and PPM waves are also shown for comparison.

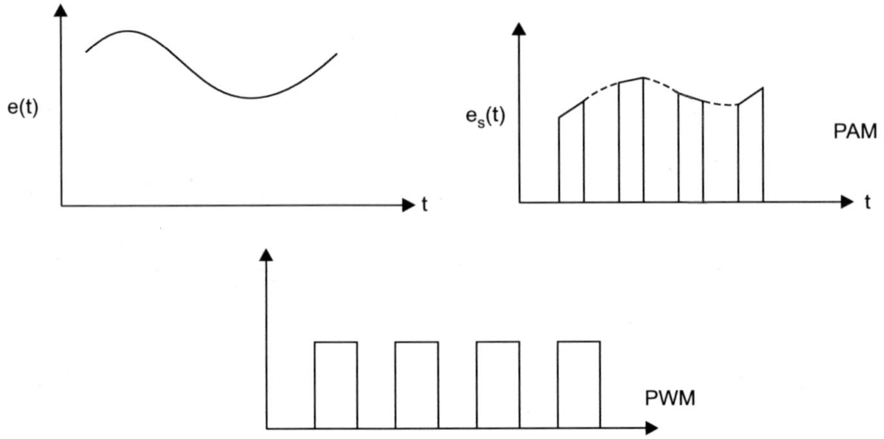

Fig. 4.5(a): Pulse Width Modulation

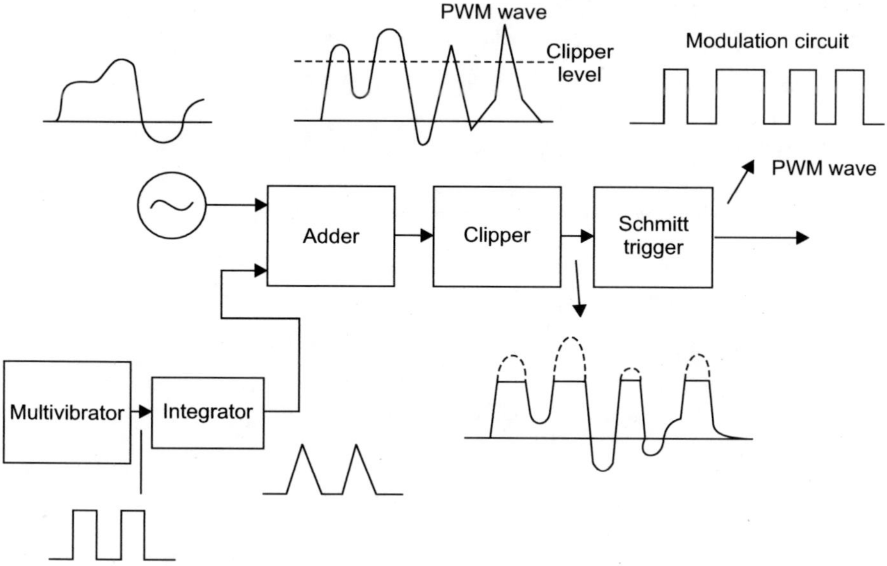

Fig 4.5(b): PWM Wave Modulation Circuit

The PWM signal can be generated by other method in which we begin by sampling a baseband signal and generating a flat topped PAM signal. Three successive samples of such PAM wave train $S_A(t)$ are shown in Fig. 4.5c.

Figure 4.5c shows (a_1) = A PAM Signal $S_A(t)$

(a_2) = An auxiliary Synchronized ramp Waveform R(t)

(a_3) = The sum $S_{(A)}$ (t) + R(t). The comparator reference level is indicated

(a_4) = A pulse duration modulated signal

(a_5) = A pulse position modulated signal.

We generate synchronously with the samples, the linear ramp-type pulse waveform R(t). These two signals [i.e., $S_A(t)$ & R(t)] are added and such is applied to comparator circuit. A comparator circuit has associated with it a reference level and an input terminal pair to which is applied in the present case, the signal S_A + R. A comparator has an output terminal whose voltage may assume only two voltage levels. One voltage level is held as long as the input is less than the reference level. The transition at the output between these two levels occurs very abruptly.

The Ramp amplitude is adjusted to be somewhat larger than the variation in amplitude of PAM samples.

4.6.3 Pulse Position Modulation (PPM)

The amplitude and width of the pulses is kept constant in this system and the position of each pulse, in relation to the position of a recurrent reference pulse is varied by each instantaneous sampled value of modulating wave. This type of PPM is shown in Fig. 4.5c.

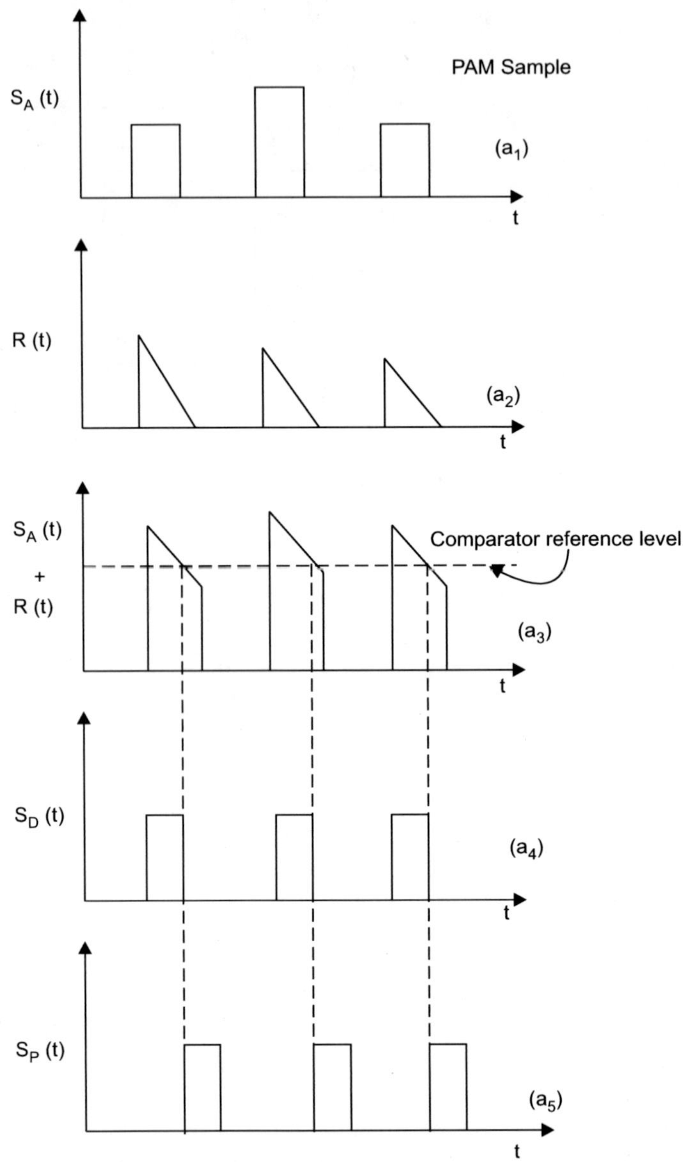

Fig 4.5(c)

The PPM can be generated from PWM with trailing edge modulation. The PWM pulses are inverted and differentiated, changing the modulated edge into position modulated positive spikes. With the little reshaping these spikes become the desired PPM wave. As a matter of fact, the principle use of PWM is for generation of PPM, because PPM is markedly superior to PWM for message transmission. It is so, because the information resides in the time location of the pulse edge, not in the pulses themselves. Thus, somewhat like the carrier frequency power of AM, the pulse power of pulse time modulation is a "wasted power" and it would be more efficient to suppress the pulses just and transmit the edge. Of course, we cannot send edges without defining the pulses. But we can send very short pulses indicating the position of edges, a process equivalent of PPM. The reduced power requirement for PPM is a fundamental advantage over PWM. Figure 4.6 shows block diagram of generating PPM-wave. The demodulation of PPM signals can be effected by passing it through low-pass filter and then integrating. However, the integration method does not take full advantage of the noise reduction properties of PPM, so the usual procedure is conversion of PAM or PWM followed by low-pass filtering.

4.6.4 Pulse Code Modulation (PCM)

PCM is entirely a different type of system than PAM, PWM or PPM. In the latter system the actual value of the sample is transmitted by varying the characteristics of a standard pulse train, in PCM the sampled signal is rounded off to the nearest value which is permitted for transmission by the system. PCM is digital process or kind of digital modulation. PCM generator produces a series of number or digits. Each of these digits, always in binary code represents the approximate amplitude of the signal sample at that instant.

In PCM system, the signal to be transmitted is sampled and each sample is approximated to the nearest allowable level as shown in Fig. 4.7a.

In this case, there are 8 allowable levels spaced 1.0 volt part. Each voltage level is now represented by certain pattern or pulses. Thus, instead of transmitting the individual samples, the corresponding pulse pattern is transmitted.

Assume that at any instant the amplitude of the signal is 6.8 volt, it is not sent as a 6.8 volt pulse, as it might have been in PAM, nor as a 6.8. μsec. wide pulse as in PWM, but simply as the digit 7, (because 7 volt is the standard amplitude nearest to 6.8) as 0111, where 0 = no pulse, 1 = pulse.

Actually it is sent as binary number back to front, i.e. as 1110 to make demodulation easy. Thus in PCM system, the signal is continuously sampled, quantized, coded and sent and each sample is converted to the nearest standard amplitude and into the corresponding back to front binary numbers.

The elements of PCM generation is shown in Fig. 4.7b. The continuous signal e(t) is first sampled to give $e_s(t)$ as in conventional pulse system. The sample values are then rounded off to the nearest predetermined discrete value. This process is called "Quantization" and the discrete values are called quantum levels.

$$f_s = \frac{1}{T} \leq 2\, f_m$$

The sampled and quantized wave $e_{sq}(t)$ which results in discrete in both time and amplitude. The encoder converts quantised samples to approximate code groups, one group for each sample and generates the corresponding digital pulsed wave forming the baseband PCM signal.

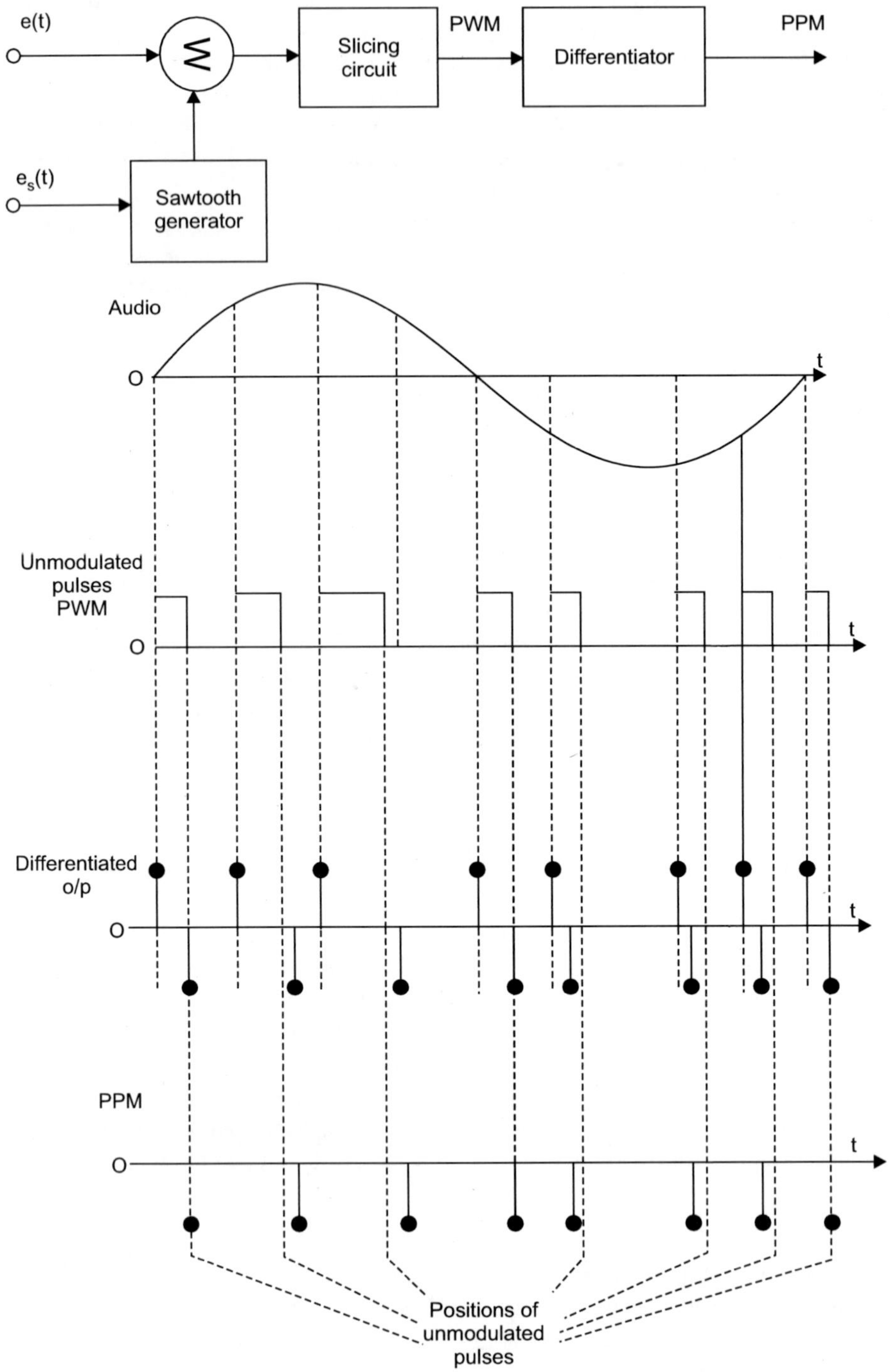

Fig. 4.6: Generation of PPM Signal

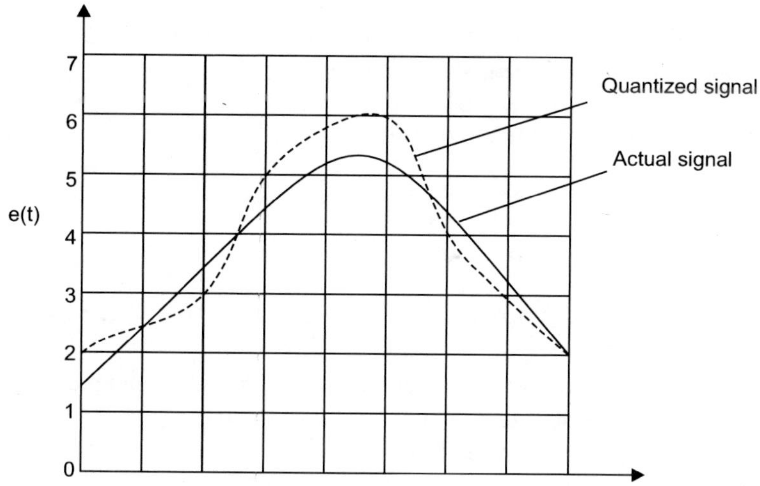

Fig. 4.7(a): Quantization (8 Level)

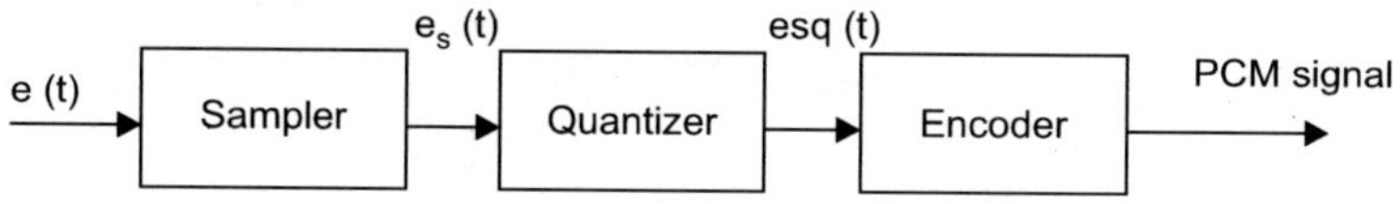

Fig. 4.7(b): Generation of PCM

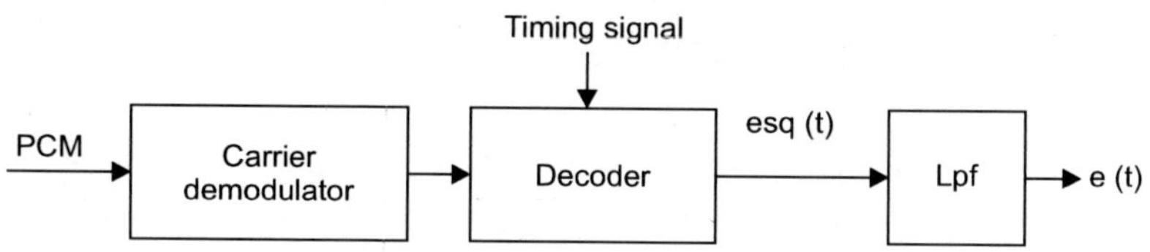

Fig. 4.7(c): Detection of PCM

As the final step of PCM generation, the baseband signal may modulate in RF carrier for radio transmission purposes. The carrier modulation can be AM (amplitude shift keying)'; Phase shift keying (PSK) or frequency shift keying (FSK).

The PCM signal can be detected as shown in Fig. 4.7c.

The receiver must decode the received signal to obtain $e_s(t)$ and construct by low-pass filtering. The first task of the decoder is recognition of the binary digits. This can be done with a decision circuit whose crossover level is set at zero volt. When the waveform is positive, the digit is taken as 1 and vice versa, clearly the decision should be made at the centre of each pulse of maximum reliability. Hence, synchronizing signals are usually supplied to bigger the decoder at the optimum time.

Bandwidth Required for PCM: The most common form of PCM is binary PCM, for which there are two possible amplitudes of pulses (0 & 1) used for coding the samples. Any one sample is represented by one code-group of these pulses. It can be shown that the number of pulses "p" in any code group is given as

$$P = \log_2 Q$$

where, Q = number of quantizing levels if Q = 8

$p = \log_2 8 = 3$

That is three pulses are needed to represent group. Similarly, Q = 16 will require 4 binary pulses in one code group.

Because several coded pulses are required for each message sample, it appears that the PCM bandwidth will be much greater than the message bandwidth.

Bandwidth can be calculated as follows:

Quantized samples occurs at a rate $f_s \geq 2f_m$ samples per second, there must be pf_s coded pulses per second. If there are no "spaces" in the code signal, the maximum permitted duration of any one pulse is $t = \frac{1}{pfs}$. On the basis of pulse resolution, we need a band, width of at least 1/2t, therefore bandwidth:

$$B \geq \frac{1}{2t} \geq \frac{pf_s}{2} \geq pf_m \qquad (1)$$

Thus, PCM bandwidth is minimum of $p = \log_2 Q$ times the message bandwidth. Of course, the carrier modulation results in the even great transmission bandwidth that of equation (1).

Quantizing Noise: If the actual signal is compared with quantized signal, it is seen that the process of quantization introduces some distortion. This is called quantizing noise or quantizing error. This noise is present because of the difference between the digit sent and actual signal at that time. If the number of levels is raised, the difference between the signal and the sent level at any instant must obviously be decreased and hence the quantizing noise is minimized. In practice Q = 128 is standard voice telephony.

Advantages and Application of PCM: As the PCM technique is so marvellous, still why are any other modulation systems used? There are three reasons for this,

(a) The other systems came first.

(b) PCM requires very complex encoding and quantizing circuitry.

(c) PCM requires very large bandwidth as compared to analog system.

PCM requires much more complex modulating procedures than analog systems. However, multiplexing equipment is very much cheaper and repeater do not have to be placed so close together because PCM tolerates much work (S/N) ratios. Specially because of large-scale integration, the complexity of PCM is no longer a problem.

Instead of one pulse per sample (as in PAM) we need to transmit a group of pulses per sample. Therefore, PCM requires more time interval for the transmission of the same information than other modulation systems. This disadvantage in PCM however is offset by its property of being more immune to noise interference. To receive PCM signal, all that is required is to know whether the pulse is present or absent, regardless of the amplitude and shape of the pulse. Thus, any external interference which may tend to introduce distortion in the height or width of the pulses has no influence whatsoever on PCM signals.

PCM is used for satellite and extra terrestrial probe communication and is being used for multichannel communication and telephone circuit over wire for short distances. Whenever PCM is used, it is always multiplexed.

4.7 OTHER DIGITAL PULSE MODULATION SYSTEM

PCM was the first digital system but by now several others have been proposed. The major ones are:

4.7.1 Differential PCM

This system is quite similar to the ordinary PCM system. Each word in the system indicates the difference in amplitude, positive or negative between any sample and previous samples. Thus, the relative value of each sample is indicated rather than the absolute value, as in normal PCM. The reason behind this system is that speech is redundant to that extent that each amplitude is related to the previous amplitude. So that large variations from one sample to the next are unlikely. Therefore, it would take fewer bits to indicate the size of amplitude change than the absolute amplitude and so a smaller bandwidth would be required for the transmission. This system has not found wide acceptance, because complications in the encoding and decoding process appear to outweigh any advantages gained.

4.7.2 Delta-Modulation

Delta-modulation is a technique by which analog signal can be encoded into binary digits(bits). Hence, delta modulation (DM) is a PCM system. However, DM has an advantage that electronic circuitry required for modulation at the transmitter and particularly for demodulation at the receiver is substantially simpler than the corresponding hardware needed for PCM.

Figure 4.8 shows block diagram for delta modulation system.

The pulse generator gives a regularly recurring train of pulses $p_i(t)$ of fixed amplitude and polarity. For simplifying the explanation, we assume that these pulses are arbitrarily narrow but still of finite area, i.e. impulses. The modulator receives these input pulses as well as signal $\Delta(t)$. The modulator output $P_0(t)$ is the input pulse train $P_i(t)$ multiplied by +1 or –1 depending on the polarity only (not the magnitude) of $\Delta(t)$. If $\Delta(t)$ is positive when $P_i(t)$ occurs, the multiplication is by +1 and if $\Delta(t)$ is negative by –1.

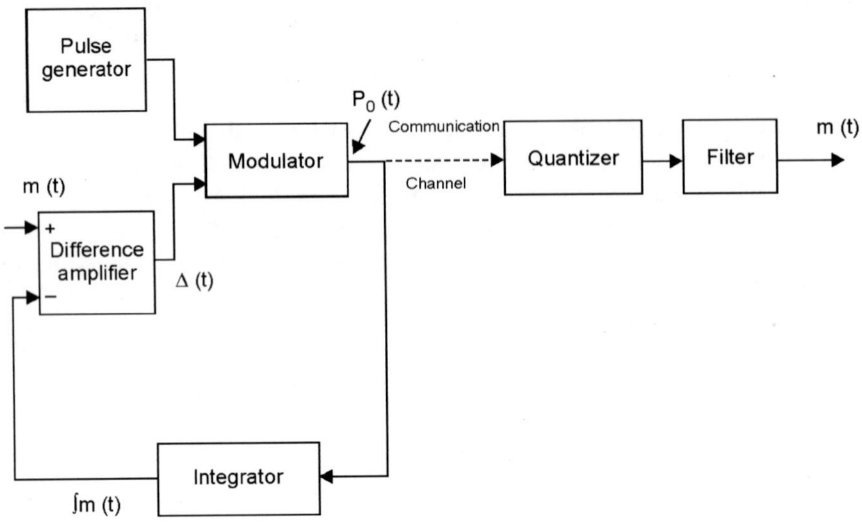

Fig. 4.8: A Delta Modulation System of Communication

The output waveform $P_0(t)$ is applied an integrator the output of which is designated as m(t).

The signal m(t) and ∫m(t) are compared in difference amplifiers. The amplifier output Δ(t) is given by [m(t) – ∫m(t)].

The quantizer performs the same function which the quantizer of a PCM receiver performs. It is hoped that if the noise added to signal in the communications channel is not overwhelming will almost always be able to distinguish a positive pulse from negative pulse.

Limitations of Delta-Modulation Due to Fixed Step Size: The fixed step in ∫m(t), imposed on DM, a limitation not encountered in other pulse modulation schemes and results in overloading when the signal changes too rapidly. All modems including delta modulation overload when the amplitude of the modulating base-band signal exceeds the range of the active devices used to process the signal.

4.7.3 Adaptive Delta Modulation

During any period of time when the changes in the signal m(t) are less than the step size, the modulator no longer follows the signal and modulator produces a train of alternating positive and negative pulses. Similarly, the modulator overloads when the slope of the signal is too high. Both these limitations can be removed by adjusting the step size in accordance with the signal being encountered. When the signal changes are small we would like the step size to be reduced, while to avoid slope-overload we would like the step size to increase. Figure 4.9 shows a delta modulation system which adjusts as step size. It is therefore known as adaptive delta modulation system. The amplifier shown has a variable gain, i.e. its gain is a

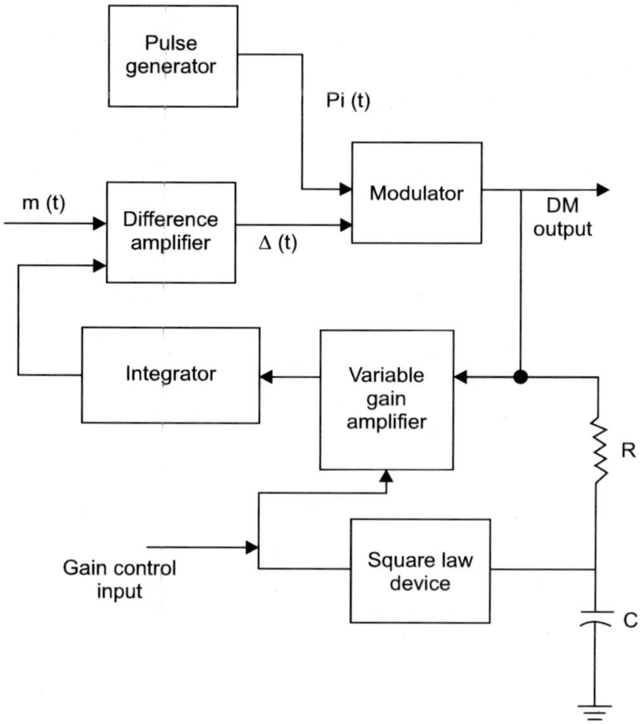

Fig. 4.9: An Adaptive Delta Modulation

function of voltage supplied at its gain control terminal. We assume that the characteristics of the amplifier are such that when the gain control voltage is zero, its gain is low is that the gain rises with increasing positive gain control voltage.

The RC combination acts as an integrator, the voltage across C being proportional to the integral of the pulse signal $P_0(t)$. The voltage across the capacitor is used to control the gain of amplifier. The square law device assures that whatever be the polarity of the voltage across the capacitor c, positive voltage will be applied to the gain control terminal of the amplifier.

Now assume that m(t) is making only small excurious so that the modulator does not follow it. The output $P_0(t)$ then consists of alternate polarity pulses. When integrated, these pulses yield an average output of almost zero. Thus, the gain control input is almost zero. The gain is low, and consequently the step-size is reduced. Consider the next case of slope overload.

If m(t) increases positively or negatively at very fast rate, then m(t) cannot follow it. The output $p_0(t)$ is then a train of all positive or negative pulses. The integrator averages and gives a large voltage output to increase the gain of the amplifier. Because of the use of a squaring circuit, the amplifier gain will increase irrespective of the polarity of the integrator voltage. The result is an increase in step size and a reduction in slope overload. It is necessary that there be an adaptive adjustment of the step size at the receiver end, as well. For this purpose in Fig. 4.9 a variable gain amplifier may be interposed before the final output filter. The gain control input of this multiple amplifier is determined, as at the receiver, by an integrator and square it, the integrated input being the received signal.

Example 4.1: A signal, $g_0(t) = 10 \cos 100 \pi t$ samples at the rate of 75 samples per second. State whether it would possible to reconstruct the signal g(t) from its samples. If not give the reasons for your answer.

Solution:

$$\text{Frequency of the signal } g(t) = \left(\frac{100\pi}{2\pi}\right) = 50 \text{ Hz}$$

$$\begin{aligned} \text{Nyquist rate} &= 2 \times \text{the highest value of frequency} \\ &= 2 \times 50 = 100 \text{ samples per second} \end{aligned}$$

Given sampling rate 75 samples per second is less than 100 samples per second and hence g(t) will not be reconstructed.

Example 4.2: A signal g(t) containing three frequencies, is given by:

$$g(t) = 10 \cos 100 \pi t + 15 \cos 150 \pi t + 5 \cos 300 \pi t$$

This signal is required to be transmitted using pulse techniques. Determine sampling rate.

If sampling is done at 200 Hz, state which part of waveform g(t) can be obtained.

Solution: The highest frequency, in the signal g(t) would be equal to

$$\left(\frac{300\pi}{2\pi}\right) = 150 \text{ Hz}$$

So, sampling rate = 2 × 150 = 300 samples per second. The other frequencies would be

$$\left(\frac{100\pi}{2\pi}\right) = 50 \text{ Hz}$$

$$\left(\frac{150\pi}{2\pi}\right) = 75 \text{ Hz}$$

With sampling frequency of 200 Hz, the highest frequency which can be reconstructed would be

$$\left(\frac{200}{2}\right) = 100 \text{ Hz}$$

So the sampler output would reproduce the 50 Hz and 75 Hz components.

Example 4.3: The signal g(t), defined by relation

$$g(t) = 10 \cos (100 \pi t) \cos (200 \pi t)$$

is required to transmitted using pulse modulation technique.
Determine Nyquist rate of sampling.

Solution: The signal g(t) can be represented as

$$g(t) = \frac{1}{2} 10 [\cos (200 + 100)\pi t + \cos (200 - 100) \pi t]$$

$$= 5 \cos 300 \pi t + 5 \cos 100 \pi t$$

So, highest frequency component in g(t) is

$$\left(\frac{300\pi}{2\pi}\right) = 150 \text{ Hz}$$

where the Nyquist rate of sampling would be (2 × 150) = 300 samples per second. Here the given signal g(t) is a bandpass signal for which sampling would be at a rate lower than the Nyqist rate.

QUESTIONS

4.1. The signals $g_1(t) = 10 \cos (100 \pi t)$ and $g_2(t) = \cos (50 \pi t)$ are both sampled at the rate of 75 times a second show that two sequences of samples thus obtained are identical. Also commented on the above.

4.2. Sketch the spectrum of signal (sin 100 πt/100 πt), if the bandwidth of signal is 50 Hz and sampling rate is 100 samples/sec.
Sketch the samples of this signal taken at a minimum sampling rate with first sample at t = 0.

PROBLEMS

4.1. Determine the sampling rate for the bandpass signal whose centre frequency f_c is 5 f_m with signal bandwidth of 2 f_m.

4.2. Find the sampling rate, if $f_c = 4.5 f_m$

4.3. What is sampling rate of following:
(i) A telephone channel limited to 4 kHz bandwidth.

4.4. Describe how spacing bias on a telegraph circuit can cause pulse shortening.

CHAPTER 5

Propagation of Waves

5.1 INTRODUCTION

Electromagnetic waves propagate in ways that depend not only on their own properties but also on those of the environment itself.

The various methods of propagation depend largely on frequency.

Wave travel in straight lines, except where the earth and its atmosphere alter their path. Thus, except in unusual circumstances, frequencies above the HF generally travel in straight lines. They propagate by means of so-called space waves. These are sometimes called tropospheric waves, since they travel in the troposphere, the portion of the atmosphere closest to the ground. Frequencies below the HF range travel around the curvature of the earth, sometimes right around the globe. The means are probably a combination of diffraction and a type of waveguide effect which uses the earth's surface and the lowest ionized layer of the atmosphere as the two waveguide walls. These ground waves or surface waves, as they are called, are one of the two original means of beyond the horizon propagation. For example, all broadcast radio signals received in daytime propagate by means of surface waves.

Waves in the HF range and sometimes frequencies just above or below it, are reflected by the ionized layers of the atmosphere and are called sky waves. Such signals are beamed into the sky and come down again after reflection returning to earth well beyond the horizon. To reach receivers on the opposite side of the earth, these waves must be reflected by the ground and the ionosphere several times. It should be mentioned that neither surface waves nor sky waves are possible in space or on airless bodies such as the moon.

Classification of Various Radio Bands

300 GHz	EHF (Extremely high frequency range) space wave, Radar and space communication.
30 GHz	SHF (Superior high frequency rang) as extremely high frequency.
3 GHz	UHF (Ultra-high frequency range)

	Space wave, Line of sight radio communication and broadcast services.
300 MHz	VHF (Very high frequency range) as UHF
30 MHz	HF (High frequency range)
	Ionospheric waves and short wave radio communication
3 MHz	MF (Medium frequency range)
	Ground waves for short distances, Ionospheric waves for longer distances, Broadcast and radio communication.
300 kHz	LF (Low frequency range)
	Surface wave, sky wave propagation is also used but less reliable radio navigation.
30 kHz	VLF (Very low frequency range)
	Surface wave and sky wave propagation to great distances
3 kHz	
300 Hz	
	ELF (Extremely low frequency range) subsurface communication
30 Hz	

5.1.1 Electromagnetic Wave

The channel in radio communication is the physical space between transmitting and receiving antennas. Antenna radiates electromagnetic waves and these waves travel outward in all directions with a velocity of 3×10^8 m/s.

These waves are transverse in nature and are therefore called 'Transverse Electromagnetic Waves (TEM Waves)'. For this type of wave, the electric field E, the magnetic field H and the direction of propagation x along which the waves travel with phase velocity V_P are mutually at right angles. These waves are little affected by surrounding atmosphere, rain, snow, etc. and are able to penetrate non-metallic objects easily, but are stopped dead by metals and fine mesh wire screen. They seem to travel best in free space and do not need a medium of propagation as do water waves.

5.1.2 Waves In Free Space

Free space by definition is that space that do not interfere with the normal radiation and propagation of radio-waves. Thus, it has no magnetic or gravitational fields, no solid body and no ionized particles. The concept of free space simplifies the theory of wave propagation.

Power density "P_d" of waves at a distance "r" due to isotropic radiator is:

$$P_d = \frac{P_t}{4\pi r^2} \text{ watts/m}^2$$

where, p_t = total power radiated.

The intensity of electric field (E) at a distance "r" is

$$E = \sqrt{\frac{30P_t}{r^2}}$$

The electric field E and power density P_d are related with each other as

$$P_d = \frac{E^2}{Z_0}$$

where, z_0 = characteristic impedance of free space.

Putting the values of P_d and E we get,

$$Z_0 = \frac{E^2}{P_d} = \frac{\left(\sqrt{\frac{30P_t}{r^2}}\right)^2}{\frac{P_t}{4\pi r^2}}$$

$$= \frac{30P_t}{r^2} \times \frac{4\pi r^2}{P_t}$$

$$= 30 \times 4\pi = 120\pi$$

$$Z_0 = 377\ \Omega$$

In free space, of course, absorption of these waves does not occur because there is nothing to absorb them. However, the picture is different in atmosphere which tends to absorb radio waves, because some fraction of electromagnetic energy is transferred to the atoms and molecules of the atmosphere. Due to this absorption, the waves are attenuated in atmosphere.

Let P_{d_1} and P_{d_2} are the power densities at distance r_1 and r_2 then attenuation constant α_p is:

$$\alpha_p = 10\log\left(\frac{P_{d_1}}{P_{d_2}}\right)$$

$$= 10\log\left(\frac{P_t / r\pi r_1^2}{P_t / 4\pi r_2^2}\right) \text{dB}$$

$$\alpha_p = 20\log\left(\frac{r_1}{r_2}\right) \text{dB}$$

Similarly, the attenuation constant for electric field intensities is

$$\alpha_E = 20\log\left(\frac{r_2}{r_1}\right) \text{ dB}$$

Example 5.1: Calculate the power density
(a) 500 m from a 500 W source and
(b) at ground from a geo-stationary satellite (stationed at 36,000 km above earth's surface) transmitting 4 kW power?

Solution: (a) Here given data

$P_t = 500$ W, $r = 500$ m.

$$\text{Power density} = P_d = \frac{P_t}{4\pi r^2} = \frac{500}{4\pi\times(500)^2}$$

$$P_d = 159.15 \ \mu \text{ W/m}^2$$

(b) Similarly,

$$P_d = \frac{P_t}{4\pi r^2}$$

$$= \frac{4\times1000\,\text{W}}{4\pi(3.6)^2\times(10^7)^2}$$

$$P_d = 2.456 \times 10^{-13} \text{ W/m}^2$$

5.1.3 Radiation

For radiation of electromagnetic waves, antennas are used. Maxwell's equations show that current flowing in an electrical wire is accompanied by a magnetic field around the wire. In case this magnetic field is variable, as is the case of an alternating current, an electric field is also found to be present. The amount of radiation depends upon the relation between the length of the wire and the wavelength of the current in the wire.

For positive a.c. cycle, nature of EM wave is as shown in Fig. 5.1(a). While for negative cycle EM waves is in Fig. 5.1(b).

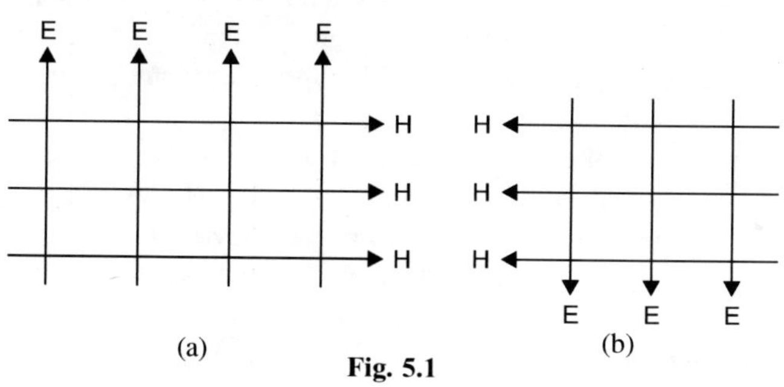

Fig. 5.1

From Fig. 5.1(a) and (b) it is seen that E and M wave changed directions but its magnitude is same. Also E and M waves are perpendicular to each other. Strength of electromagnetic waves depend upon amplitude of a.c. signal and its frequency dependant upon frequency of a.c. signal.

5.1.4 Polarization

Electromagnetic waves are transverse and its electric and magnetic fields are mutually perpendicular. Since the magnetic field surrounding the wire is perpendicular to it, the electric field becomes parallel to it. This configuration also exists after the electromagnetic waves has been radiated by the wire.

Polarization defines the physical orientation of the radiated waves in space. Waves are said to be linearly polarized if they have the same alignment throughout, in space. Mostly all antennas radiates linearly polarized waves.

Incoherent sources, such as sun or other heavenly bodies radiate wave vectors with haphazard alignment and are said to randomly polarized.

5.2 ENVIRONMENTAL FACTORS ON WAVE PROPAGATION

Propagation of radiowaves near the earth, causes a number of changes in electromagnetic wave. A wave near earth is susceptible to reflection, refraction, interference, diffraction, etc. Reflection is caused due to earth surface, mountains, buildings and any other similar structures. While passing through atmospheric layers of different densities, electromagnetic waves get refracted.

Diffraction may cause due to massive structures, such as mountains, buildings, etc. Waves also interfere with each other, when they belong to the same source but travels through different paths and finally cross at same common point.

5.2.1 Reflection

Reflection of electromagnetic waves is similar to that of light waves.

The reflection of electromagnetic waves can be explained by the following experiment.

Just stand before a mirror. Now place another mirror at your back and look into the first mirror.

What do you find in this mirror, of course, a number of your images. You will note that the brightness of the images goes on decreasing progressively. This indicates that after each reflection, a part of the light waves are absorbed and lost. With radio waves the situation is exactly same.

To define the amount of absorption in each reflection, we define a parameter called reflection coefficient, which we denote by "P". Reflection coefficient is defined as the ratio of the electric intensity of reflected wave to that of the incident wave. In case of perfect conductor (a perfect conductor reflects an electromagnetic waver completely, i.e. 100%), "P" is unity and for other practical cases, it is always less than unity.

5.2.2 Refraction

Refraction in radio waves is quite similar to its optical counterpart. It occurs when an electromagnetic wave passes from one propagating medium to another having different density. The wave velocity of the wave also changes, while it changes the medium.

5.3 INTERFERENCE OF ELECTROMAGNETIC WAVES

Interference occurs when two waves from the same source meet at a point while travelling through separate paths. This type of interference mostly occurs in high frequency sky-wave propagation or in microwave space-wave propogation. In case of microwave propagation, the antenna is located near the ground, and the waves reach the receiving point not only directly but also after being reflected from the ground.

Figure 5.2 shows the path of two waves.

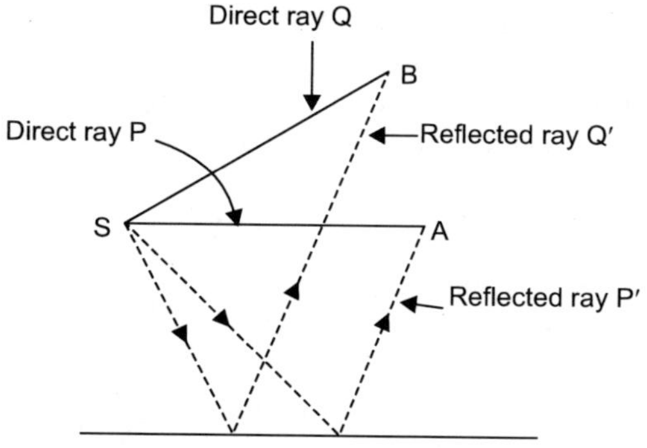

Fig. 5.2: Interference between the Direct and Ground Reflected Rays

As can be seen, direct path is shorter than the path with reflection. If in any case, the frequency and height of antenna is such that the difference between the direct and the reflected ray exactly equals half wavelength the two waves completely cancel at receiving point A. Provided the ground is a perfect reflector. However, if the ground does not act as a perfect reflector, partial cancellation of waves occurs. If the two paths have a path difference of exactly one wavelength as between the paths Q and Q, reinforcement of the received waves take place.

A succession of such points may be found lying one above the other giving an interference pattern having alternate cancellations and reinforcements.

5.3.1 Diffraction of Radio Waves

Diffraction is concerned with the behaviour of electromagnetic waves, as affected by the presence of small slits in a conducting plane or sharp edges of obstacles.

If a plane wave is considered as in Fig. 5.3a the question that arises immediately is why the wavefront continues as plane, instead of spreading out in all directions. The answer is that an infinite plane wave has been considered and mathematics shows that cancellation of the secondary wavelets will occur in all directions other than the original direction of the wavefront, thus the wavefront does continue as a plane.

When a finite plane wave is considered, the cancellation in spurious directions is no longer complete so that some divergence or scattering will take place. For this to be noticeable, however, the waveform must be small, such as that obtained with the aid of the slot in a conducting plane, as Fig. 5.3b. It is seen that instead of being 'squeezed through' as a single ray, the wave spreads out past the slot, which now acts as Huygen's point source on a wavefront and radiates in all directions. The radiation is maximum in front of the slot and diminishes gradually away from it.

Figure 5.3 shows what happens when a plane wave meets the edge of an obstacle. Although a sharp shadow might have been expected, diffraction takes place once again for precisely the same reasons as before. If two nearby points on the wavefront P and Q are again considered sources of wavelets, it is seen that radiation at angles away from the main direction of propagation is obtained. Thus, the shadow zone receives radiation. If the obstacle edge had not been there, this side radiation would have been cancelled by other point sources on the wavefront.

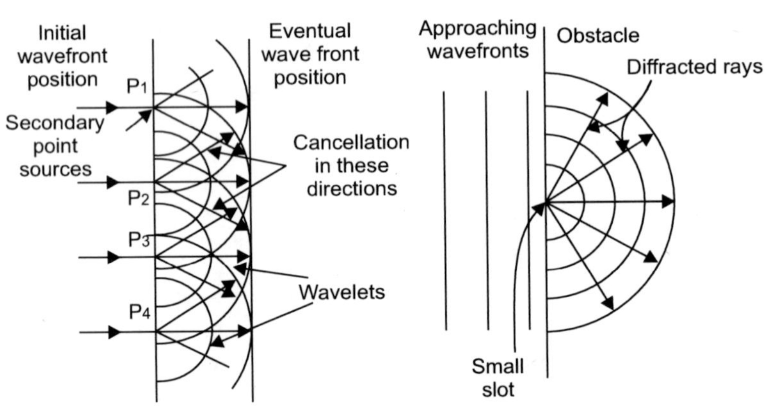

Fig. 5.3: Diffraction (a) Of a Plane Wavefront (b) Through Small Slot

Radiation once again dies down away from the edge, but not so gradually as with a single slot because some interference takes place, this is the reason why two point sources on the wavefront were shown. Given a certain wavelength and point separation, it may well be that rays a and a′, coming from P and Q respectively, have a path difference of a half wavelength so that their radiations cancel. Similarly, the path difference between the rays b and b′ may be a whole wavelength, in which case reinforcement takes place in that direction. When all the other point sources on the wavefront are taken into account, the process becomes less sharp. However, the overall result is still a succession of interference fringes as one moves away from the edge of the obstacle.

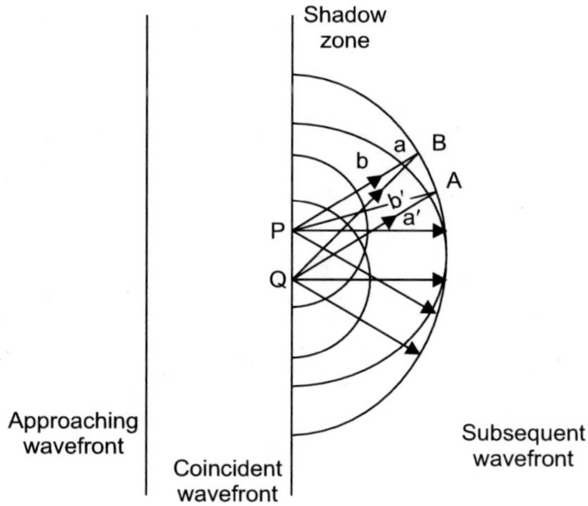

Fig. 5.4: Diffraction Around the Edge of an Obstacle

This type of diffraction is of importance in two practical situations. First, signals propagated by means of the space wave may be received behind tall buildings, mountains and other similar obstacles as a result of diffraction. Second in the design of microwave antennas, diffraction plays a major role in preventing the narrow pencil of radiations which is often desired by generating unwanted side lobes.

5.4 PROPAGATION OF WAVES

There are number of mechanisms by which radio waves may travel from a transmitting to receiving antenna. The more important of these are designated by the terms ground waves, sky waves and space or tropospheric waves.

The ground wave also some times called surface wave, can exist when the transmitting and receiving antennas are close to the surface of the earth and are vertically polarized. These waves, supported at its lower edge by the presence of the ground, are of practical importance for braodcast at lower frequencies. Thus, all broadcast signals received in the daytime are through ground waves. The sky wave represents energy that reaches the receiving antenna as a result of a bending of the wave path introduced by the ionization of upper atmosphere.

This ionized region, terms the ionosphere which begins about 80 km above the earth's surface accounts practically for all very long distance radio communication. Figure 5.5 shows details of wave propagation.

The space-wave represents energy that travels from the transmitting to the receiving antenna in the earth troposphere, the portion of the earth's atmosphere in the first 15 km. Adjacent to the earth's surface. The space wave commonly consists of at least two components. One of these rays is that travels directly from transmitter to receiver, while the other is a ray that reaches the receivers a result of reflection

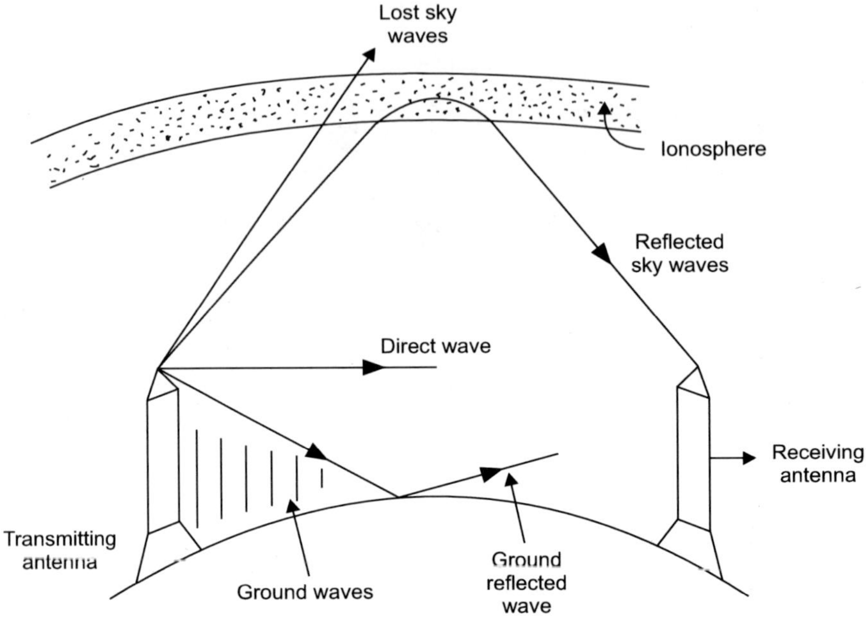

Fig. 5.5: Wave Propagation

from the surface of the earth. Space wave energy may also reach the receiver as a result of reflection or refraction produced by variations in the electrical characteristic of troposphere and by diffraction around the curvature of earth, hills, etc. Radio transmission at frequencies above that 30 MHz is generally by space wave propagation. It means that television transmission, frequency modulation and radar, etc. are systems of this method of propagation.

5.4.1 The Ground Waves

These waves progress along the surface of the earth and must be vertically polarised to prevent short-circuiting of the electric component. The ground wave induces charges in the earth, which travel with the wave and so constitute a current. As the wave passes over the surface of the earth, it is weakened as a result of energy absorbed by the earth in order to supply the power loss resulting from the induced current, flowing through the earth's resistance.

There is another way in which the surface wave is attenuated, because of diffraction the wavefront (Electric vector) gradually tilt over. As the wave propagates, it tilts more and more and increasing tilt causes greater short-circulating of the electrified component of the wave and hence field strength reduction. Eventually at some distance from antenna, the waves lies down and dies.

This means ground wave is generally useful only at very low frequencies (10–30 kHz) and low frequencies (30–300 kHz). Even below 500 kHz reliable communication can be made over distances upto 1000 km by the ground waves.

AM radio broadcast in medium frequency band (300–3000 kHz) are transmitted primarily via the ground waves but at high frequencies (3–30 MHz) and very high frequencies (30–300 MHz) used for FM, TV and radar, etc. the ground wave is so reduced in strength by absorption as to virtually becomes useless beyond an area of a few kilometres around the transmitter.

The medium wave (550–1600 kHz) broadcast service normally utilizes the surface wave. Some energy will be transmitted into ionosphere, where during day-time conditions, it is completely absorbed in D-layer.

During nighttime conditions, an appreciable portion of the ionospheric wave is returned to earth, which extends the service area of the station well beyond that covered by the surface wave. There will also be a zone in which both the surface wave and the ionospheric wave are of the same order of magnitude and the signal strength is the phasor sum of the two. Unfortunately, fluctuations in the ionosphere produce fluctuations in the phase of the reflected wave relative to surface wave, resulting in the severe fading of the combined wave.

5.4.2 Tropospheric Propagation

The Mode of Propagation

The region which is 10 km above the earth's surface is known as tropospheric region. In this region free space conditions are modified by: (i) The surface of earth. (ii) The earth's atmosphere.

For simplicity, assume earth as a flat surface as shown in Fig. 5.6. Transmitting antenna of height h_T and receiving antenna of height h_R are separated by a distance d. If wave is propagated by space wave propagation from transmitting antenna to

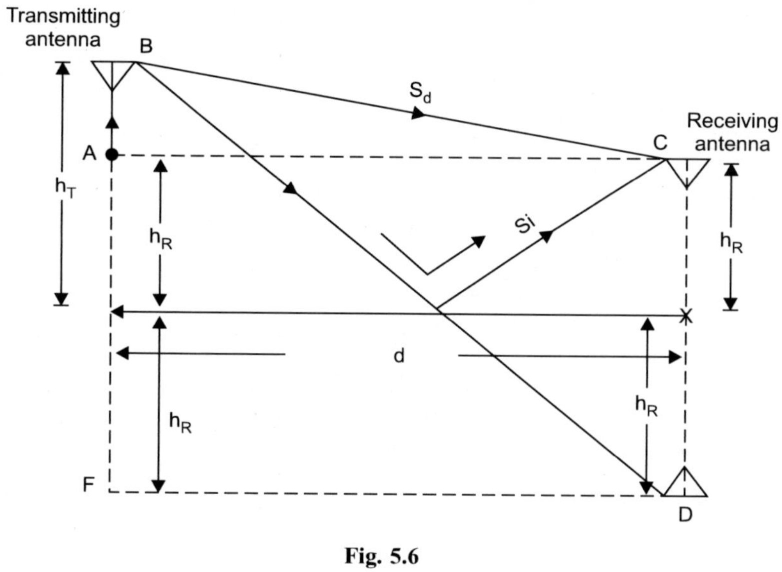

Fig. 5.6

receiving antenna within tropospheric region. Then space wave reaching at receiving antenna has two components: (i) Direct wave and (ii) Ground reflected wave.

Direct wave follows path S_d, while ground reflected wave travels greater distance than direct wave. Resultant wave formed by these two components whose amplitude is negligibly affected, while phase is affected by significant amount. Let Δs be the path difference between these two components. Then phase change corresponding to ΔS is

$$\phi_s = \frac{2\pi}{\lambda} \times \Delta S$$

Now our aim is to calculates ΔS

From AFBD $\quad S_i^2 = (h_T + h_R)^2 + d^2$

and from ΔABC, we have $\quad S_d^2 = (h_T - h_R)^2 + d^2$

$$\therefore \quad S_i^2 - S_d^2 = [(h_T + h_R)^2 + d^2] - [(h_T - h_R)^2 + d^2]$$

$$\therefore \quad S_i^2 - S_d^2 = 4h_T\, h_R \qquad \text{(L)}$$

We can expand $(S_i^2 - S_d^2)$ as

$$\therefore \quad S_i^2 - S_d^2 = (S_i - S_d)\,(S_i + S_d)$$

$$= (S_i + S_d) \cdot \Delta S$$

For most practical purpose$S_i \cong Sd \cong d$

$$\therefore \quad S_i^2 - S_d^2 \cong (2d) \cdot \Delta S \qquad \text{(M)}$$

Comparing (L) and (M), we have

$$(2d)\, \Delta S = 4h_T\, h_R$$

$$\Delta S = \frac{2h_T h_R}{d}$$

From this equation it is seen that path difference is directly proportional to antenna heights and inversely proportional to the distance between two antennas.

We know that $\quad \phi s = \frac{2\pi}{\lambda} \times \Delta S$

$$\therefore \quad \phi_s = \frac{2\pi}{\lambda} \times \frac{2h_T\, h_R}{d}$$

$$\therefore \quad \phi_s = \frac{4\pi h_T h_R}{\lambda d}$$

We can conclude that ground reflected wave lags by an angle $\phi_s = \frac{2\pi\, h_T\, h_R}{\lambda d}$ with respective to direct wave since S_i covers larger distance as compared to S_d.

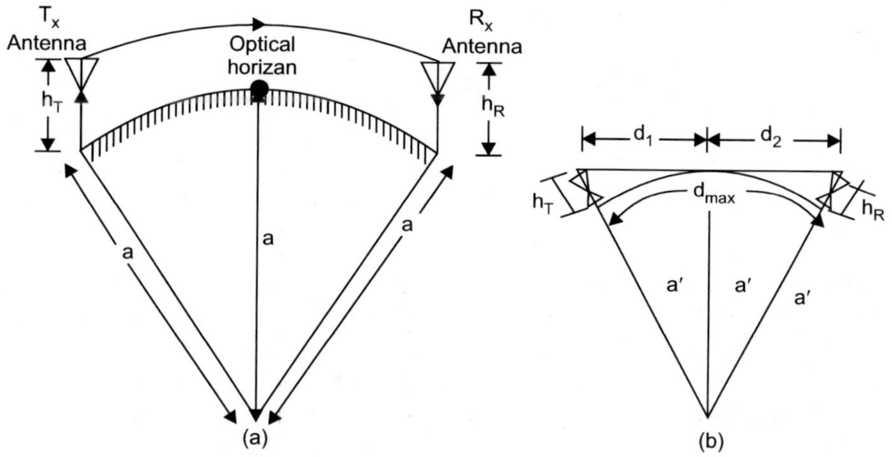

Fig. 5.7

Space Wave Propagation: Due to the curvature of earth, the range of transmission is limited, Fig. 5.7(a) shows practical diagram for transmission of radio signal. Optical horizon exists in Fig. 5.7(a) due to reflection or refraction of radio signal by earth whose actual radius is denoted by a Fig. (5.7b) is simplified version of Fig. 5.7(a). In Fig. 5.7(b) earth is considered a flat and is imaginary radius is a′. Under standard, atmospheric condition a and a′ is related by $a' = \frac{4}{3}a$. Above mentioned phenomena, exists in space wave propagation. The space waves (more or less) travels in a straight line. The range of transmission by space wave is limited by curvature.

From Fig. 5.7(b), we have:

$$(a' + h_T)^2 = d_1^2 + (a')^2 \qquad (1)$$

Similarly

$$(a' + h_R)^2 = d_2^2 + (a')^2 \qquad (2)$$

$$a'^2 + 2a' h_T + h_T^2 = d_1^2 + a'^2$$

$$a'^2 + 2a' h_R + h_R^2 = d_2^2 + a'^2$$

Since $a' >> h_T$ then $\quad d_1^2 \cong 2a'h_T$

Similarly, $a' >> h_R$ $\quad d_2^2 \cong 2a'h_R$

$$d_{max} = d_1 + d_2 = 2a' h_T + 2a' h_R$$

$$d_{max} = \sqrt{(2a'h_T) + (2a' \times h_R)} \qquad (1)$$

From equation (1) it is seen that maximum radio range is depended upon transmitting and receiving antenna heights since a′ is constant.

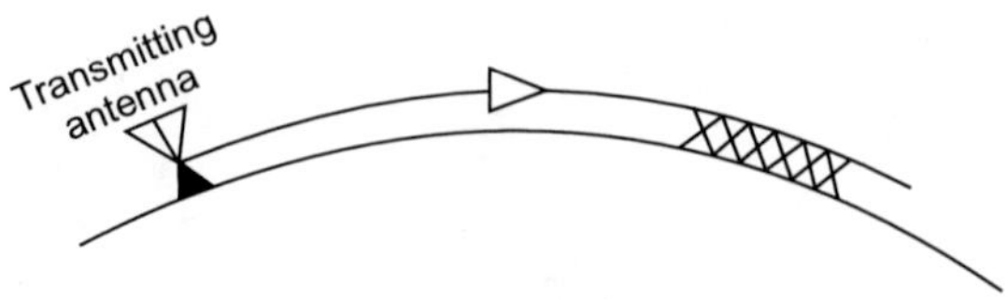

Fig. 5.8

Broadcast Fading Zone

The medium wave broadcasting service (550 to 1600 kHz) normally occur due to surface wave; but some energy of its transmitts into ionosphere, where during daytime conditions it is almost completely absorbed in D layer.

During nighttime conditions, considerable portion of the ionospheric wave returns to earth. Due to this, the service area of a station extends well beyond that covered by the surface wave. Due to this phenomena, zone is produced by surface wave and ionospheric waves. These two waves has same magnitude resultant signal strength is the phasor sum of two. Unfortunately, fluctuations in the ionosphere produce fluctuations in the phase of the reflected wave relative to surface wave. Due to these combinations of surface wave and flucated ionospheric wave fading exist. The area in which this occurs is known as broadcasting fading zones as shown in Fig. 5.9

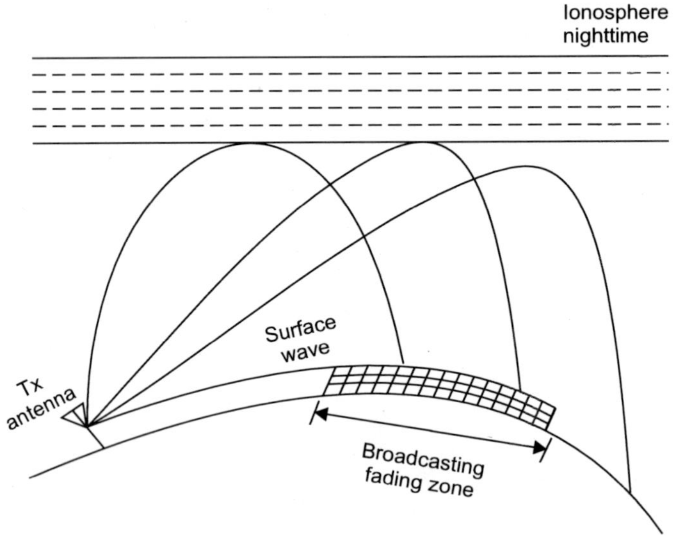

Fig. 5.9

5.4.3 Sky Waves

Energy radiated in directions above the horizons travel through space until it reaches the ionized region in the upper atmosphere. If conditions are favourable there, the path of the wave bends down back. The waves reflected in such a way from the ionosphere returns to the earth at very great distances from the transmitter. These types of waves are called sky waves and are very helpful for long distance rapid communication.

The reason for reflection of radio waves from top of the atmosphere is because of the positive and negative ions formed due to dissociation atmosphere molecules by energy rays from sun. The molecules, so ionized, remain in the form for long period.

The fact that makes long distance transmission possible is the existence of a series of electrically ionized layers in the ionosphere. These layers act as a radio-mirror to bounce the sky waves back to earth.

5.5 IONOSPHERE

This is also called as ionosphere propagation since radio waves propagates into space with the help of ionosphere. The ionosphere is the upper portion of the atmosphere, which continually absorbs large quantities of radiant energy from the sun and thereby becomes heated and ionized. A state is reached where the free electron density is maintained almost constant, the ionization rate being balanced by the recombination rate of electrons with positive ions. Clearly, the electron density will vary between day and night conditions and will also show variation between winter and summer as the ionisation rate is dependent on solar radiation.

Various peaks are observed in electron density corresponding to the heights at which various gases settle in the upper atmosphere. The layers follow a meteorological classification, being known as the C-layer, D-layer, E-layer and F_1 and F_2 layers. The last two layers F_1 and F_2 combine at night to form one single layer.

Due to variation in physical properties of the atmosphere such as temperature, composition density, etc. the ionosphere becomes stratified in distribution. The α, β and γ and the ultraviolet radiations from sun are mainly responsible for ionisation. Figure 5.10a shows the different layers along with the electron density.

D-layer is the lowest one and it lies between about 50 and 90 km. above earth's surface.

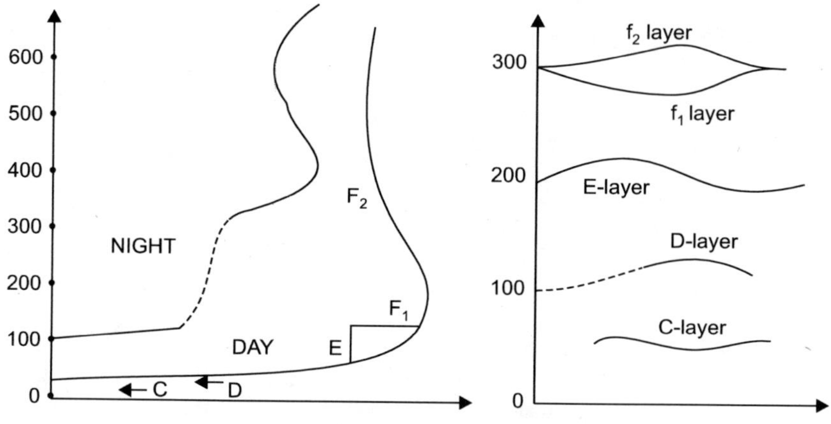

Fig. 5.10(a): Electron Density

It has an average thickness of about 10 km. The degree of ionization depends on the sun and so it disappears at night. The exact structure of D-layer is not known with certainty. It reflects some VLF and LF waves and absorbs MF and HF waves to certain extent. HF communication is not at all help D-layer.

The E - layer is next to D - layer and is located at about 110 km w thickness of about 25 km. Like the D - layer, D - layer also disappears at due to the deionization of the ions in the absence of the sun. The main role played by this layer in communication is to aid the MW surface wave propagation and to reflect HF-waves to some extent in daytime.

The F_1 layer exists at a height of about 180 km. in daytime with a thick of about 20 km. At night F_1 combines with F_2. F_1 reflects some of the waves but most of these get past this layer and are reflected by F_2-layer, the main effect of F_1-layer is to provide attenuation for HF-waves.

The F_2-layer is by far the most important layer and its height ranges from 250 to 350 km with a possible thickness upto 200 km at daytime. At night-time-layer falls to a height of about 300 km where it combines with F_1-layer height and ionization density varies tremendously and it depends on the average ambient temperature, local time of day and the sun spot cycle. The important property of F-layer is that it persists at night, unlike the other.

The actual mechanism of sky waves can be understood by Fig. 5.10b.

Sky wave 1 radiated at small angles is reflected from the F-layer return to the earth at a great distance from the transmitter.

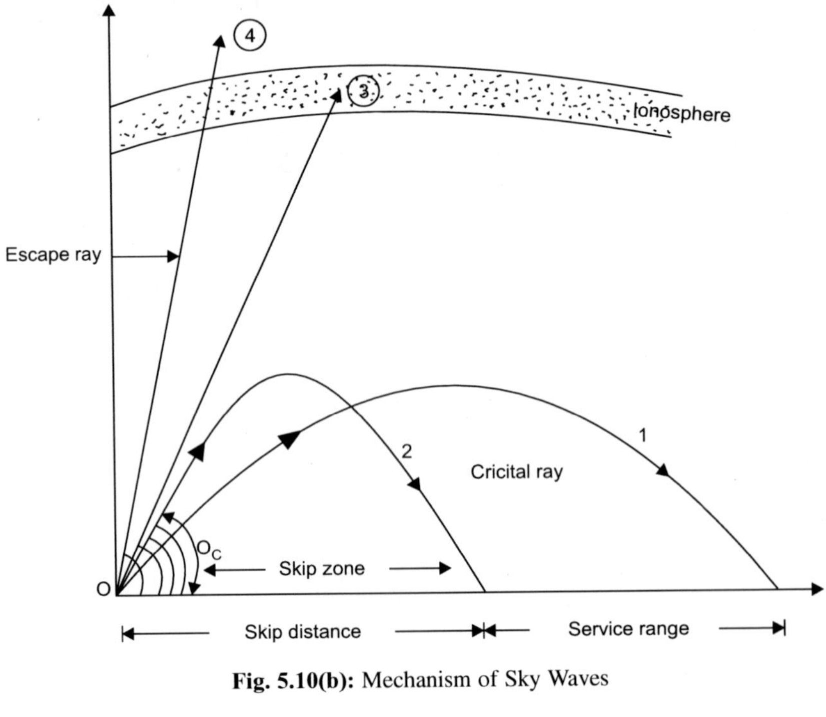

Fig. 5.10(b): Mechanism of Sky Waves

Sky wave 2, leaves the antenna at a somewhat greater vertical angle is reflected back to the earth at a smaller distance.

Skip Distance: The distance between transmitting antenna and the point where the sky wave is first received after returning to the earth is called "skip distance".

Skip Zone: Right near the transmitter, the signal is heard via ground waves. As we leave the ground wave region, we encounter a zone of silence called "skip zone".

Critical Angle (θ_c): As the vertical angle of the sky wave with respect to earth is increased the ionospheric layers are no longer capable of reflecting the wave back to the earth. The angle above which the sky wave no longer returns to the earth, but travels outward into space is called "Critical angle" (θ_c).

Critical Ray: The wave beamed at θ_c is called critical ray.

Multiple Hops: The service range of a transmission of a given frequency is determined by the critical ray at the nearest point and the glancing ray at the farthest point. The maximum possible range is reached when the critical ray coincides with the glancing ray. The range can be increased by using multiple hops in which wave is reflected from the earth after the first hop, to be reflected from the ionosphere once again farther on. It is shown in Fig. 5.10c.

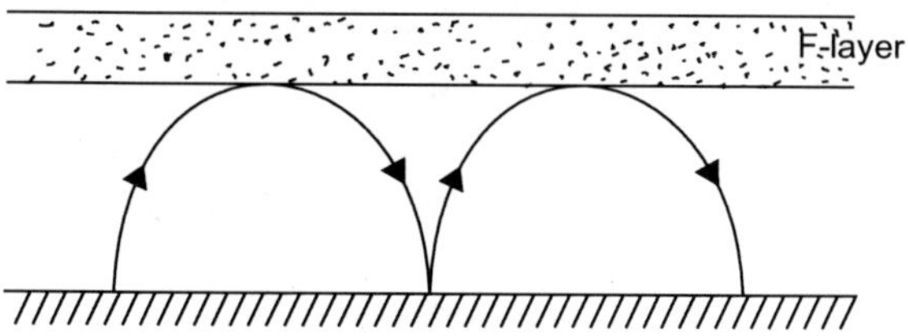

Fig. 5.10(c): Multiple Hops

Virtual Height: Actually the radio waves are not reflected but refracted. As the wave reaches a given layer, ionization density increases and this reduces the refractive index (n) of the layer. The incident wave is gradually bent farther and farther away from the normal and eventually becomes parallel to the layer and then bends downward, finally emerging from the ionized layer at an angle equal to the angle of incident. However, below the ionized layer, the incident and refracted rays follows path which are exactly the same as they would have been if reflection had taken place from the surface located at a height more than the actual height of the layer. Fig. 5.10d shows the virtual height.

Virtual height varies with angle of incidence and is important characteristics of any given layer.

Maximum Usable Frequency (MUF): As we know that the waves will be reflected from the ionosphere if its angle with the ground surface is less than the critical angle θ_c. The wave beamed at θ_c is called critical ray.

Suppose a wave of frequency "f" is incident at an angle (θ) less than critical angle (θ_c). Keeping the angle fixed, the frequency of the wave is increased upto a

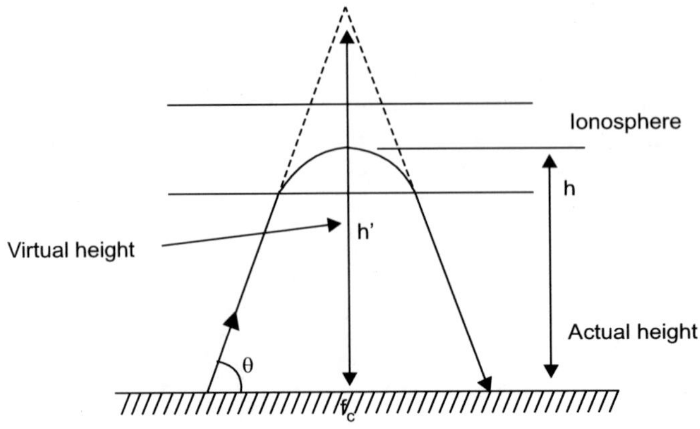

Fig. 5.10(d): Virtual Height

value when the given angle becomes the critical angle for that wave. If the frequency is increased further, the wave will not be reflected back, but will escape into space. This maximum limiting value of the frequency at a given angle with respect to the ground is called "Maximum Usable Frequency (MUF)".

Optimum Working Frequency (OWF): Frequencies about 15% lower than the maximum usable frequency should be used. This frequency is called "Optimum Working Frequency (OWF).

5.6 SPACE WAVE PROPAGATION

This type of propagation is also called tropospheric propagation. The troposphere is the region of the earth's atmosphere immediately adjacent to the earth's surface and extending upward for some tens of kilometres. In this region, the free space conditions are modified by (1) the surface of the earth and (2) the earth's atmosphere.

When the earth's curvature can be neglected, space wave propagation takes place in the manner shown in Fig. 5.11.

Here, the energy reaches the receiver in two ways (1) by a ray travelling directly between transmitting and receiving antennas over a path S_d and (2) by a ray travelling over path S_i that involves a reflection from the surface of the earth.

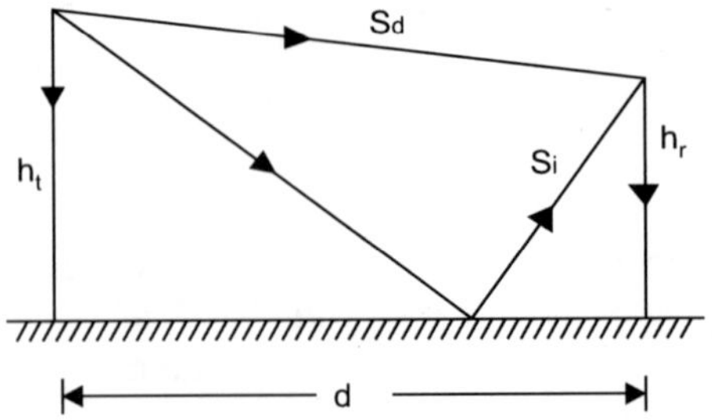

Fig. 5.11: Space Waves

The field strength at the receiving antenna is a vector sum of the fields represented by these two rays. It can be shown that field strength at the receiving antenna is given by

$$E_R = \frac{2E_0}{d} \sin \frac{2\pi h_t h_r}{\lambda_d}$$

where E_0 = field strength produced at unit distance by the transmitting antenna in the desired direction when earth is absent.

d = distance from transmitter to receiver.

λ = wavelength, same unit as d.

For large values of d, the above equation is modified as:

$$E_R = \frac{2E_0}{d} \sin \frac{2\pi h_t h_r}{\lambda_d} = E_0 4\pi h_t h_r / \lambda_d.$$

This equation shows the importance of antenna heights, for example, doubling the height of the base station antenna will double the field strength at the receiving point. Note also that the field strength is proportional to E_0, which in turn is proportional to the square root of the transmitted power Doubling the power results in only a $\sqrt{2}$ increase in the field strength.

Radio Horizon: (Effect of Earth's Curvature): As the distance "d" increases, it becomes necessary to take into account the curvature of the earth. Earth curvature has a more important effect in that it presents a horizon that limits the range because the effect of the earth's atmosphere is to cause a bending of the radio wave, which carries it beyond the horizon. The path of the ray can be considered straight by assigning a greater radius to the earth, than it actually has.

5.7 MICROWAVE LINKS

There are basically three types of microwave communication systems. These can be classified as:

(a) Line of sight. (LOS)

(b) Troposcatter and Diffraction systems.

(c) Satellite Communication System.

Majority of the systems used in India are LOS - microwave systems. Troposcatter systems are used usually to get over difficult terrain conditions or to cover large distances where LOS - microwave links may not be feasible or attractive. Most of the microwave communications of today work in the range of 1 to 12 GHz.

5.7.1 LOS Links

LOS–microwave systems are used to carry telephony television and data signals. Majority of the systems, carry multichannel telephone signals.

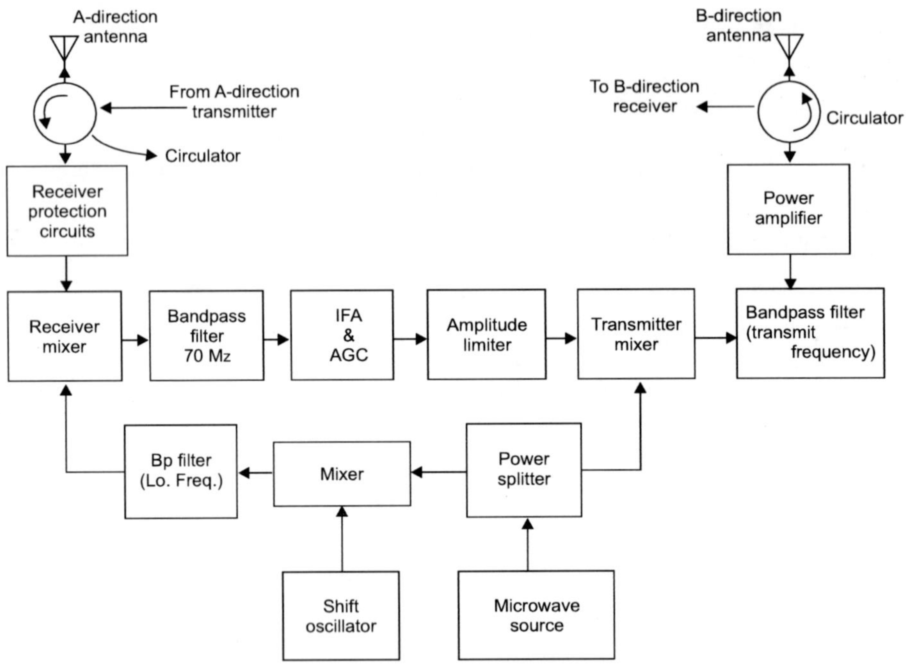

Fig. 5.12: Simplified Block Diagram of Microwave Link Carrier Chain, Shown Receiving from a Direction B and Transmitting in A Direction

A microwave link terminal has a number of similarities to a coaxial cable terminal.

Figure 5.12 shows block diagram of typical microwave repeater.

The repeater receives a frequency modulated microwave signal from one repeater and transmits to the next one, an identical chain is provided for working in the other direction. The only difference here is that transmissions in the two directions are somewhat different in frequency to avoid interference, the frequency difference typically being a few hundred MHz at 4 or 6 GHz operating frequencies.

The amplification of the received signal can be carried out at RF, IF or baseband frequency. However, it has been found that amplification at intermediate frequency gives best results. This IF is mostly 70 MHz and it is at this frequency that the bulk of amplification takes place in the link. The reason for conversion of RF to IF is noise reduction. It is easier to produce a very low noise amplifier at 70 MHz than at 4 or 6 GHz.

The antennas most commonly used are those with parabolic reflectors and hoghorn antennas.

The circulator ensure a connection between the adjoining ports in the direction of arrow but not between any other ports. This means that the transmitter is connected to the antenna and antenna to the receiver but the transmitter has no direct connection to the receiver input. If this were not ensured, the receiver mixer would be burned out very fast. The mixer is safeguarded further by protection circuits from overloads, caused by any transmission, mostly generated by transmitters connected to the same antenna.

These days, almost always the receiver mixer is a Schottky barrier diode. This is very low noise mixer diode. The mixer is followed by bandpass filter, usually at frequency of 70 MHz and having bandwidth of about 12 MHz. The filter is used to provide selectivity of the system, ensuring that signals belonging to the other carriers in the system are rejected adequately. IF amplifier provides most of the gain of the repeater.

IF is followed by amplitude limiter. This helps to prevent spurious amplitude modulation.

After this stage a transmitter mixer stage is employed to bring the IF output upto the transmitting microwave frequency. Varactor diodes are mostly used in this mixer. This mixer is followed by a bandpass filter to prevent any over duration by FM -signal and thus straying into unauthorized portion of the frequency spectrum or causing interference to the other carriers in the link.

The older systems usually had a klystron oscillator as a microwave source, but nowadays, a VHF transistor crystal oscillator, with varactor multiplier chain is employed.

The towers used by microwave link antennas range in height from 10 to 100 m, depending on the terrain, length of the particular link and location of the tower itself.

5.7.2 Tropospheric Scatter Links

A troposcatter link terminal is quite similar to microwave link terminal and in fact a typical block diagram is sufficiently like Fig. 5.11 so that separate block diagram is not shown. The main difference lies in the very much higher output powers and lower receiver noise figures in troposcatter links. Typical output powers range from 1 to 10 kW, but powers as high as 100 kW have been used for broadcast links.

Diversity is almost invariably used in troposcatter links. It may be frequency or space diversity or a combination of the two where fading is particularly severe. This results in greatly improved reliability, although the cost and the complexity increases at the terminals.

Most of the troposcatter links in the world are single span but others with even upto 20 spans are working. In fact, this depends on circumstances a point-to-point link over in accessible terrain is likely to be single span with length of upto 1000 km.

A typical broadband link may carry upto 192 two way telephone channels i.e. three supergroups plus one group. Sometimes upto 5 supergroups, i.e. 300 telephone channel capacities are also designed.

The capital cost of troposcatter links in rupees per circuit kilometre works out to about four times that of coaxial cable, making it about 12-times that of microwave links. Troposcatter links are used where special considerations so demand, rather than interchangeably with the other broadband transmission media.

5.8 USE OF SATELLITES

Communication through satellite requires three processes:

(a) Multiplexing of the individual telephone channels into a single composite baseband signal.

(b) Modulation of a radio frequency carrier by the baseband signal, and

(c) Multiple access of the satellite transponder by several earth stations.

In case of single channel working, multiplexing is not required. To identify various types of systems it is usual to give the initials of the three processes in above order. For example: FDM-FM-FDMA implies frequency division multiplexing, frequency modulation and frequency division multiple access.

There are three ways of multiplexing telephone channels into a common baseband signal. These are

(a) Frequency Division Multiplexing (FDM).

(b) Time Division Multiplexing (TDM).

(c) Code Division Multiplexing (CDM).

The ground segment of a satellite communications system in a broad sense consist of (1) the ground terminal that controls and monitors the satellite and (2) the various earth stations that operates with each other via satellite.

5.8.1 Description of an Earth - Station

Figure 5.13 shows block diagram of an earth - station.

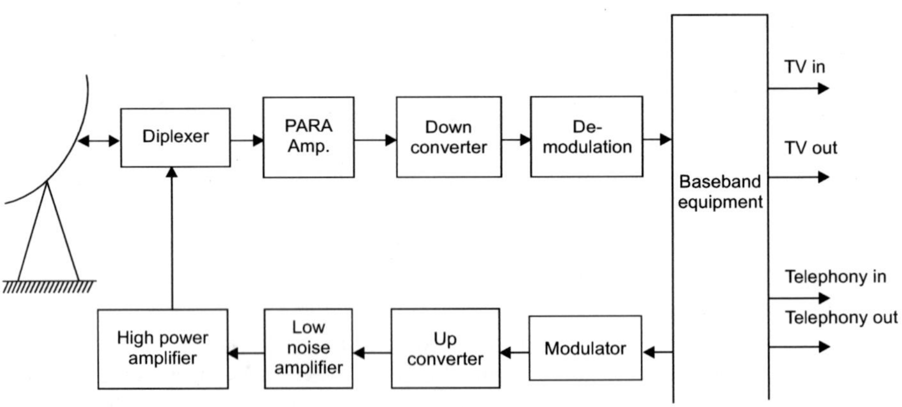

Fig. 5.13

The major subsystems are:

(1) Antenna and feed,

(2) Tracking and Servo control.

(3) Low noise amplifier

(4) High power amplifier

(5) UP and down converter.
(6) Modulators and demodulators.
(7) Baseband and multiplex equipment.
(8) Power supply system

(1) Antenna and Feed: This is one of the costliest subsystem in the earth station, costing as much as one-third to one half of the cost of installation. Mostly parabolic reflectors with suitable feed systems are used as earth station antennas.

The most important parameters that should be taken into account while designing antenna system for satellite communication earth stations are:

(1) Type of mount,
(2) Sterrability
(3) Gain and rms surface accuracy.
(4) Beam-width
(5) Side lobe levels
(6) Pointing accuracy
(7) Wind-speed

Thee are three types of mounting systems available for earth station antennas. They are:

(1) Azimuth - elevation mount.
(2) X-Y - mount, and
(3) Equatorial or Hour angle-declination mount of these last one finds extensive use in radio astronomy work.

(2) Tracking and Servo System: The absolute pointing accuracy depends on many factors such as the alignment, wind velocity, structural rigidity, etc. for tracking the satellite there are several methods available.

(1) Manual tracking
(2) Auto tracking
(3) Programme tracking
(4) Step tracking, etc.

In manual tracking mode the antenna position is manually adjusted as per the orbit predictions.

Auto tracking involves generation of an error signal for each axis, in a monopulse or conical scan tracking feed and receiver system and then feeding this error signal to the servo control system to null the error.

The automatic tracking gives the minimum pointing error. Programme tracking depends upon the accuracy of orbit predictions. In this system, the antenna is controlled by a computer to which are fed the orbit parameters.

In step tracking system, the antenna is stepped intermittently in Azimuth and elevation.

(3) High Power Amplifier System: One of the critical and expensive item in an earth station transmit chain is the high power amplifier system. The transmitter

power required varies anywhere between few watts in the case of thin route telephony to a few kilowatts typically 3 kW in case of TV signal transmission.

Typical high power amplifiers in an earth station accept power input of the order of few milliwatts and amplify them to power level of a few kilowatts. The main design consideration of these high power amplifiers are:

(1) Gain
(2) Input and output power levels
(3) Bandwidth
(4) Efficiency
(5) Tunability
(6) AM - PM conversion
(7) Cooling
(8) Monitoring control and alarm provisions.
(9) Reliability
(10) Cost.

A travelling wave tube (TWT) is used as an amplifier which can operate over the full bandwidth of 500 MHz.

(4) Low Noise Amplifier: The signal received from communication satellites are very weak and hence extremely low noise temperature receivers are employed in earth stations. The noise temperature of a receiver is defined as the equivalent temperature at which the receiver input has to kept to produce the same noise power at the output.

(5) Up and Down Converters: The up converters are used for translating the 70 MHz modulated signal to the 6 GHz transmitting band. Up conversion is normally done in two stages, first the 70 MHz is up converted to IF in the range between 500 and 3000 MHz and then this is further up converted to 6 GHz band.

Down converters can also be double conversion type but because the filtering requirements are not very stringent in this case, single conversion type down converters are increasingly being used. Compact size and reduced costs are the main advantages.

(6) Modulators and Demodulators: The baseband signal after suitable multiplexing modulates the 70 MHz carrier. The type of modulation depends upon the signal format and access scheme. TV signal is normally frequency modulated using a large deviation. Telephony channels are either frequency multiplexed and then frequency modulated or TDM and then phase shift keyed.

Similarly, demodulation process depends on the access scheme and signal format.

5.8.2 Satellite Space Craft Systems

A complete satellite consists of several subsystems, but most important of these are:

(1) Power supply subsystem.
(2) Altitude and Orbit Control.
(3) Telemetry and Telecontrol
(4) Communication Subsystem.

Figure 5.14 shows block diagram of a satellite system.

Description:

(1) Power Supply Subsystem: This section provides electric power for the operation of various mechanical, electrical and electronic systems incorporated in a satellite. This subsystem is divided into three parts:

(1) Solar array, (2) Battery, (3) Power Supply Control Circuit.

Thousands of minute solar cells are placed on the surface of the satellite. If this surface area is inadequate then separate projections are used. These so-called solar panels, convert sun energy into electrical energy. Sunlight is not available to the solar pannels for all the 24 hours. Thus, a battery is floated in parallel with the solar cells.

During the daytime, i.e. when the satellite solar cells are exposed to sunlight, these cells generate sufficient energy so as the recharge the battery and also provide for the satellite working. When no sunlight is available, the charged batteries supply the necessary electrical power for proper control of this power and for protection purposes, power supply control circuit is provided.

(2) Altitude and Orbit Control: This subsystem consists of five major parts:

(1) Sensors, (2) Propulsion system, (3) Altitude control, (4) Antenna control (5) Orbit control.

Altitude and orbit control is used to control the orbit of the satellite besides helping to maintain stabilization and its position. The control can be effected from the satellite itself (automatically) and from ground control.

The sensors sense the various changes in parameters of the orbit of the satellite and the error signals generated by the sensors are fed to altitude control, orbit control and antenna control, to suitably adjust these three. The three parameters can also be controlled by ground control using telecontrol. The sensor signals are also transmitted to the ground with the help of telemetry telecontrol section.

(3) Telemetry and Telecontrol: Error signals generated from the sensors are fed to a telemetry encoder block. This block encodes the error signals in a predetermined fashion and feeds it to the telemetry transmitter block. This stage suitably amplifies the signal and beams back it to earth through a diplexer and an antenna. The ground control station analyzes this signal and a suitable command is transmitted which is received by the satellite antenna and is fed to a command receiver. This receiver amplifies the telecontrol command signal and feeds it to the block labelled command decoder. The output of command decoder is used to control the altitude the orbit and the antenna.

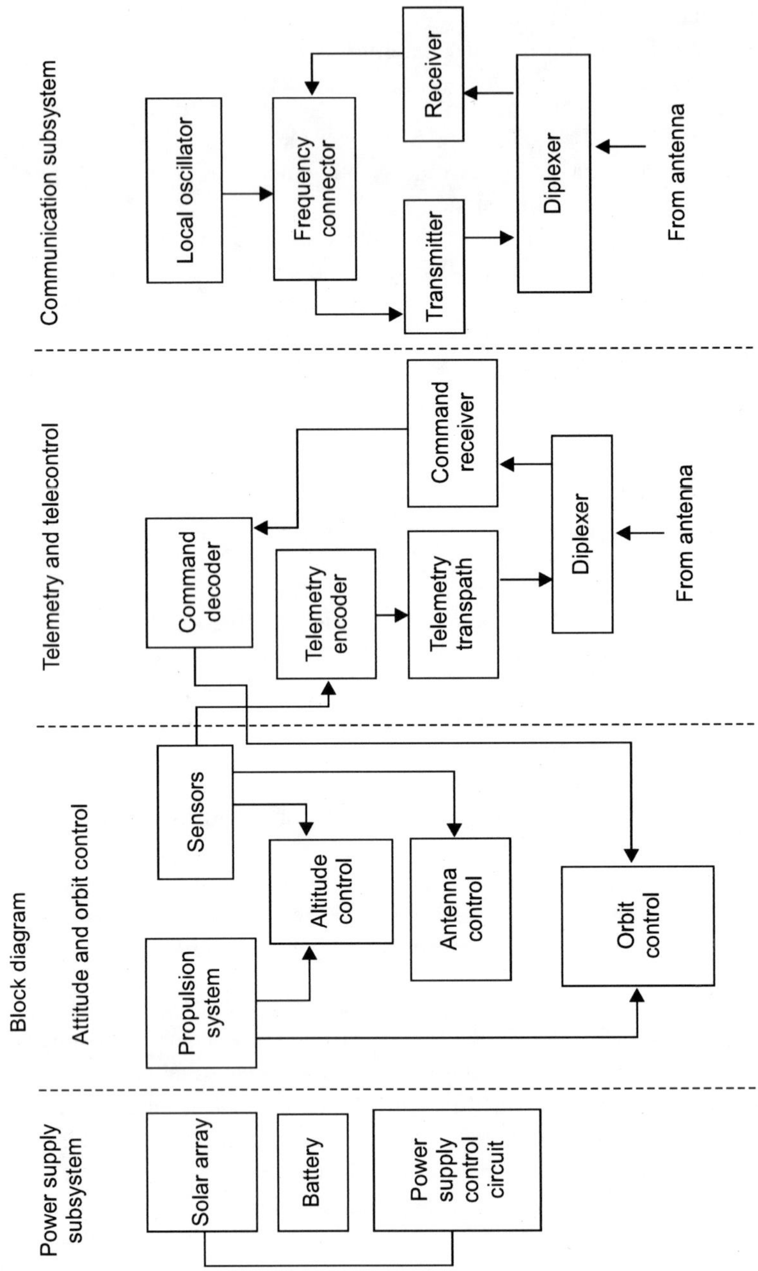
Block diagram
Power supply subsystem
Solar array
Battery
Power supply control circuit
Attitude and orbit control
Propulsion system
Sensors
Altitude control
Antenna control
Orbit control
Telemetry and telecontrol
Command decoder
Telemetry encoder
Telemetry transpath
Command receiver
Diplexer
From antenna
Communication subsystem
Local oscillator
Frequency connector
Transmitter
Receiver
Diplexer
From antenna

Fig. 5.14

(4) Communication-Subsystem: The signal beam from the ground station is intercepted by the satellite antenna and is fed to a receiver. This receiver has a tunnel diode front end r.f. amplifier chain to amplify the received signal. Signals from earth stations are received over the frequency range from 5.93 to 6.42 GHz in parallel transponder. The output of receiver is fed to a frequency converter which is usually down converter. The received signal thus down converted to the range of 3.705 to 4.19 GHz and is fed to TWT power amplifier. The overall amplification achieved lies in the range of 125 to 150 dB. The amplified output is beamed back to the earth using high-gain antennas.

Numericals

Example 5.2: The maximum electron density of E layer of ionosphere at a particular time is found to be $5 \times 10^{10}/m^3$ at virtual height of 100 km. This layer is to be used for commonly between two points on earth separated by 1000 km distance. Assuming a flat earth, find maximum frequency which can be used for communication.

Solution:

$$N = 5 \times 10^{10}/m^3$$

$$\frac{5\times10^{10}}{(100\times cm)^3} = 5 \times 10^4/cm^3$$

$$MUF = \frac{f_c}{\cos\phi_0}$$

$$\therefore f_{muf} = f_c \sec \phi_0$$

$$f_c^2 = 81\ N$$

$$= 81 \times 5 \times 10^4\ (kHz)^2$$

$$= 400 \times 10^4 = 4 \times 10^6$$

$$\therefore f_c = 2.01\ MHz$$

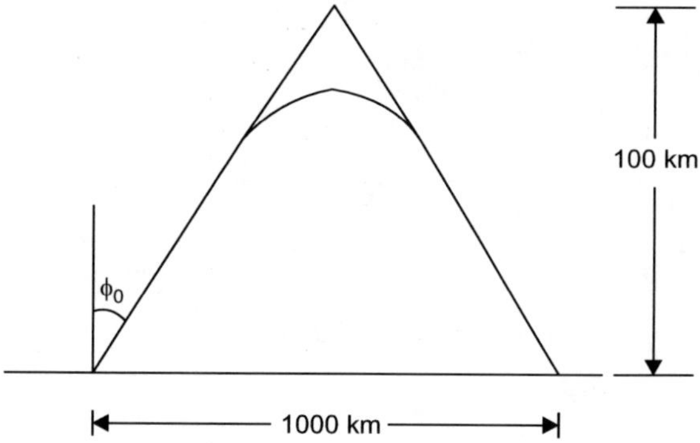

$$\sin\left(\frac{\pi}{2}-\phi_0\right) = \frac{100}{\sqrt{100^2+500^2}}$$

$$\cos\phi_0 = \frac{100}{\sqrt{10000+250000}}$$

$$\cos\phi_0 = \frac{100}{\sqrt{260000}}$$

$$\sec\phi_0 = 5.099$$

$$\text{MUF} = f_c \sec\phi = 10.2 \text{ MHz.}$$

Example 5.3: Find the attenuation function, field at the receiver and transmission loss for $P_1 = 15$ W, $\lambda = 35$ cm, $G_1 = 100$ (or 20 dB) $G_2 = 100$ (or 20 dB), $h_t = 80$ m, $h_2 = 20$ m and $r = 8$ km, if for a transmission over a dry ground $\varepsilon = 4$, $\sigma = 0.001$ S/m, $\gamma = 0.08$ rad and $\lambda = 35$ cm for which $R = 0.91$ and $\theta = 180°$ in the case of horizontal polarization and $R = 0.66$ and $\theta = 180°$ in the case of vertical polarization.

Solution: For horizontal polarization attenuation function is given by

$$f = \sqrt{1+2R\cos\left[\theta+\left(\frac{2\pi}{\lambda}\right)\Delta r\right]+R^2} = 0.74$$

$$f_{dB} = 20\log_{10}(0.74) = -2.6 \text{ dB}$$

and the rms value of the resultant field is given by

$$E_{r.m.s.}(h) = \frac{173\sqrt{P_1 k_w G_1}}{\gamma_{km}}\sqrt{1+2R\cos\left[\theta+\left(\frac{2\pi}{\lambda}\right)\Delta r+R^2\right]}\ \text{mv/m}$$

$$= \frac{173\sqrt{(15\times10^{-3}\times1000)}}{8}\times0.740$$

$$= 19.6 \text{ mV/m.}$$

Hence, the transmission loss is given by

$$L_{dB} = 10\log_{10}L$$

$$= Lf_s\,\text{dB} - 20\log_{10} f$$

$$= 20\log_{10}(4\pi r) - 20\log_{10}(\lambda) - P_{dB} - G_1\,\text{dB} - G_2\,\text{dB}$$

$$= 100 + 9.1 + 2.6 - 20 - 20$$

$$= 71.7 \text{ dB}$$

For vertical polarization, the r.m.s. value of the resultant field is given by

$$E_{r.m.s.}\ (v) = \frac{173\sqrt{(15\times10^{-3})\times100}}{8}\times0.71$$

$$= 18.9 \text{ mV/m.}$$

and $$L_{dB} = 100 + 9.1 + 2.9 - 20 - 20$$

$$= 72.0 \text{ dB}$$

Example 5.4: Compute the value of field intensity due to a wave freely propagated in a fog (with a visibility of about 30 m) at a distance of 5 km from the transmitter with P_1 = 1 W, G_1 = 600 and λ = 2 cm, if the corresponding attenuation δ is 0.25 dB/km.

Solution: Here

$$F = 10\frac{(-\delta_{(dB/km)}l_{(km)})}{20} = 0.87$$

and $$E_{r.m.s.} = \left[\frac{173\sqrt{P_1\,kW\,G_1}}{\gamma_{km}}\right]e^{-\delta/(km)}\text{mV/m.}$$

$$= \frac{173\sqrt{(10^{-3}\times600)}}{5}\times e^{-1.25} = 24.28 \text{ mV/m}$$

$$\because r = l = 5 \text{ km}$$

Example 5.5: Find the required transmitter power to ensure continuous working during 99% of a time along an ionospheric path 1800 km long at 60 MHz (X = 5 m), if the transmitting and receiving antenna are identical with a gain of 30 dB, the received [power is to be 5 × 10^{-15} W and $F_{99\%}$ = – 110 dB.

Solution: Now the received power is given by

$$P_2 = \frac{P_1 G_1 G_2 \lambda^2 F_1^2}{(4\pi r)^2}\ W_1$$

here F_1 is the power attenuation function.

Also $F_1 = 10^{(-110/20)} = 10^{-5.5} = 3.16 \times 10^{-6}$

where substitution for G_1, G_2 λ and r gives

$$P_1 = 10.5 \times 10^3 \text{ W} = 10.5 \text{ kW.}$$

It will be noted that ionospheric scattering of metric waves can be profitably utilized on distances upto 2000 km for transmission of signals with a bandwidth upto 6 kHz and that transmission loss may be off set by increase in transmitter power and use of highly directional antennas.

Example 5.6: Find how far from a 1 kW F.M. station can a 100 MHz signal be picked up with a half wave dipole, with a 6 dB noise figure for the receiver and an isotropic transmitting antenna.

Solution:

Now Noise Power $= S_N$

$$= k\,T\,\Delta f = \left(\frac{kT}{q}\right) q\,\Delta f$$

where $\left(\frac{kT}{q}\right) = \left(\frac{1}{40}\right) V$

$$q = 1.6 \times 10^{-19}\ C.$$

$$\therefore S_N = 10^{-14.4}\ W/MHz \text{ or } -114\ dBm/MHz$$

Example 5.7: A 150 m antenna transmitting at 1200 kHz has an antenna current of 8 Amps. Determined the voltage received by a receiving antenna of 2 m height and place away at a distance of 40 km from the transmitting antenna.

Solution:

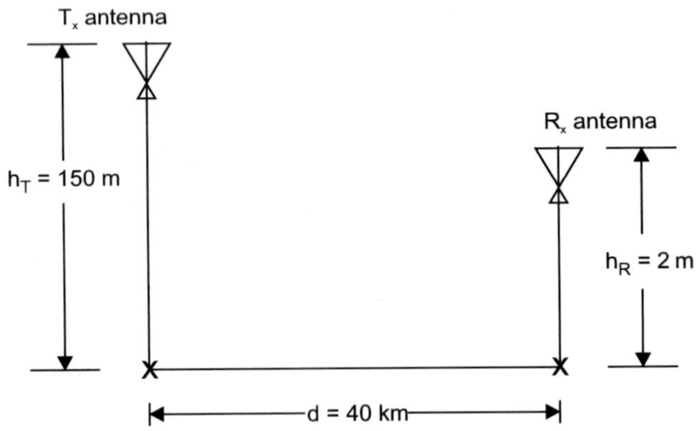

We know $f = C\lambda$

$$\therefore \quad \lambda = \frac{C}{f} = \frac{3\times10^8}{1200\times10^3}$$

$$\therefore \quad \lambda = 250\ m$$

Also we have $V = \dfrac{120\times\pi\times h_T\times h_R\times I}{\lambda d}$

$$\therefore \qquad V = \frac{120 \times \pi \times 150 \times 2 \times 8}{250 \times 40 \times 10^3}$$

$$= \frac{9047 \times 78.68}{10000 \times 10^3}$$

$$= 90.5 \times 10^{-3}$$

∴ Voltage received by receiving antenna

$$V = 90.5 \text{ mV}$$

Example 5.8: Determine maximum range for a tropospheric transmission, with antenna heights being 100 ft and 60 ft.

Solution: $d_{max\ (miles)} = \sqrt{2h_T (f_{t)}} + \sqrt{2h_R\ (f_t)}$

$$= \sqrt{200} + \sqrt{120}$$

$$d_{max\ (miles)} = 25.1 \text{ miles}$$

Example 5.9: In a VHF mobile radio system, the base station transmits 100 W at 150 MHz with an antenna 20 m above the earth.

Estimate the field strength at a receiving antenna, of height 2 m at a distance of 40 km.

Solution:

Here $\lambda = \dfrac{300 \times 10^6}{150 \times 10^6} = 2\,\text{m}$

$$E_0 = \sqrt{30 P_T G_T} / d \text{ V/m}$$

$$\therefore \qquad E_0 = \sqrt{30 \times 100 \times 1.64}, \text{ at, } d = 1 \text{ m}$$

Since the angle $[2\pi\ h_T\ h_R/\lambda d] << 0.5$ rad,

We have, $E_R = E_0 \left(\dfrac{4\pi h_T h_R}{\lambda d^2} \right)$

$$= 70 \left[\frac{4 \times 3.14 \times 20 \times 2}{2(40 \times 10^3)^2} \right]$$

$$\therefore \qquad E_R = 11\ \mu\text{V/m}$$

QUESTIONS

5.1. Enumerate the different modes of propagation and associated maximum range that can be achieved for e.m. waves in different frequency ranges.

5.2. Write technical notes on:
 (a) Virtual height
 (b) Critical frequency
 (c) Maximum usable frequency
 (d) Optimum working frequency and
 (e) Skip distance.

5.3. Explain the term 'ducting'. State the factors which would account for ducting.

5.4. Explain why vertical polarization is preferred to horizontal polarization in MW broadcasting.

5.5. Briefly describe the ionosphere and also the e.m. wave propagation through the ionosphere.

PROBLEMS

5.1. If the field intensity, of 25 mV/m, develops a p.d. of 2.7 V in a certain aerial, what would be the effective height of the aerial? **[Ans.: 108 m]**

5.2. Compute the power density, of an e.m. wave reaching the sun's surface, from a 2 MW high power pulse transmitter situated on the earth's surface, if the aerial gain is 60 dB and the distance of the sun from the earth, is 1.5×10^8 km. **[Ans.: 7.1 pW/m²]**

5.3. Evaluate the attenuation, of an e.m. wave in decibels, between two aerials with gains of 20 dB and spaced 50 km apart, if the operating frequency is 300 MHz. **[Ans.: 75.96 dB]**

5.4. Find the ground wave range for which the maximum usable frequency would be 10 MHz, assuming reflection to take place at a height of 400 km, with the maximum ionosphere density corresponding to a refractive index of 0.9 at the given frequency and neglecting the curvature of the earth within the required ground range. **[Ans.: 1652 km]**

CHAPTER 6

Fundamental of Information Theory

6.1 INTRODUCTION

In fundamental of information theory, the terms like information, noise frequently come hence there is a need to define terms like information and noise.

6.2 DEFINITION

6.2.1 Information

The thing which attribute some sense or meaning is known as information.

6.2.2 Noise

The thing which does not attribute any sense or no meaning is known as noise.

6.3 NEED OF SOURCE CODING

For efficient communication from source to destination, there is need of source coding. Convolution code, Huffman's code, Cycle code, Linear block code and Hamming code are examples of source coding.

On the other hand, suppose we have to do reliable communication over noisy channel, from source to destination there is necessity of error control coding. Examples of error control coding are parity check bit, error correction code and forward error correction.

6.4 COMMUNICATION SYSTEM

The purpose of communication system is to carry information bearing baseband signal from one place to another place over a communication channel.

6.5 INFORMATION THEORY

Information theory not only related to communication but also related to computer science. Computer science understood written information in coded form.

Information theory related statistical physics, statistical interference. Statistical interference is measured in the form of electromagnetic interference and electromagnetic compatibility. Information theory plays vital role in probability and statistics.

Information theory deals with mathematical modelling and analysis of communication rather than physical sources and channel. Examples of physical sources are spoken message, written message and light sources. Examples of channels are open wire, coaxial cable, twisted pair of wires and fibre optics cable. Open wire is nothing but electromagnetic waves propagated in space.

6.6 ENTROPY

Probabilistic behaviour of source of information is known as entropy. Information is measured in terms of entropy.

6.7 CHANNEL CAPACITY

Ability of a channel to convey information is known as channel capacity. It is naturally related to noise characteristics of a channel. Noise characteristics depends upon internal noise present within cable. If entropy of a source is less than channel capacity there is error-free communication taking place over the channel.

6.8 UNCERTAINTY, INFORMATION AND ENTROPY

Suppose discrete source of information emits κ symbols during unit time, that is one second. These κ symbols emitted by source are $s_0, s_1, s_2, \ldots s_{k-1}$ within a period of one second with probability

$$P(s = s_\kappa) = P_\kappa \text{ where } \kappa = 0, 1, \ldots, \kappa - 1 \quad (6.1)$$

Then source output modulated as discrete random variable s whose set is given by

$$\phi = \{s_0, s_1, s_2, \ldots, s_{\kappa-1}\} \quad (6.2)$$

This set of probability must satisfy the condition – $p_0 + p_1 + p_2 + \ldots + p_{\kappa-1} = 1$. That is,

$$\sum_{\kappa=0}^{\kappa-1} P_\kappa = 1 \quad (6.3)$$

Experiment is probabilistic symbols emitted during one second are stastically independent. The source having property given by eq. (6.3) is called discrete memoryless source. Memoryless source means present output is not dependent upon previous input but present output depends upon present input.

6.9 UNCERTAINTY, SURPRISE AND INFORMATION

Uncertain amount of information emitted by source.

Suppose there are two events that are event $s = s_i$ and event $s = s_\kappa$.

Event $s = s_i$ emitting i symbols $s_0, s_1, s_2, \ldots s_{i-1}$ with probability $p_0, p_1, p_2, \ldots, p_{i-1}$ and event $s = s_\kappa$ emitting κ symbols $s_0, s_1, s_2, \ldots s_{\kappa-1}$ with probability $p_0, p_1, p_2, \ldots p_{\kappa-1}$ where $\kappa = 0, 1, 2, \ldots \kappa - 1$.

Case I: Clearly, if $p_\kappa = 1$ and $p_i = 0$ for $i \neq \kappa$, it means that within communication system there is only noise and there is no information.

Case II: When symbols s_κ emitted, then $p_\kappa = 0$ and $p_i = 0$ for $i = \kappa$, it means that within communication system or information source there is no noise and there is only information.

Case III: When source occurs with different probabilities like p_k, p_i, p_l and when $p_\kappa = 0$, $p_i = 1$ and $p_l = 2$ and when p_κ is low, it means that within information source there is no noise and there is more information. Thus, uncertainty, surprise and information are interrelated.

Case (a): Before the event $s = k_\kappa$ occurs there is amount of uncertainty.

Case (b): When the event $s = s_\kappa$ occurs there is amount of surprise.

Case (c): After occurrence of event $s = s_\kappa$ there is gain in information.

Amount of information is equal to inverse of probability of occurrence.

6.10 RELATIONSHIP BETWEEN INFORMATION, PROBABILITY AND ENTROPY

When experiment is done, and when event $s = s_\kappa$ observed, source emits κ symbols $s_0, s_1, s_2, \ldots, s_{\kappa-1}$ with probabilities $p_0, p_1, p_2, \ldots, p_{\kappa-1}$ then amount of information gained are $I(s_0), I(s_1), I(s_2), \ldots I(s_{\kappa-1})$ respectively.

Then information and probability are related by

$$I(s_\kappa) = \log\left(\frac{1}{p_\kappa}\right) \tag{6.4}$$

Definition given by eqn. (6.4) gives following important properties:

Property 1: $I(s_\kappa) = 0 \text{ for } p_\kappa = 1$ (6.5)

From this condition it is understood that there is no information gained from source and source contains only noise.

Property (2): $I_{s\kappa} \geq 0 \text{ for } 0 \leq p_\kappa \leq 1$ (6.6)

From this condition it is understood that there is some information or no information gained from source and also source contain noise.

Property (3): $I(s_\kappa) > I(s_i) \text{ for } p_\kappa < p_i$ (6.7)

From this condition it is understood that more information gained from source and source contain negligible noise practically.

Property (4): $I(s_\kappa \cdot s_l) = I(s_\kappa) + I(s_l)$ (6.8)

From this condition or property, it is understood that event s_κ and event s_l are statistically independent.

Equation (6.4) can be written as

$$I(s_\kappa) = \log_2\left(\frac{1}{p_\kappa}\right) = -\log_2(p_\kappa)$$

where $$\kappa = 0, 1, \ldots \kappa - 1 \tag{6.9}$$

When $p_\kappa = 1/2$ we have $I(s_\kappa) = -\log_2(1/2) = \log_2(2)$

$$I(s_\kappa) = 1 \text{ bit} \tag{6.10}$$

Since we gain one bit of information from the source. It means that information $I(s_\kappa)$ is positive and probability (p_κ) is negative.

$$p_\kappa \uparrow \therefore I_\kappa \downarrow < \text{one bit}$$

6.11 DERIVATION FOR ENTROPY

Suppose source $s = s_\kappa$ emits κ symbols $s_0, s_1, s_2, \ldots s_{\kappa-1}$ with probability $p_0, p_1, p_2, \ldots p_{\kappa-1}$ and information $I(s_0)$, $I(s_1)$, $I(s_2)$, …, $I(s_{\kappa-1})$, respectively; where $\kappa = 0, 1, 2, \ldots, \kappa - 1$.

Then mean of information $(I_{s\kappa})$ over source alphabet ϕ is given by:

$$H(\phi) = E[I_{(s\kappa)}]$$

$$\therefore \quad H(\phi) = \sum_{\kappa=0}^{\kappa-1} p_\kappa I_{(s\kappa)} = p_0 I(s_0) + p_1 I(s_1) + \ldots + p_{\kappa-1} I(s_{\kappa-1}) \tag{6.11}$$

$$\therefore \quad H(\phi) = \sum_{\kappa=0}^{\kappa-1} p_\kappa \log_2\left(\frac{1}{p_\kappa}\right)$$

$$\therefore \quad \text{Entropy} = \sum_{\kappa=0}^{\kappa-1} p_\kappa \log_2\left(\frac{1}{p_\kappa}\right) \tag{6.12}$$

From this eq. (6.12), it is understood that entropy is proportional to probability (p_κ).

Some Properties of Entropy: Consider a discrete memoryless source whose mathematical model defined by equation $\phi = \{s_0, s_1 \ldots, s_{\kappa-1}\}$ and $p(s = s_\kappa) = p_\kappa$ where $\kappa = 0, 1, \ldots, \kappa - 1$. Then entropy $H(\phi)$ of such a source is bounded as follows:

$$0 \le H(\phi) \le \log_2 \kappa \tag{6.13}$$

where κ is number of symbols of the alphabet ϕ of source.

Then there are two properties of entropy

(1) $H(\phi) = 0$ if and only if the probability $p_k = 1$, probability $p_i = 0$ and probability $p_l = 0$. From this property it is understood that entropy contains no uncertainty that is more information gained from source.

(2) $H(\phi) = \log_2 \kappa$ if and only if $p_\kappa = 1/\kappa$ for all κ. From this property it is understood that entropy contains maximum uncertainty that is source emits only noise.

6.12 CHANNEL CODING

6.12.1 Discrete Memoryless Channels

Up to this, we have seen discrete memoryless source for information generation. Now we will see discrete memoryless channel for reliable information transmission.

For reliable information transmission, consider statistical model of discrete memoryless channel. The channel has input X and output Y, here input X corrupted by noise. Both X and Y are random in nature.

Within one second, channel accepts an input symbol X from alphabet X and channel emits an output symbol Y from an alphabet Y. Channel is 'discrete' since both X and Y have finite sizes. Channel is said to be memoryless because output depends upon current input and not on previous input.

Figure 6.1 shows discrete memoryless channel.

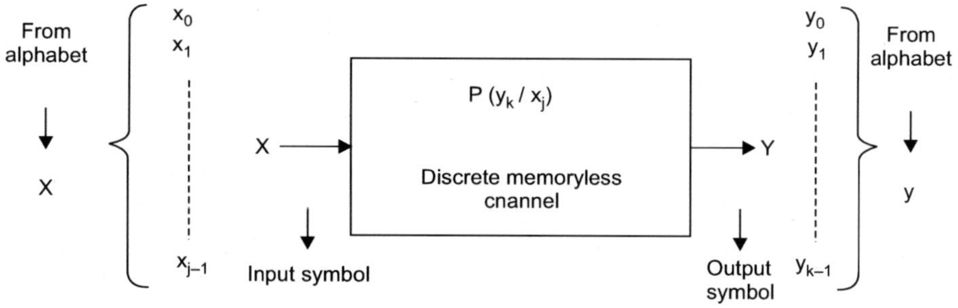

Fig. 6.1: Discrete Memoryless Channel

The channel is described in terms of an input alphabets:

$$x = (x_0, x_1, \ldots, x_{j-1}) \tag{6.14}$$

and output alphabets

$$y = (y_0, y_1, \ldots, y_{k-1}) \tag{6.15}$$

and a set of transition probabilities is given by:

$$P(y_k/x_j) = P\,(y = y_k \,/\, x = x_j) \text{ for all j and k} \tag{6.16}$$

Naturally, we have $0 \le P(y_\kappa/x_j) \le 1$ for all j and κ.

(1) Input alphabet x and output alphabet y need not have the same size.
(2) For example, in channel coding the size k of output alphabet y may be larger than the size J of the input alphabet x; thus $\kappa \geq J$.
(3) Also $\kappa \leq J$

For convenience, a discrete memoryless channel is also arranged the various transition probabilities of channel in the form of a matrix.

$$P = \begin{bmatrix} P(y_0/x_0) & P(y_1/x_0) & \cdots & P(y_{\kappa-1}/x_0) \\ P(y_0/x_1) & P(y_1/x_1) & \cdots & P(y_{\kappa-1}/x_1) \\ \vdots & \vdots & & \vdots \\ P(y_0/x_{j-1}) & P(y_1/x_{j-1}) & \cdots & P(y_{\kappa-1}/x_{j-1}) \end{bmatrix}$$

The J by κ matrix P is called channel matrix or transition matrix.

Fixed channel input of transition matrix are rows i.e. $(x_0, x_1, \ldots, x_{j-1})$ and fixed channel outputs of transition matrix are columns that is $(y_0, y_1, y_{\kappa-1})$.

(1) Fundamental property of channel matrix P: Sum of elements along any row of the matrix is always equal to one.

Mathematically, it can written as:

$$\sum_{\kappa=0}^{\kappa-1} P(y_\kappa / x_j) = 1 \text{ for all } j \tag{6.17}$$

(2) Joint probability distribution: If X is a channel input and Y is a channel output. The joint probability distribution of the random variable X and Y is given by

$$\begin{aligned} P(x_j, y_\kappa) &= P(x = x_j; y = y_\kappa) \\ &= P(y = y_\kappa/x = x_j)\, P(x = x_j) \\ P(x_j, y_\kappa) &= P(y_\kappa/x_j)\, p(x_j) \end{aligned} \tag{6.18}$$

(3) Marginal probability distribution: Marginal probability distribution of output Y (random variable) is obtained by averaging out the dependance P (x_j, y_κ) on x_j.

$$P(y_\kappa) = \sum_{j=0}^{J-1} P(y_\kappa / x_j) P(x_j) \text{ for } \kappa = 0, 1, \ldots, \kappa - 1 \tag{6.19}$$

6.12.2 Binary Symmetric Channel

Binary symmetric channel have theoretical and practical importance. Binary symmetric channel is special case of discrete memoryless channel when $J = \kappa = 2$. The channel has two input symbols ($x_0 = 0$ and $x_1 = 1$) and two output symbols ($y_0 = 0$ and $y_1 = 1$). If probability of receiving 1 if 0 is sent is same as probability of receiving 0 if 1 is sent then channel is known as binary symmetric channel.

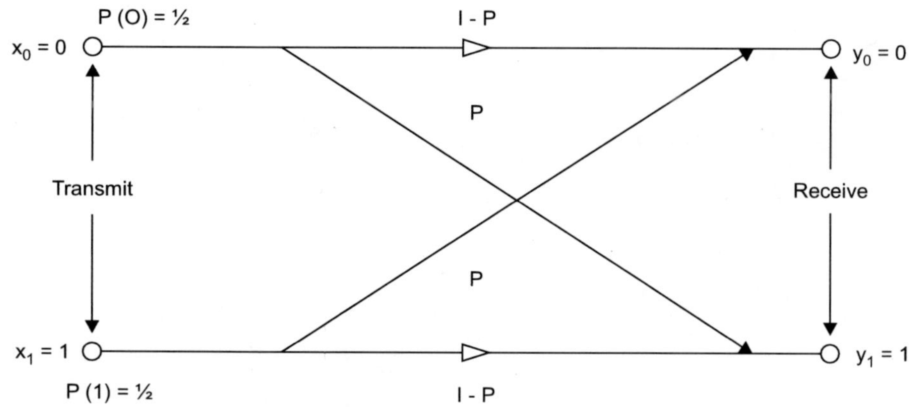

Fig. 6.2: Binary Symmetric Channel

Mathematically, $P(0) = P(1) = 1/2$.

Hence, channel is binary symmetric.

There is probability P of an error in transmission. Technically, we can say when bit 0 is transmitted then bit 1 is received or when bit 1 is transmitted then bit 0 received.

The probability of correct transmission is $1 - P$. Technically we can say, when 0 is transmitted then 0 received or when 1 is transmitted then 1 received.

It is great interest to relate transitional probability with conditional probability. Conditional probability of error P_{10} and P_{01} that were determined for the pulse code modulation receiver. For the case when binary symbols 0 and 1 are equiprobable, then we can show that the optimized values of these two error probabilities are equal. Technically $P_{10} = P_{01}$

We have,
$$P_{10} = P_{01}$$
$$P_{10} = P(y = 1/x = 0)$$
and
$$P_{01} = P(y = 0/x = 1)$$

We immediately see that for the pulse code modulation receiver
$$P_{10} = P_{01} = P$$

6.13 MUTUAL INFORMATION

Suppose channel between input X and output Y is noisy channel. Input X selected from alphabets $x_0, x_1, \ldots, x_j, x_{j-1}$ and output Y selected from alphabets $y_0, y_1, \ldots, y_K, y_{k-1}$.

Entropy H(x) is a measure of uncertainty about X before observing output Y. Now we have to measure amount uncertainty in X after observing output Y.

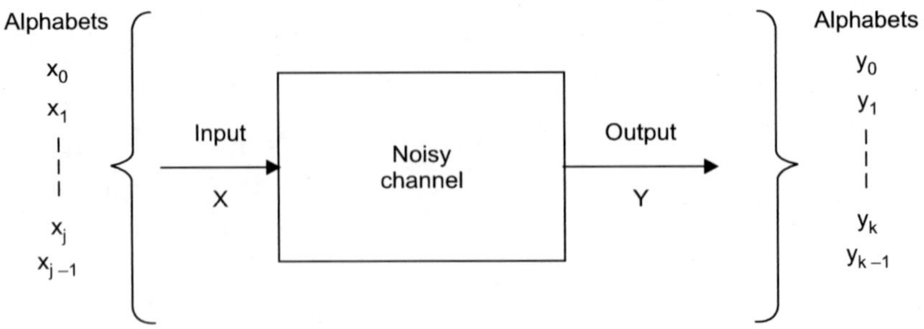

Fig. 6.3: Noisy Channel

To understood this, we should know conditional entropy of X selected from alphabets x_0, x_1, ... x_j, x_{j-1} given that $Y = y_\kappa$.

Specifically, we write:

$$H(x/Y = y_\kappa) = \sum_{j=0}^{J-1} P(x_j/y_\kappa)\log_2\left[\frac{1}{P(x_j/y_\kappa)}\right] \tag{6.20}$$

This quantity $H(x/Y = y_\kappa)$ is itself a random variable that takes on values $H(x/Y = y_0)$, ..., $H(x/Y = y_{\kappa-1})$ with probabilities $P(y_0)$, $P(y)$, $P(y_{\kappa-1})$ respectively.

The mean of entropy $H(x/Y = y_\kappa)$ over the output alphabet Y is therefore given by:

$$H(x/y) = \sum_{\kappa=0}^{\kappa-1} H(x/Y = y_\kappa)\, P(y_\kappa)$$

$$= \sum_{\kappa=0}^{\kappa-1}\sum_{j=0}^{j-1} P(x_j/y_\kappa)P(y_\kappa)\log_2\left[\frac{1}{P(x_j/y_\kappa)}\right] \tag{6.21}$$

$$H(x/y) = \sum_{\kappa=0}^{\kappa-1}\sum_{j=0}^{j-1} P(x_j, y_\kappa)\log_2\left[\frac{1}{P(x_j/y_\kappa)}\right] \tag{6.22}$$

where $P(x_j, y_\kappa) = P(x_j/y_\kappa)\, P(y_\kappa)$

The quantity H(x/y) is called conditional entropy and H(x/y) represents amount of uncertainty; remaining about the channel input after the channel output has been observed.

(1) Entropy H(x) represents uncertainty about the channel input before observing channel output.

(2) Conditional entropy H(x/y) represents uncertainty about the channel input after observing channel output.

(3) Difference H(x) – H(x/y) must represent uncertainty about the channel input that is resolved by observing the channel output.

The important quantity H(x) – H(x/y) is known as mutual information of the channel. This mutual information is denoted by I(x, y). Thus, I(x, y) can be written as

$$I(x, y) = H(x) - H(x/y)$$

Similarly, we may write:

$$I(y, x) = H(y) - H(y/x)$$

where H(y) is entropy of channel output and H(y/x) is conditional entropy of channel output given the channel input.

6.13.1 Properties of Mutual Information

Property (1): The mutual information of a channel is symmetric.
Mathematically,

$$I(x, y) = I(y, x)$$

where I(x, y) is known as measure of uncertainty about the channel input that resolved by observing the channel output and I(y, x) is known as measure of uncertainty about the channel output that is resolved by sending the channel input.

Property (2): Mutual information is always non-negative
Mathematically,

$$I(x, y) \geq 0$$

Property (3): The mutual information of a channel is related to the joint entropy of the channel input and channel output. Mathematically,

$$I(x_1\ y) = H(x) + H(y) - H(x, y)$$

where I(x, y) is known as mutual information and H(x, y) is known as joint entropy of channel input and channel output.

6.14 SHANNON'S HARTELY LAW FOR ADDITIVE WHITE GAUSSIAN NOISE (AWGN) CHANNEL

Suppose AWGN channel has input X and output Y and AWGN channel is corrupted by noise (n), then AWGN channel output Y is given by:

$$Y = X + n$$

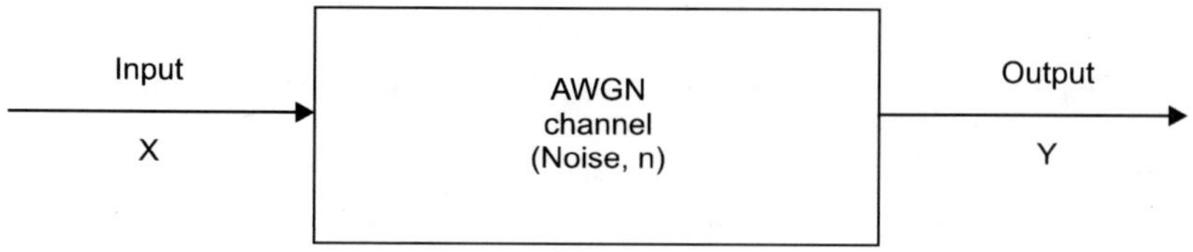

Fig. 6.4: AWGN Channel

where n is an additive white Gaussian noise present in bandlimited signal.

Then channel capacity C_s of an AWGN channel is given by:

$$C_s = \max_{f_x(x)} . \ I(x, y)$$

or $$C_s = \frac{1}{2}\log_2\left(1+\frac{S}{N}\right) \text{ bits/sample}$$

where S/N is signal to noise ratio at channel output.

If channel bandwidth B is fixed, then output Y is a bandlimited signal which is sampled by Nyquist rate according to Nyquist theorem. Nyquist theorem is stated as follows if signal has maximum frequency f_{max} then signal is sampled by Nyquist sampling theorem. Mathematically, it is

$$\text{sampling rate} \geq 2\, f_{max}.$$

Then the channel capacity (c) bits/sec of AWGN channel is given by:

$$C = 2BC_s = 2B \times \frac{1}{2}\log_2\left(1+\frac{S}{N}\right)$$

$$\therefore \quad C = B\log_2\left(1+\frac{S}{N}\right) \text{bits/sample} \tag{6.23}$$

From eqn. (6.33) it is seen that capacity of AWGN channel is directly proportional to bandwidth over which noise is measured and signal to noise ratio.

6.15 SHANNON'S THEOREM

Shannon's theorem is a fundamental to the theory of communication. It relates to rate of transmission of information over a communication channel.

6.15.1 Statement (1) of Theorem

Suppose source emits m equal message and if m >> 1. Suppose source generating information at rate R and given a channel with channel capacity C

Then if $$R \leq C$$

That is, Rate of information generated ≤ channel capacity

Statement (1): Rate of information generated from source less than or equal to channel capacity then output of source transmitted over the channel with a probability of error.

6.15.2 Statement (2) of Theorem

Suppose source emits m equal messages and if m >> 1. Suppose source generating information at rate R and given channel with channel capacity C.

Then if mathematically $R > C$

This is, Rate of information generated > channel capacity

Statement (2): Rate of information generated by source is greater than channel capacity then theorem states that if information rate R exceeds a specified value of channel capacity c then error probability will increases towards unity.

6.16 LINEAR BLOCK CODE

In linear block code, a code word has the form

$$a_1a_2a_3 \ldots a_\kappa c_1c_2 \ldots c_r \tag{6.24}$$

where κ is number of information bits and r is number of parity bits and the code word $n = \kappa + r$ = number of bits. There are $M = 2^\kappa$ valid code words

Then uncoded word is

$$\bar{A} = [a_1a_2 \ldots a_\kappa] \tag{6.25}$$

For generation of linear block code starts with a selection of the number r of parity bits to be added and thereafter with specification of an $\bar{H}$ matrix

$$\bar{H} = \begin{vmatrix} h_{11} & h_{12} & \ldots & h_{1\kappa} & 1\ 0\ 0 \ldots 0 \\ h_{21} & h_{22} & \ldots & h_{2\kappa} & 0\ 1\ 0 \ldots 0 \\ \ldots & \ldots & \ldots & \ldots & \ldots \\ h_{r1} & h_{r2} & & h_{r\kappa} & 0\ 0\ 1 \ldots 0 \end{vmatrix} \tag{6.26}$$

$(\bar{h})$ $r \times \kappa$ submatrix $r \times r$ $(\bar{I})$ identity matrix

From matrix (6.26) it is seen that $\bar{H}$ consists of submatrix $(\bar{h})$ whose rows are κ and columns are r and identity matrix $(\bar{I})$ have r rows and r columns.

From eqns. (6.24) and (6.25), it is seen that logical variable a'^s C'^s have possible values 0 or 1. Also from matrix (6.26) it is seen that h are constant logical variable whose values are given 0 or 1.

The merits of the block code depends on the status with which h'^s have been selected.

In linear block code case since we deal with logical variable, it is seen that the arithmetic product is replaced by the logical product. Logically, the AND operation is replaced by Exclusive-OR operation.

We rewrite that these logical operations are defined as follows:

AND	Exclusive - OR
$Y = A \cdot B$	$Y = A \oplus B = \bar{A}B + A\bar{B}$
$0 = 0 \cdot 0$	$0 = 0 \oplus 0 = 1 \times 0 + 0 \times 1$
$0 = 0 \cdot 1$	$1 = 0 \oplus 1 = 1 \times 1 + 0 \times 1$
$0 = 1 \cdot 0$	$1 = 1 \oplus 0 = 0 \times 1 + 1 \times 1$
$1 = 1 \cdot 1$	$0 = 1 \oplus 1 = 0 \times 1 + 1 \times 0$

The transpose of matrix $\bar{H}$, is $\bar{H}^T$ is derived by simply interchanging rows and columns is:

Hence,

$$\bar{H}^T = \begin{vmatrix} h_{11} & h_{21} & \cdots & h_{r1} \\ h_{12} & h_{22} & \cdots & h_{r2} \\ \hline h_{1\kappa} & h_{2\kappa} & \cdots & h_{r\kappa} \\ 1 & 0 & \cdots & 0 \\ 0 & 1 & \cdots & 0 \\ 0 & 0 & \cdots & 0 \\ 0 & 0 & \cdots & 1 \end{vmatrix} \begin{matrix} \kappa \times r\ \bar{h}^T \text{ sub matrix} \\ \\ \\ \\ \\ \\ r \times r \text{ identity matrix} \end{matrix} \tag{6.27}$$

Now, transposed matrix $\bar{H}^T$ consists of an r × r identity matrix and a submatrix $\bar{h}^T$ which is transpose of $\bar{h}$.

To generate a code word $\bar{T}$ from the uncoded word $\bar{A} = [a_1, a_2, \ldots, a_\kappa]$. We have to form a generator matrix G where

$$\bar{G} = \begin{vmatrix} 1 & 0 & 0 & \cdots & 0 & h_{11} & h_{21} & \cdots & h_{r1} \\ 0 & 1 & 0 & \cdots & 0 & h_{12} & h_{22} & \cdots & h_{r2} \\ \hline 0 & 0 & 0 & \cdots & 1 & h_{1\kappa} & h_{2\kappa} & \cdots & h_{r\kappa} \end{vmatrix} \tag{6.28}$$

$$\underbrace{\qquad\qquad}_{\kappa \times \kappa} \qquad \underbrace{\qquad\qquad}_{\kappa \times r}$$

where $\bar{G}$ have identity matrix having κ rows and κ columns and second matrix is the submatrix $\bar{h}^T$ having κ columns and r rows.

We have to verify that $$\bar{G}\,\bar{H}^T = 0 \tag{6.29}$$

Thus, if for example, we form $(\bar{G}\bar{H}^T)_{11}$ we get $(\bar{G}\bar{H}^T)_{11} = h_{11} = 1 \oplus 1$. That is, $h_{11} = 0$.

Similarly, all other elements of $\bar{G}\bar{H}^T$ are zero.

To generate matrix $\bar{G}$ have been formed, the code word $\bar{T}$ corresponding to each uncoded word $\bar{A}$ is $\bar{T} = \bar{A}\,\bar{G}$.

Most importantly, it is seen that the coded words so formed have the property that according to eqn. (6.29)

$$\bar{T}\bar{H}^T = \bar{A}\bar{G}\bar{H}^T = 0$$

That is, $\bar{A}\bar{G}\bar{H}^T = 0$

(1) Linear block code is suitable for systematizations, organization and comparison of codes.

(2) It is also used for purpose of clarifying mechanics of encoding and decoding.

6.17 CONVOLUTIONAL CODING

6.17.1 Code Generation

A convolutional code is generated by combining the outputs of a κ stage shift registers through the employment of exclusive-OR logic summers. Such a coder is shown in Fig. 6.5 for the case κ = 4 and v = 3.

In Fig. 6.5, m_1, m_2, m_3 and m_4 are shift registers or flip-flops or one-bit memory. Output of m_1 given to input of m_2, output of m_2 given to input of m_3, output of m_3 given to input of m_4. Actual input b_i given to m_1. In Fig. 6.5 there are

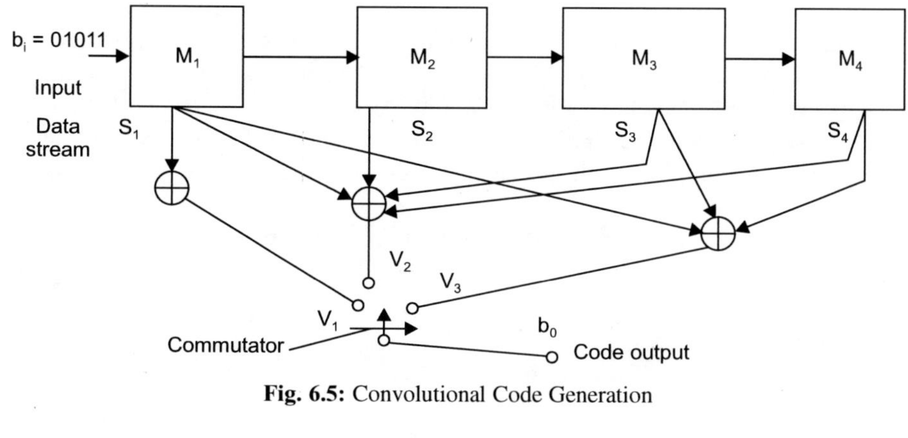

Fig. 6.5: Convolutional Code Generation

three exclusive-OR gates. Output v_1 has input s_1, output v_2 have inputs s_1, s_2, s_3 and s_4. Output v_3 have inputs s_1, s_3 and s_4.

Code output, b_0 have commutator. Commulator connects output b_0 to v_1 or v_2 or v_3 depending upon our desiration.

Depending upon clock input b_i = 01011 shifts into m_1, m_2, m_3 and m_4. For first clock pulse bit 1, that is least significant bit (L.S.B.) shifted to m_1, for second clock L.S.B. bit 1 shifts to m_2 and bit 1 shift m_2. For third clock L.S.B. bit 1 shifts m_3, 1 into m_2 and 0 into m_3. For fourth clock, L.S.B. bit 1 shifts to m_4, 1 into m_3, 0 into m_2 and 1 into m_1.

Now m_1, m_2 m_3 and m_4 have bit stream 1011 respectively. The outputs v_1, v_2 and v_3 of the adders shown in Fig. 6.5 are $v_1 = s_1$

$$v_2 = s_1 \oplus s_2 \oplus s_3 \oplus s_4 \text{ and } v_3 = s_1 \oplus s_3 \oplus s_4$$

We assume that initially shift register is a clear. The first bit of the input data is entered to m_1. During this message bit internal the commutator samples and hence the adder output v_1, v_2 and v_3. In present case output is three coded. The encoder is therefore have sampling rate 1/3. The next message bit then enters m_1, while the bit initially in m_1 transfer to m_2 and the commutator again samples all the v adder output. This process continues until the last bit of message has been entered into m_1. 0 are added to the message to transfer the last message bit through m_4 and hence out of shift register. The shift register then finds itself in its initial "clear" condition.

6.18 HUFFMAN CODING

Huffman's code is optimum code. It has maximum (highest) efficiency.

(1) Arrange the source symbols in orders of decreasing probability.

(2) Add the probabilities of the two symbols having the lowest probability and reorder the resultant probabilities, this step is called reduction I. The same procedure is repeated until there are two ordered probabilities remaining.

(3) Start encoding with the last reduction, which consists of exactly ordered probabilities. Assign 0 as the first digit in the code words for all source symbols associated with first probability. Assign 1 to the second probability.

(4) Now go back and assign 0 and 1 to second digit for the two probabilities that were combined in the previous reduction step retaining all assignments made in step 3.

(5) Keep regressing this way until the first column is reached.

Numerical 6.1:

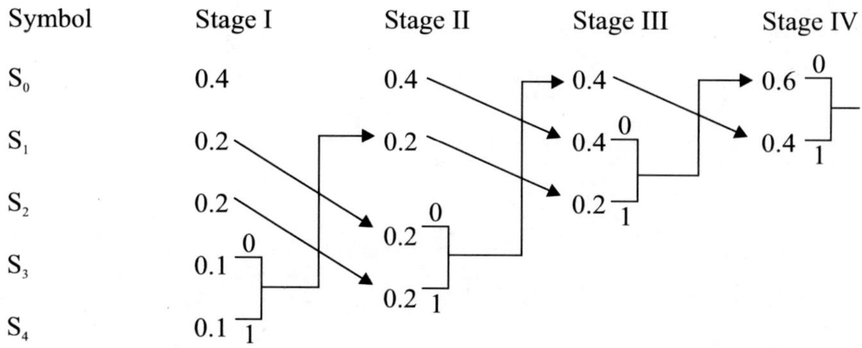

Symbol	Probability	Codeword
S_0	0.4	00
S_1	0.2	10
S_2	0.2	01
S_3	0.1	010
S_4	0.1	011

The average code word length = $\bar{L}$ = 0.4(2) + 0.2(2) + 0.2(2) + 0.1(3) + 0.1(3)

$$\bar{L} = 2.2$$

$$H(x) = 0.4\log_2\left(\frac{1}{0.4}\right) + 0.2\log_2\left(\frac{1}{0.2}\right) + 0.2\log_2\left(\frac{1}{0.2}\right) + 0.1\log_2\left(\frac{1}{0.1}\right) + 0.1\log_2\left(\frac{1}{0.1}\right)$$

$$H(x) = 0.52877 + 0.46439 + 0.46439 + 0.33219 + 0.33219$$

$$H(x) = 2.12193 \text{ bits}$$

Numerical 6.2:

x_i	$p(x_i)$	Code
x_1	1/2	0
x_2	1/4	10
x_3	1/8	110
x_4	1/8	111

Solution: $I(x_1) = -\log_2 1/2 = 1 = n_1$

$I(x_2) = -\log_2 1/4 = 2 = n_2$

$I(x_3) = -\log_2 1/8 = 3 = n_3$

$I(x_4) = -\log_2 1/8 = 3 = n_4$

We know that,

$$H(x) = \sum_{i=1}^{4} P(x_i)I(x_i) = \frac{1}{2}\times 1+\frac{1}{4}\times 2+\frac{1}{8}\times 3+\frac{1}{8}\times 3$$

$$H(x) = 1.75$$

$$L = \sum_{i=1}^{4} P(x_i)n_i = \frac{1}{2}\times 1+\frac{1}{4}\times 2+\frac{1}{8}\times 3+\frac{1}{8}\times 3$$

$$L = 1.75$$

$$\eta = \frac{H(x)}{L} = \frac{1.75}{1.75} = 1\times 100 = 100\%$$

Numerical 6.3: A discrete memoryless channel x has five symbols x_1, x_2, x_3, x_4 and x_5 with probabilities $p(x_1) = 0.4$, $p(x_2) = 0.19$, $p(x_3) = 0.16$, $p(x)_4 = 0.15$ and $p(x_5) = 0.1$. Calculate $H(x) = ?$; $L = ?$, and $\eta = ?$

Solution:

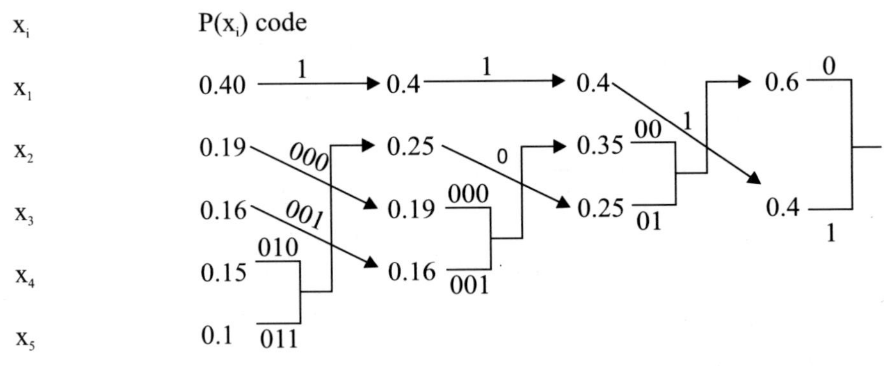

x_i	$p(x_i)$	codeword
x_1	0.4	0
x_2	0.19	01
x_3	0.16	10
x_4	0.15	110
x_5	0.1	111

$$L = \sum_{i=1}^{5} P(x_i)n_i = 0.4(1)+0.19(3)+0.16(3)+0.15(3)+0.1(3)$$

$$L = 2.2$$

$$H(x) = \sum_{i=1}^{5} P(x_i)n_i = 0.4(1) + 0.19(2) + 0.16(2) + 0.15(3) + 0.1(3)$$

$$H(x) = 2.15$$

$$\eta = \frac{H(x)}{L} = \frac{2.15}{2.2} = 0.977$$

$$\%\eta = 0.977 \times 100 = 97.7\%$$

6.19 REED-SOLOMON CODE (RS CODE)

Specifically on RS (n, κ) code is used to encode m bit symbols into block of $n = (2^m - 1)$ symbols that is $m(2^m - 1)$ bits where $m \geq 1$.

m bits symbols converted into n $(2^m - 1)$ bit symbols or n $(2^m - 1)$ bits by RS (n, k) code where $m \geq 1$.

Thus, the encoding algorithm expands a block of κ symbols to n symbols by adding (n – κ) redundant symbols.

κ symbols by adding (n – k) redundant symbols by using encoding algorithum converted into n symbols.

When m is an integer power of two, that is 2^m, the m bit symbols are called bytes. A popular value of m is s that is 2^8. Indeed, 8-bit RS code are extremely powerful.

A t error correcting RS code has a few parameters:

(1) A t error correcting RS coding has a block length n equal to $(2^m - 1)$ symbols.

(2) A t error correcting RS code has a message size of κ symbols.

(3) A t error correcting RS code has a parity check size of (n – k) equal to 2t symbols.

(4) A t error correcting RS code has a minimum distance $d_{minimum}$ is equal to 2t + 1 symbols.

Reed–Solomon codes is efficient for decoding.

6.20 PARITY CODING

The simplest method of coding is a parity coding. Parity bits are added to message bit at the time of transmission. At the reception end, receiver checks these parity bits. Error is detected if the expected pattern of parity bit is not received. It may be noted that parity coding is used only to detect the errors.

6.20.1 Even Parity

The parity of a binary word is known as 'even' if it contains even number of 1's. For example, 01101001 has even parity because there are four number of 1's in the word.

6.20.2 Odd Parity

The parity of binary word is known as 'odd' if it contains odd number of 1's. For example, 10010001 has odd parity because there are three number of 1's in this word.

6.20.3 Vertical Redundancy Check (VRC)

In this method, the parity check is done of each character separately. This method is always used in ASCCI code. In ASCCI code, each character has seven binary digit. One redundancy digit is added to each character to make the parity odd or even. As an example, let us coding shown in Table 6.1.

Table 6.1

Character	ASCCI Code of Character								Additional Parity bit
K	b_1	b_2	b_3	b_4	b_5	b_6	b_7	b_8	
	1	1	0	1	0	0	1	0	Even number of 1 in code ∴ parity zero
O	1	1	1	1	0	0	1	0	Odd number of 1 in code ∴ parity 1
I	1	0	0	1	0	0	1	0	Odd number of 1 in code ∴ parity 1
A	1	0	0	0	0	0	1	0	Even number of 1 in code ∴ parity zero

As shown in Table 6.1, the message consists of total 8 bits. Bit b_1 to b_7 are used as ASCCI code of character and bit b_8 is used as parity bit.

For example, the ASCCI code of k is 1101001. This consists of even number of 1's therefore 8th parity bit has '0' value. Hence, transmitted message k will be 11010010 which has even parity.

The ASCCI code is 11110010. This contains odd that is five number of 1's, hence, the parity has value '1'. Hence transmitted message for 0 is 11110011, which has even parity.

During the transmission if single error is created, it will make number of 1's odd. For example, let us consider that the received message for character κ is as shown below in Table 6.2.

Table 6.2

	b_1	b_2	b_3	b_4	b_5	b_6	b_7	b_8
Transmitted message for character →	1	1	0	1	0	0	1	0
κ by even parity received message	1	1	0	1	①	0	1	0

↑ This digit is an error

Table 6.2: Transmitted message block

	Character	STx.	N	A	I	T	I	S	0	K	ETB	BCC
Message bits	b_1	0	0	1	1	0	1	1	1	1	1	1 column
	b_2	1	1	0	0	0	0	1	1	1	1	0 parity
	b_3	0	1	0	0	1	0	0	1	0	1	0 bit
	b_4	0	1	0	1	0	1	0	1	1	0	1
	b_5	0	0	0	0	1	0	1	0	0	1	1
	b_6	0	0	0	0	0	0	0	0	0	0	0
	b_7	0	1	1	1	1	1	1	1	1	0	0
	b_8 ↑ Row parity bit	1	0	0	1	1	1	0	1	0	0	

Hence, as shown in Table 6.2, bit b_5 has error and it makes number of 1's odd that is 5. The receiver detect this error. However, it is not possible for the receiver to detect exactly which digit is in error. The receiver will only detect there are odd number of 1's and thus received message contains some error. Thus, receiver has to send repeat request to the transmitter to retransmit the message. If two digits are in error, the number of 1's in the received message is even. In this case, receiver will detect no error. In fact, this is the major drawback of vertical redundancy check.

6.20.4 Longitudinal Redundancy Check (LRC)

In vertical redundancy check (VRC) only single error are detected. On the other hand, in longitudinal redundancy check (LRC) large message blocks are divided into several character. Now parity check is applied to both rows and columns. Thus, it is also possible to correct single error with LRC. Block check character (BCC) is added at the end of block of information. Block check code can be even or odd parity. An even parity of block check code may be obtained by using exclusive OR Gate.

Figure 6.6 shows structure of information block transmitted with block check character.

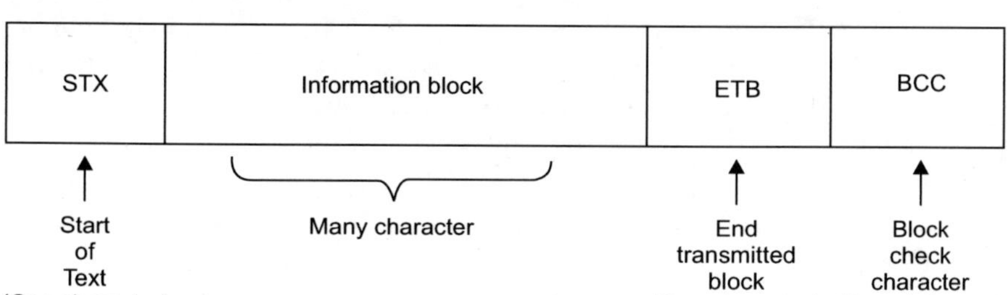

Fig. 6.6: Structure of Information Block

In Fig 6.6 given above, STx is one character in ASCCI. After this STx, there is multiple character information block. After information block there is single character; end of transmitted block. After ETB there is single character block check character.

As shown in Fig. 6.6 above, consider an example, let us consider transmitted and received message block as shown in Table 6.2 which uses LRC.

LRC method is used with ASCCI code only. Table 6.2 shows transmitted message block and Table 6.3 shows detected message block.

Table 6.3: Detected message block

Message bits		N	A	I	T	I	S	O	K		
b_1	0	0	1	1	0	1	1	1	1	1	1
b_2	1	1	0	0	0	0	1	1	1	1	0
b_3	0	1	0	0	1	0	0	1	0	1	0
b_4	0	1	0	1	(1)	1	0	1	1	0	1x
b_5	0	0	0	0	1	0	1	0	0	1	1
b_6	0	0	0	0	0	0	0	0	0	0	0
b_7	0	1	1	1	1	1	1	1	1	0	0
b_8 ↑ Row paity bit	1	0	0	1	1x	1	0	1	0	0	1

x implies an error in the parity bit

The first row of table shows character which are transmitted. It may be observed that first character transmitted is STx. The ASCII value of STx is shown in its column. The ASCII value of STA is $(b_1b_2b_3b_4b_5b_6b_7) = (0100000)$.

The last row (for b_8) shows the even parity bit is added for each character. Hence, $b_8 = 1$ for STx because ASCII code of STx contain only one that is odd binary 1. Hence, transmitted character for STx is $(b_1b_2 \ldots b_8) = (01000001)$.

The sequence of character N.A.I.T. etc. is transmitted following ST_x. This is known as information block. It may be noted that second last character transmitted is ETB. The ASCII value of ETB $(b_1, b_2, \ldots, b_7)$ (1110100) is shown in its column. The parity bit b_8 is zero. After ETB, BBC is transmitted. The value of B.B.C. in particular row is obtained by performing Ex-OR operation.

For example, consider the row b_1. This is

STx	N	A	I	T	I	S	O	K	ETB1	BBC
$b_i = 0$	0	1	1	0	1	1	1	1	1	1

At this point, it may be observed that the row for b_1 contains odd that is seven number of 1's. The Ex-OR operation will provide a value of 1. Hence, in row b_1,

BBC has value 1. Note that complete row of b_1 has even parity. Similarly, in row b_2 BBC is zero. It has also got even parity. BBC is thus obtained by parity coding. BBC also provided parity check in vertical and horizontal direction. As shown in Table 6.3 the encircled digit is in error. This error is easily located in rows and columns parity check. Hence, single error can be detected and corrected.

Advantage of LRC over VRC

(1) LRC method detect single error and also correct single error.

(2) By LRC method, double and triple error can be detected.

6.21 LEMPEV-ZIV (LZ) CODING

The drawback of the Huffman's code is that it requires knowledge of statistical model of source. Unfortunately, in practice, source statistics are not always known as before.

To overcome these practical limitations, we may use the Lempev-ZIV (L–Z) algorithms which is adaptive and simple to implement than Huffman's coding.

Numerical 6.4: Consider a discrete memoryless source with source alphabets $\phi = \{s_0, s_1, s_2\}$ with respective probabilities $p_0 = 1/4$; $p_1 = 1/4$ and $p_2 = 1/2$. Determine entropy $H(\phi)$.

Solution:

$$H(\phi) = \sum_{k=0}^{k-1} p_K I(s_K) = p_0 \log_2\left(\frac{1}{p_0}\right) + p_1 \log\left(\frac{1}{p_1}\right) + p_2 \log_2\left(\frac{1}{p_2}\right)$$

$$\therefore H(\phi) = \frac{1}{4}\log_2(4) + \frac{1}{4}\log_2(4) + \frac{1}{2}\log_2(2)$$

$$\therefore H(\phi = \frac{3}{2} \text{ bits}$$

Numerical 6.5: Symbols of ϕ^2

Probability (σ_i)	σ_0	σ_1	σ_2	σ_3	σ_4	σ_5	σ_6	σ_7	σ_8
i = 0, 1, 2, …, 8	$\frac{1}{16}$	$\frac{1}{16}$	$\frac{1}{8}$	$\frac{1}{16}$	$\frac{1}{16}$	$\frac{1}{8}$	$\frac{1}{8}$	$\frac{1}{8}$	$\frac{1}{4}$

Determine $H(\phi)^2$ = ?

Solution:

$$H(\phi^2) = \sum_{i=0}^{8} P(\sigma_i)\log_2 \frac{1}{P(\sigma_i)}$$

$$\therefore \quad H(\phi)^2) = \frac{1}{16}\log_2(16) + \frac{1}{16}\log_2(16) + \frac{1}{8}\log_2(8) +$$

$$\frac{1}{16}\log_2(16) + \frac{1}{16}\log_2(16) + \frac{1}{8}\log_2(8) +$$

$$\frac{1}{8}\log_2(8) + \frac{1}{8}\log_2(8) + \frac{1}{4}\log_2(4)$$

$$\therefore H(\phi^2) = 3 \text{ bits}$$

Numerical 6.6: In a certain system the (S/N) power ratio is 10 and bandwidth is 10,000 Hz. Find the maximum permissible information rate. What will be the effect on the system if S/N falls to a value of 5.

Solution: Here the maximum permissible information rate is given by:

$$R_{max} = B\log_2\left(1+\frac{S}{N}\right) = 10{,}000 \log_2(1+10)$$

$$\therefore R_{max} = 10{,}000 \times \log_2(11) = 10{,}000 \times \frac{\log_{10} 11}{\log_{10} 2}$$

$$\therefore R_{max} = 10{,}000 \times 3.45943$$

$$R_{max} = 34594.3 \text{ bits/sec.}$$

If S/N becomes 5 then the channel capacity or maximum rate R_{max} becomes 10,000 $\log_2(6)$ = 25849.6 bits/sec.

Thus, the rate of signalling would exceed the channel capacity causing excessive error to appear.

Numerical 6.7: What will be the system bandwidth for 30 dB signal to noise ratio, if it is desired to transmit digital data at a rate of 30,000 bits/sec?

Solution: The channel capacity expressed as:

$$C = B\log_e\left[\left(1+\frac{S}{N}\right)\right]$$

Assuming a channel with B.W. (B) = 30 kHz and (S/N) = 1

then $C = 30 \times 10^3 \times \log_e (1 + 1) = 30 \times 10^3 \log e^2$

$$C = 30 \times 10^3 \times 0.693147$$

$$\therefore C = 20794.4 \text{ bits/sec.}$$

So for standard voice telephone circuits if power is increased by a factor of 1000 (corresponding to a (S/N) ratio of 1000 or 30 dB), the bandwidth can be reduced to

$$B = 3 \text{ kHz}$$

Numerical 6.8: Find the minimum (S/N) power ratio in a system with bandwidth of 3 kHz and information transmitting rate of 5 bits/second.

Solution: The rate of information transfer (R_m) is given by

$$\left(\frac{R_m}{B}\right) \leq \frac{S+N}{N} \leq \log_2(S+N) - \log_2(N) \text{ bits}$$

So, $5B \leq B\log_2\left(1+\frac{S}{N}\right)$

$$\therefore \left(1+\frac{S}{N}\right) \geq 32$$

$$\therefore \left(\frac{S}{N}\right) \geq 32-1$$

$$\therefore \left(\frac{S}{N}\right) \geq 14.9 \text{ dB}$$

6.22 DIGITAL MODULATION TECHNIQUE

Digital and numerical modulation techniques are broadly classified as

(1) Amplitude shift keying
(2) Frequency shift keying
(3) Phase shift keying

6.22.1 Amplitude Shift Keying (ASK)

If modulating signal is m and carrier is cos ω_ct. By this digital data when unmodulated carrier get modulated as amplitude shift keyed (A.S.K.) then for logic 1, output of A.S.K. modulated signal is m(t) cos ω_ct and for logic 0, output of ASK modulated signal will be 0.

6.22.2 Frequency Shift Keying (FSK)

When modulating signal is m(t) that is digital data and unmodulated carrier is cos ω_ct. By frequency shift keying (F.S.K.) then for logic 1 output frequency of FSK modulated signal is to be ωc_1. On the other hand, for logic 0 output frequency of F.S.K. modulated signal is to be ωc_0.

6.22.3 Phase Shift Keying (PSK)

When modulating signal is m(t) that is digital data and unmodulated carrier is cos ω_ct. By phase shift keying (PSK) at leading or rising edge and trailing or falling edge there is 180° change in phase. At falling edge of modulating data phase changes from p(t) cos ω_ct to p(t) cos (ω_ct + π). On the other hand, at rising edge of data phase changes from p(t) cos (ω_ct + π) to p(t) cos ω_ct.

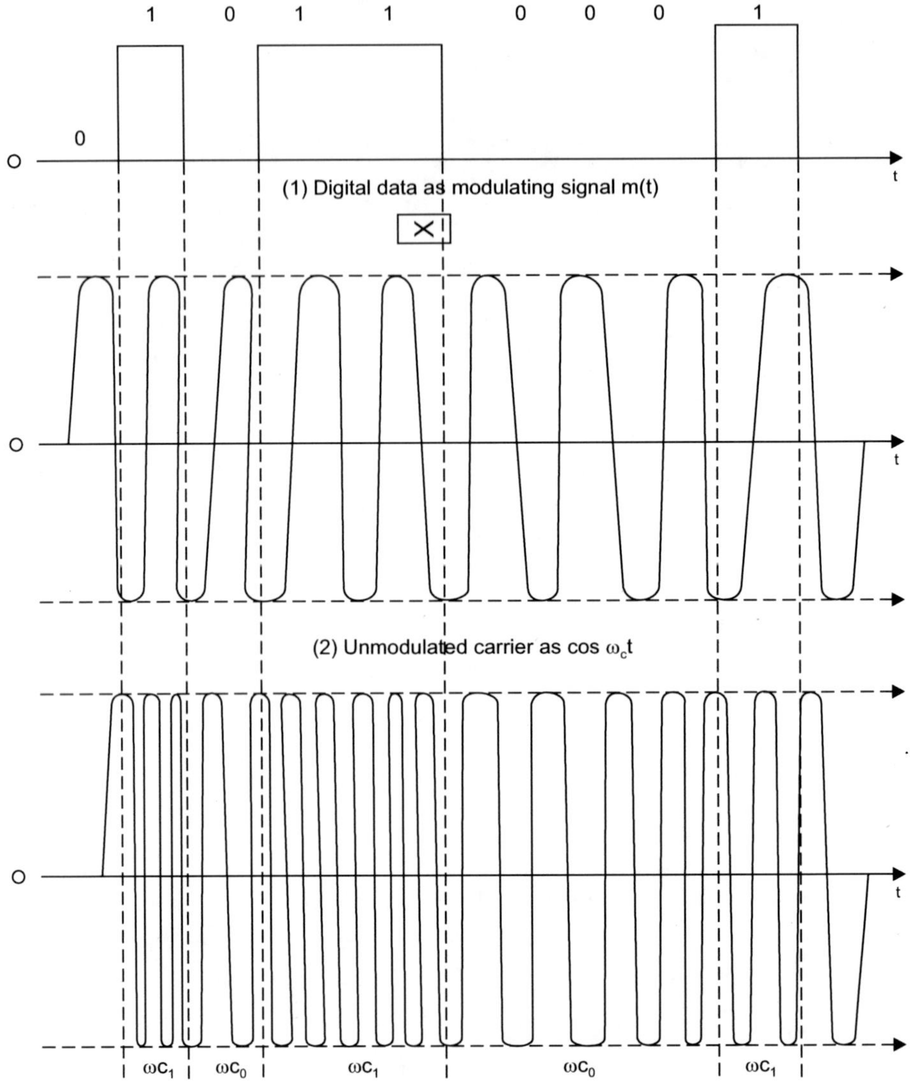

(3) Frequency shift keying output

Fig. 6.7

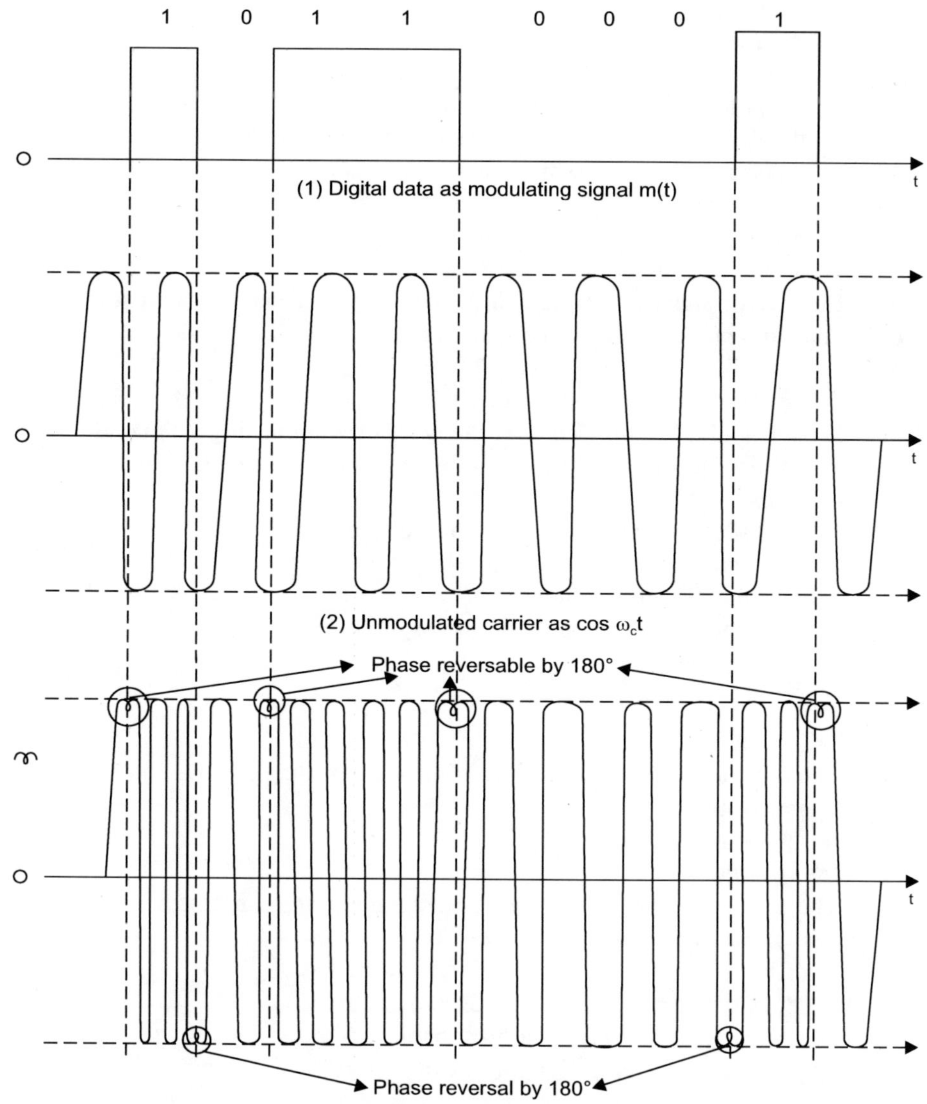

Fig. 6.8

QUESTIONS

6.1. Draw waveforms of amplitude shift keying.

6.2. Define the term entropy.

6.3. Define source and channel coding.

6.4. A source generates four messages. The entropy of the source will be maximum

(a) all probabilities are equal.
(b) one of the probabilities equal 1 and 2 others are zero.
(c) the probabilities are 1/2, 1/4 and 1/2.
(d) the two probabilities are 1/2 each and other zero.

6.5. The spectral density of white noise is:
(a) exponential (b) uniform
(c) random (d) Gaussian

6.6. What is the information rate of source?

6.7. Define term signal and explain difference between analog and digital signal.

6.8. State and prove sampling theorem in time domain.

6.9. What is Nyquist rate and Nyquist interval?

6.10. Classify the digital modulation technique.

6.11. What is multiplexing? What are the types of multiplexing technique?

6.12. What are the types of transmission media used for data and signal transmission?

6.13. What is additive white Gaussian noise channel?

6.14. What is binary symmetric channel?

6.15. What is mutual information?

6.16. Define conditional entropy.

6.17. What is channel matrix?

PROBLEMS

6.1. The capacity of a communication with a bandwidth of 4 kHz and SNR 15 is approximately
(a) 20 kbps (b) 16 kbps
(c) 10 kbps (d) 8 kbps

6.2. The entropy of a message source generating four messages with probability 0.5, 0.25, 0.125 and 0.125 is
(a) 1 b/message (b) 1.75 b/message
(c) 3.32 b/message (d) 5.936 b/message

6.3. A source generates four equiprobable symbols. If source coding is empoloyed, the average length of code for 100% efficiency is:
(a) 6 bits/symbol
(b) 4 bits/symbol
(c) 3 bits/symbol
(d) 2 bits/symbol

6.4. The channel capacity under the Gaussian noise environment for a discrete memoryless channel with a B.W. of 4 MHz and SNR of 31 is:
(a) 20 Mbps (b) 4 Mbps
(c) 8 kbps (d) 4 kbps

6.5. Find the Nyquist rate and the Nyquist interval for the signal $x(t) = \frac{1}{2\pi} \cos(4000\,\pi t) \times \cos(1000\,\pi t)$.

CHAPTER 7

Telephone Exchange

7.1 OVERVIEW OF AUTOMATIC TELEPHONE NETWORK EXCHANGE

Figure 7.1 shows overview of automatic telephone network. Automatic telephone network consists of following components:

1. Subscriber loop
2. Signalling
3. Control, and
4. Switching

(1) Subscriber Loop: Subscriber has plane old telephone instrument, it is connected to local exchange by twisted pair of wires hence loop gets formed between subscriber and local exchange which is known as subscriber loop.

(2) Signalling: Basic function of the signalling equipment is to monitor the incoming lines. Forward the appropriate status of D.C. Signalling detects ON hook, OFF hook to the telephone exchange to which the subscriber is permanently connected; status of subscriber plane old telephone. When calling party lifts his handset by

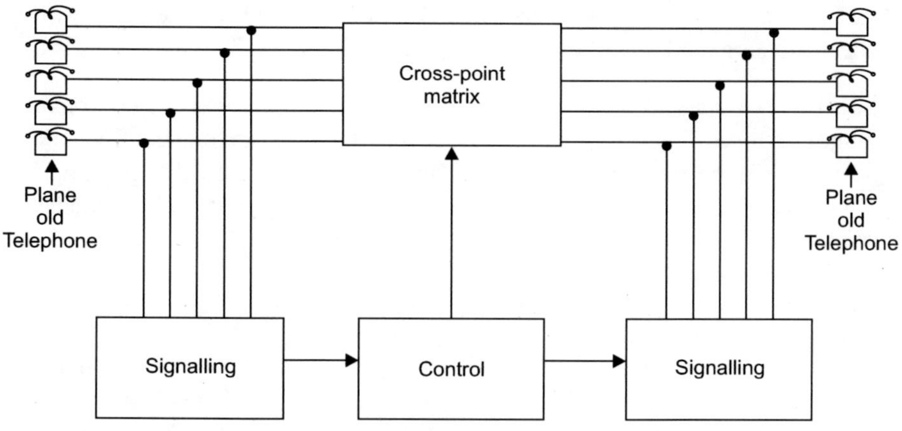

Fig. 7.1: Overview of Automatic Telephone Network Exchange

going off hook he sends request to exchange. Exchange in turn process the request and send a dial tone. The calling subscriber is then ready to transmit the address of a called subscriber. On receiving ringing the called subscriber lifts the handset and make a voice call. When the called subscriber goes on hook, the calling subscriber receivers a busy tone indicating the termination of a call.

A link is established between calling subscriber to called subscriber by circuit switching in the exchange. A path is established between calling and called subscriber for duration of call.

(3) Control: Function of control is scan subscriber lines and monitor the activity and do the call processing, maintenance and operation. Event monitoring is first priority, call processing second priority and operation, maintenance, and billing last priority. Control takes a each function as per assigned priority.

(4) Cross-point matrix: Calling subscriber and called subscriber are connected to each other through a cross point which is allocated by control for duration of call. This cross point is released on termination of the call. There are electromaganical, solid state cross points used. Cross points are fabricated using integrated circuit technology in modern times. Cross points are designed, based on multistage space and time switching in two dimensions or three dimensions. E1OB exchanges currently in operation uses time space time (TST) three-dimensional switching. Multidimensional switching provide flexibility in design that provide alternative path. It is cost-effective.

7.2 INTRODUCTION TO TELEPHONE EXCHANGE

(1) Subscriber Telephone Unit: It has given unique identification number. Early time exchange was pulse dialling type. Nowadays it is of dual tone multifrequency type. All subscriber near about thousands in local geographical area are connected to local exchange by twisted pair of wires. The loop formed between subscriber and local exchange is known as subscriber loop.

(2) Local Exchange and Main Exchange: Local exchange is connected to main exchange by trunks. Trunks is group of wires over which we can transmit frequency division multiplexed signal or synchronous time division multiplexed signal. Early days in local and main exchange, switching was step-by step switching. Nowadays switching is crossbar switching along with stored programme control is known as electronics switching.

(3) Trunk: Exchange are connected to exchange by a trunk over which frequency division multiplexed or synchronous time division multiplexed signals are transmitted. When telecommunication traffic is high, then instead of coaxial cable optical fibre cable is used. Generally, optical fibre cable used between local exchange to main exchange or main exchange to another main exchange.

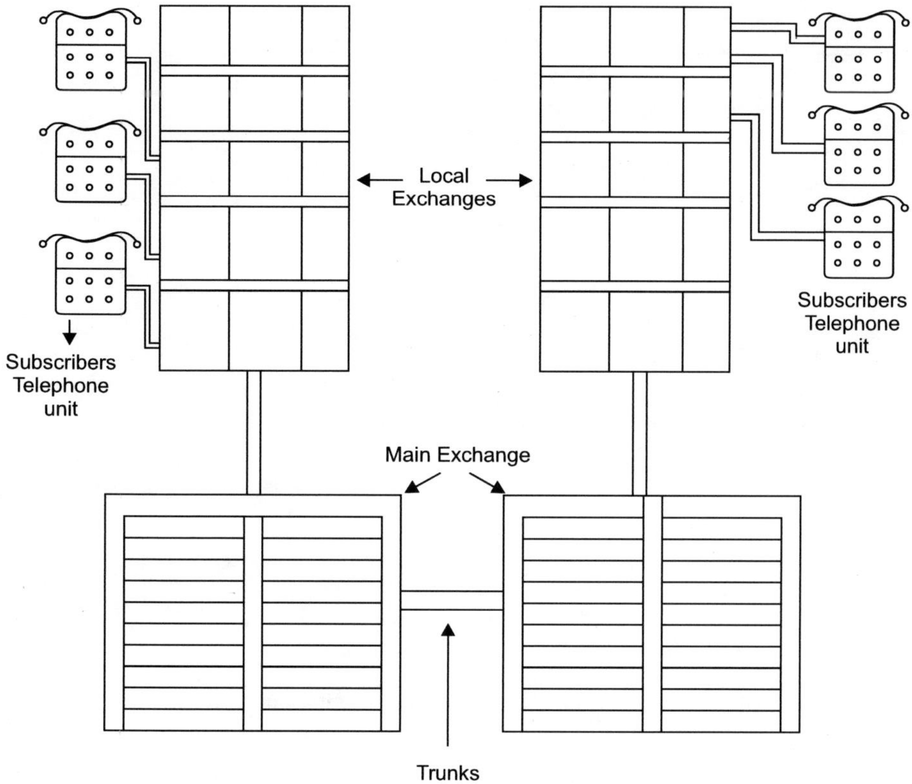

Fig. 7.2: Architecture of Telephone Exchange

7.3 CLASSIFICATION OF TELEPHONE SWITCHING SYSTEM

In any telephone exchange system, switching system is heart of telephone exchange. Switching system is broadly classified as follows:

(1) Manual switching system: In earlier days, switching system were manual and operator oriented. Drawback of these switching systems recognized and eliminated. Hence, automatic exchange came into existence.

(2) Automatic switching system: These switching systems can be classified as electromechanical and electronics switching system. Electromechanical switching system subdivided as step-by-step switching system and crossbar switching system. The step-by-step switching system is also known as strowger switching system.

Electromechanical switching classified as strowger or step-by-step switching and crossbar switching. Stored program control switching are further classified as space division switching and time division switching. Time division switching subdivided as analog switch and digital switch. Digital switch classified as space

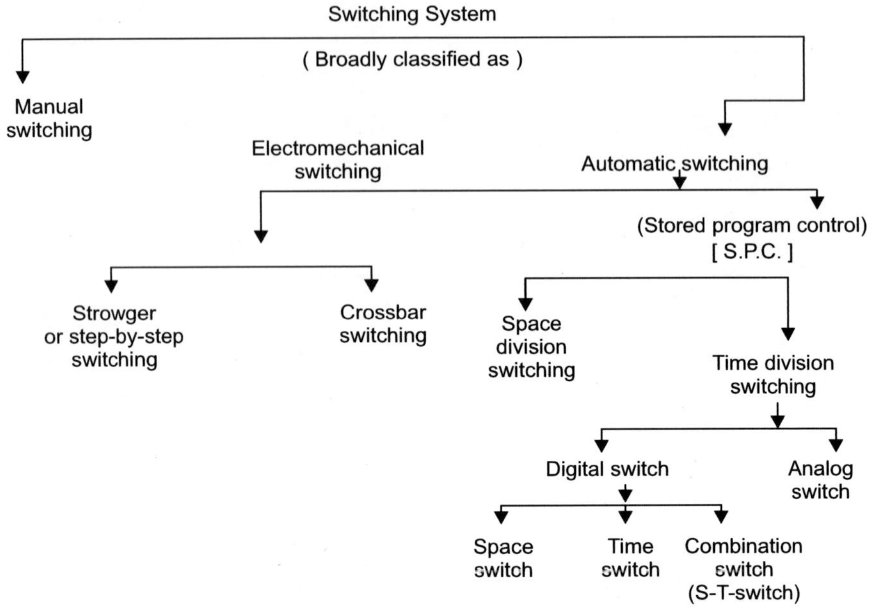

Fig. 7.3: Classification of Switching System

switch, time switch and combination switch, i.e. combination of space and time switch.

7.4 CROSSBAR SWITCHING SYSTEM

Crossbar switching systems are hard-wired control subsystem which uses relays and latches.

7.4.1 Principle of Crossbar Switching

If electromagnets M_2 and M'_3 are energised, a contact is established at the cross point 6 such that subscriber B is connected to the subscriber C. In telephone conversation, the one who initiates the call is referred to as the calling subscriber and one for whom the call is destined is the called subscriber.

In electronics switching system, the control functions are performed by a computer or processor. Hence, these systems are called stored program control (S.P.C.) system. The switching scheme used by electronics switching system may be either space division switching or time division switching. In space division switching a dedicated path is established between calling and called subscribers for the entire duration of call. Space division switching is also the technique used in strowger switch. A crossbar switching system with stored program control qualifies an electronics exchange.

In time division switching samples of speech signals are transferred at fixed interval. Time division switching may be analog or digital. In analog switching the sampled voltage levels are transmitted as they are whereas in digital switching they

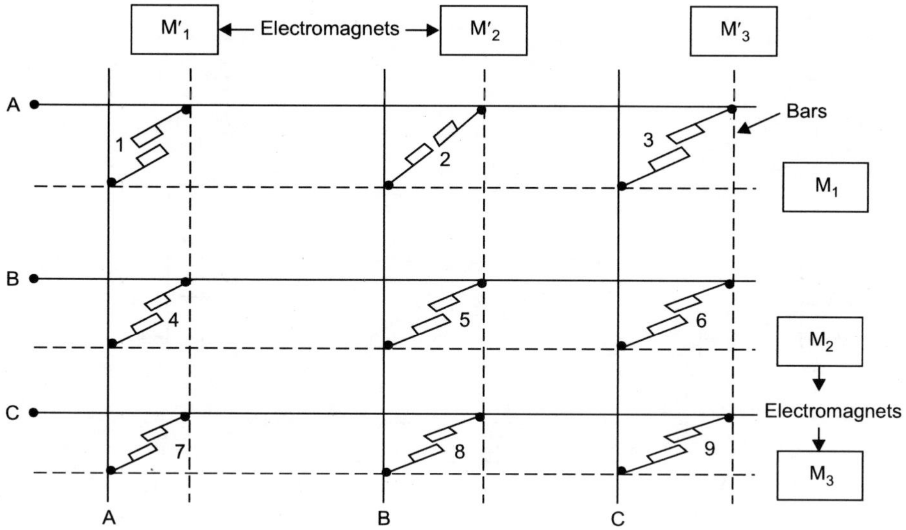

Fig. 7.4: 3 × 3 Crossbar Switching

are binary coded that is 011101 and transmitted. If coded values are transmitted during the same time interval from input to output, the technique is called space switching. If the values are stored and transferred at a later time interval, the technique is called time division digital switching. A time division digital switch may also be designed by using a combination of space and time switching technique.

7.5 A TELECOMMUNICATION NETWORK

Figure 7.5 shows a telecommunication network. The lines that run between the switching systems are called trunks. Trunks are group of wires over which frequency division multiplexed or synchronous time division multiplexed signals are transmitted and those that run to the subscriber promises are known as subscriber lines. These lines are twisted pair of wires.

Drawbacks of electromechanical switching

(1) It has limited capacity.

(2) Impossible to modify them to provide additional functionalities.

7.6 MAIN DISTRIBUTION FRAME

The complexity of the telephone network managed using a hierarchical structure. Generally four levels of cabling are used as shown in Fig. 7.6.

(1) **Drop wires:** Drop wires comes from subscriber end and terminate at distribution point. The drop wires are nothing but twisted pair of wires that run into subscriber premises.

(2) **Distribution cable (DC):** Many distribution cable from nearby geographical locations are terminated on a feeder point.

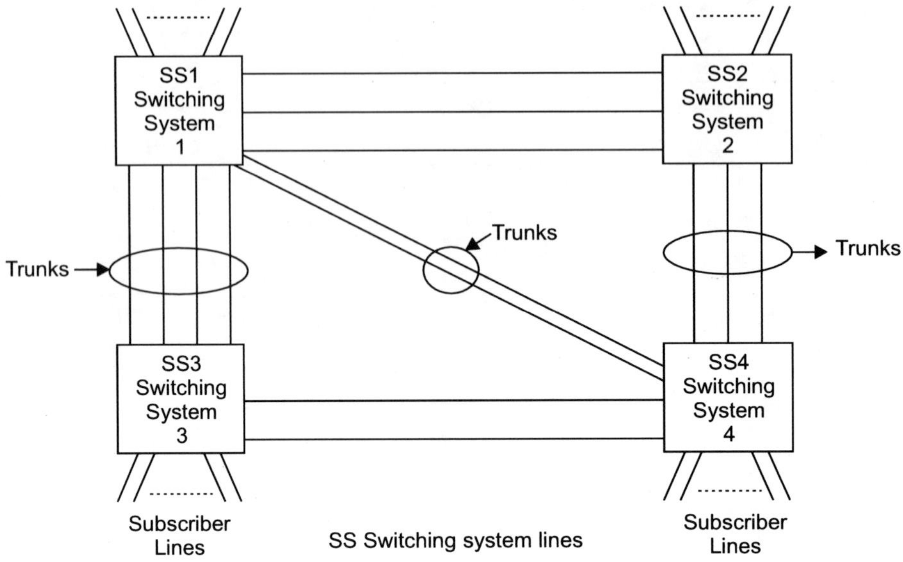

Fig. 7.5: Telecommunication Network

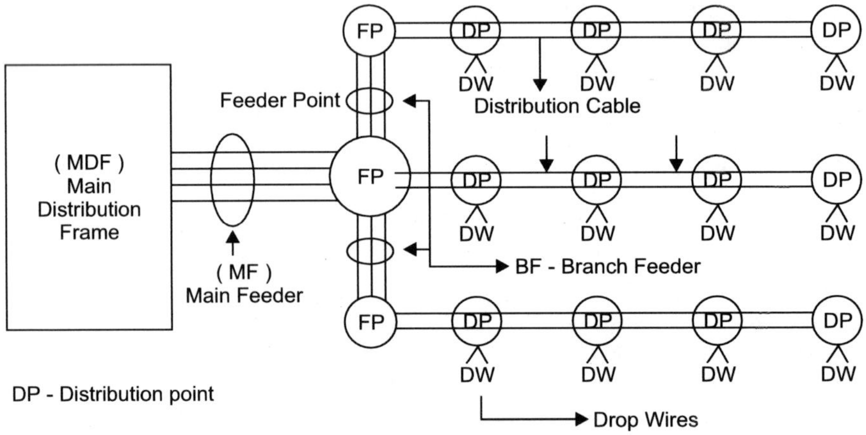

Fig. 7.6: Cable Hierarchy for Subscriber Loops

(3) **Branch feeder cable (BF):** In between feeder points whatever cable used is branch feeder cable.

(4) **Main feeder cable (MF):** In between feeder points to main distribution from whatever cable used is main feeder. The main feeder cables contains 100–200 wire pairs. Subscriber pair and exchange pair are interconnected at the main distribution frame by means of jumpers. The main distribution frame thus provides a flexible interconnection mechanism which is very useful in reallocating cable pair and subscriber number.

7.7 SWITCHING HIERARCHY AND ROUTING

Within a local exchange and main exchange type of switching used is either step-by-step switching or crossbar switching or software switch known as packet switching. Telephone network require some form of interconnection of switching exchanges to route traffic effectly and economically. Exchanges are interconnected by group of trunk lines. Two trunk groups are required between any two exchanges.

When calling subscriber lifts his handset then dial tone goes as a control signal, then called subscriber receives dial tone then he also lifts his handset. Call goes from calling subscriber local exchange to main exchange and from main exchange to called subscriber handset. Once call established between calling subscriber to called subscriber then conversation takes place between calling and called subscriber.

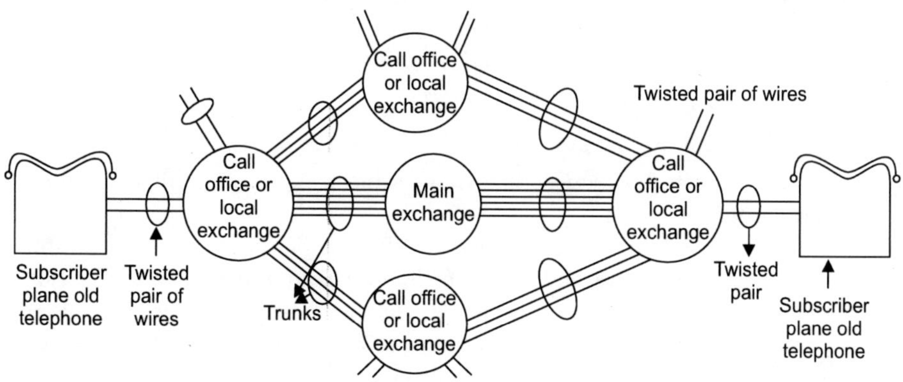

Fig. 7.7: Switching Hierarchy and Routing

7.8 TELEPHONE NETWORK TOPOLOGIES

Three topologies are adopted for interconnection of exchange.

(1) Mesh topology
(2) Star topology
(3) Hierarchical network

(1) **Mesh topology:** Mesh topology is fully connected network that is there is no blocking of calls. Mesh network is as shown in Fig. 7.8.

In mesh topology entity 1, entity 2, entity 3, entity 4 and entity 5 interconnected directly by 10 links.

Suppose n number of subscribers directly interconnected then number of wires or cables or links required are

$$L = \frac{n(n-1)}{2}$$

For example, take n = 1000 then number of links required

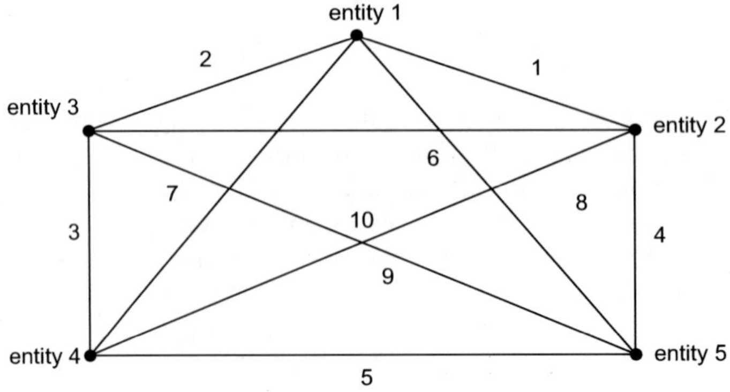

Fig. 7.8: Mesh Topological Network

$$L = \frac{n(n-1)}{2} = \frac{1000(1000-1)}{2}$$

Number of links required, is 4,99,500.

For 1000 subscriber directly connected by mesh topology number of links required are 4,99,500.

To install 4,99,500 links are not physically possible because cost of cable becomes too high. Hence, we have to go for hierarchical structure or topologies. Three basic topologies come into existence in telephone exchange. Mesh topology is fully connected network. Technically there is no blocking probability, hence mesh topology used for heavy traffic in metropolitan area.

(2) **Star topology:** Star topology used in telephone exchange. A star exchange requires intermediate exchange called a tandem exchange.

Tandem exchange is at centre and this tandem exchange local exchanges are connected. Figure 7.9(a) and (b) shows single level and two level star topological exchanges.

(a) *Subscriber:* Twisted pair of wires runs from subscriber plane old telephone of its nearby geographical location to local exchange.

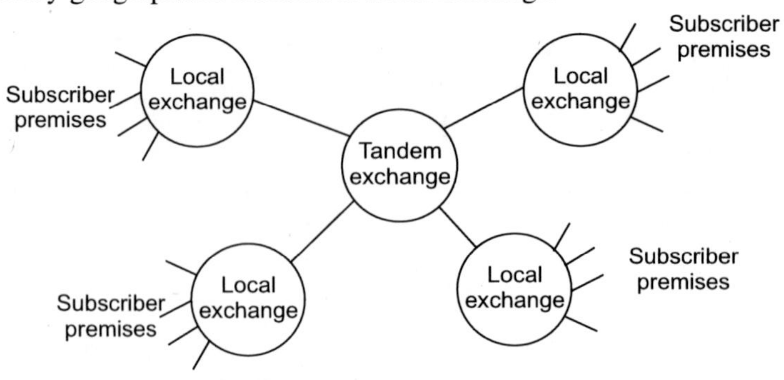

Fig. 7.9(a) Single Level Star Topology

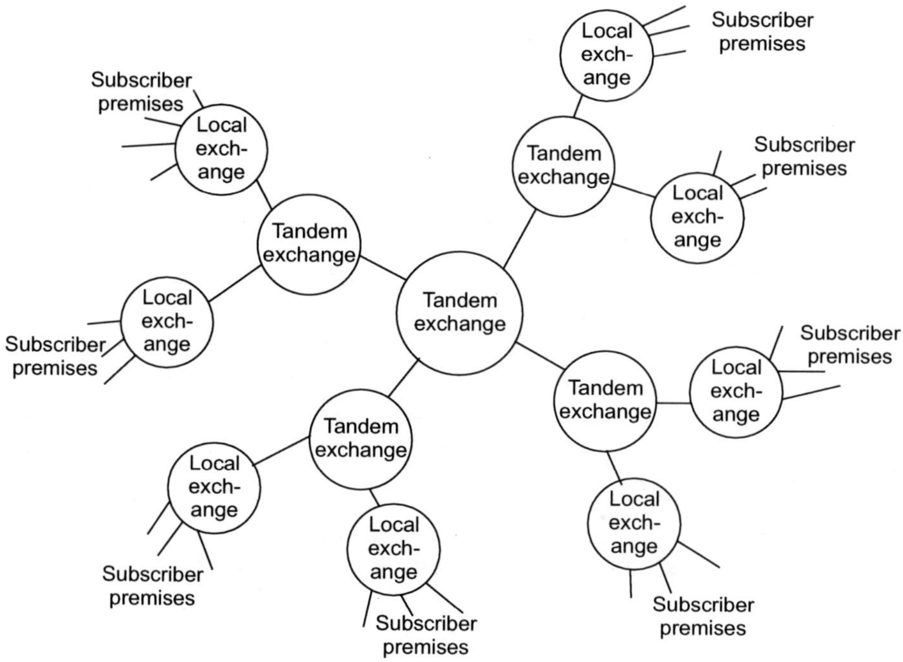

Fig 7.9(b) Two-level Star Topology

(b) *Local Exchange:* Local exchange connected to tandem exchange by trunks. Trunk is a group of wires over which frequency division multiplexed or synchronous time division multiplexed signals are transmitted.

(c) *Tandem Exchanges:* Tandem exchanges are connected to main tandem exchange by trunks.

(d) *Main tandem exchange:* All telephone calls go through main tandem exchange. Billing facility is also done within main tandem exchange.

(3) **Hierarchical Network:** Hierarchical network are capable of handling heavy traffic requirement and at the same time use minimum number of trunk groups.

7.9 TRANSMISSION PLANE

Transmission systems are broadly classified as radio system, coaxial cable system and optical fibre communication.

Radio systems are subdivided as (1) Sky wave propagation (2) Line of sight microwave communication (3) Tropospheric scatter communication and (4) Satellite communication.

(1) **Coaxial Cable Transmission:** Coaxial cable is a two-wire transmission line with a unique arrangement of conductors and shielded in concentric configuration by a dielectric. One conductor is grounded. Outer conductor provides

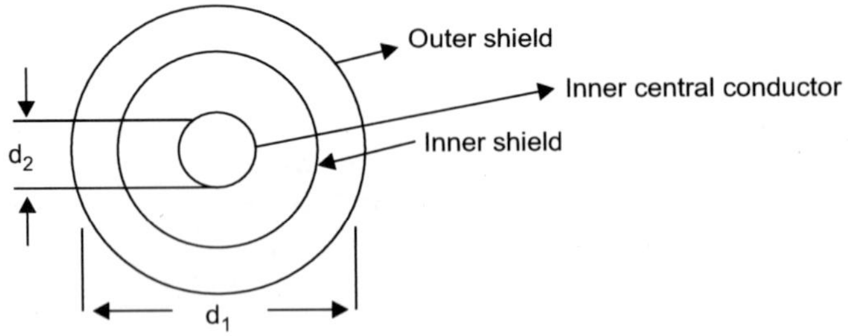

Fig. 7.10: Coaxial Cable

electromagnetic shield. Reduces pick up of unwanted interference. Coaxial cable is basically an unbalanced transmission line. Inner conductor or central conductor carries the signal. Outer and inner shield provide protection to conductor radiation losses, crosstalk is avoided by shield.

(2) **Optical fibre communication:** In optical fibre communication either single mode fibre or multimode fibre is used. Single mode fibre is used for long distance communication since in single mode fibre losses are minimum but cost of fibre is high. Multimode fibre is used for short distance communication because losses are maximum in multimode fibre but cost of fibre is low.

For converting electrical signal into light signal within transmitter source used is either edge emitting LED or laser. This light signal passes through fibre. At receiving end, photodiode is used for converting light signal into electrical signal.

(3) **Radio System:** As explained above, radio system is classified as (a) Sky wave (b) Line of sight microwave communication (c) Topospheric scatter communication and (d) Satellite communication.

(a) *Sky wave propagation:* For sky wave propagation source contain large power along with transmitting antenna. Transmitting antenna radiate electromagnetic waves. Mechanism of electromagnetic waves depends upon angle of incidence and operating frequency of rays. Figure 7.11(a) shows mechanism of sky wave depending upon angle of incidence and Fig. 7.11(b) shows mechanism of sky-wave propagation depending upon angle of incidence.

When angle of incidence $\theta <$ critical angle θ_c then ray 2 strikes the ionosphere and comeback to earth at larger distance. When angle of incidence $\theta = \theta_c$ then ray 1 strikes ionosphere and comeback to earth at lesser distance. When angle of incidence $\theta >$ critical angle (θ_c) then ray 3 penetrate into ionosphere and escaped into space. Hence, ray 3 is known as escaped ray. Sky wave propagation is used for long distance communication.

Figure 7.11(b) shows mechanism of sky wave propagation depending upon operating frequency.

When operating frequency (f) is equal to maximum usable frequency ($f_{M.U.F.}$) then ray 1 strikes the ionosphere and come back to earth at lesser distance.

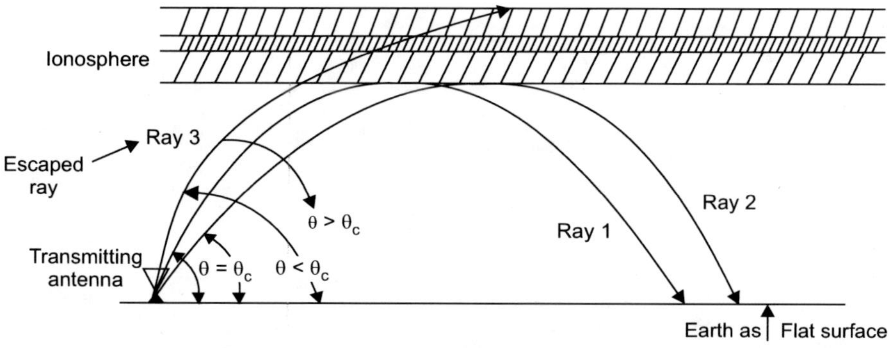

Fig. 7.11(a): Mechanism of Sky Wave Propagation Depending upon Angle of Incidence

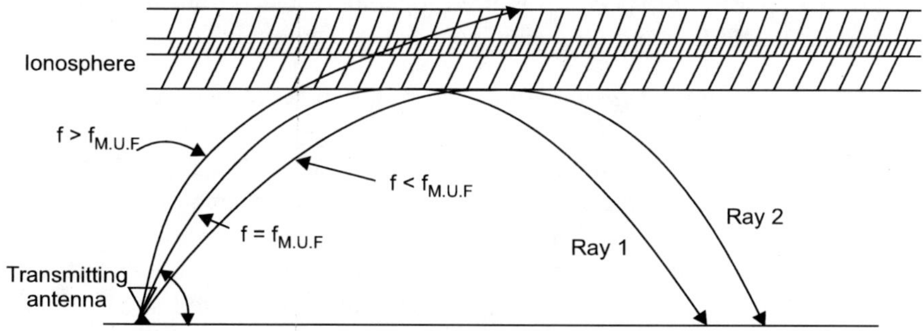

Fig. 7.11(b): Sky Propagation Depending upon Operating Frequency

When operating frequency (f) is less than maximum usable frequency ($f_{M.U.F}$) that is $f < f_{M.U.F.}$ then ray 2 strikes the ionosphere and come back to earth at larger distance. When operating frequency (f) greater than maximum usable frequency ($f_{M.U.F.}$) that is $f > f_{M.U.F.}$, then ray 3 penetrate into ionosphere and escapes from ionosphere into space. Sky wave propagation used for long distance communication.

(b) *Line of sight microwave communication:* Line of sight microwave communication made up of two terminal sites and usually one or more repeater sites as shown in Fig. 7.12.

(i) *Transmitting terminal site A:* At transmitting terminal site A, voice channels or spoken messages are frequency division multiplexed into a baseband signals which is then modulated into RF carriers.

(ii) *Receiving terminal site B:* At receiving terminal site B, the RF carrier is demodulated and the resulting baseband signal demultiplexed into individual voice channels.

(iii) *Repeater (R):* A repeater site R is characterised by two antennas for the two directions. It receives, amplifies and retransmits the RF signal to the next site in sequence. Some repeaters may add and drop of few lines serving some local connection.

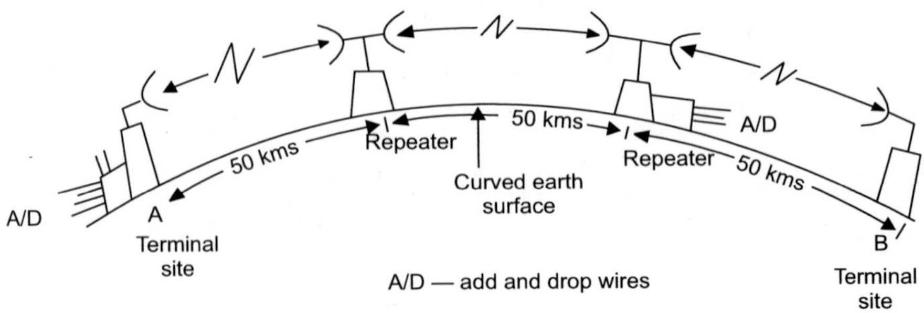

Fig. 7.12: Light of Sight Microwave Communication

The 4 GHz portion of spectrum is used for long distance transmission. The 6 GHz band is useful for both long and short distance communication.

(c) *Tropospheric scatter communication:* Tropospheric scatter communication shown in Fig. 7.13.

Tropospheric scatter communication also known as troposcatter or forward scatter propagation. Tropospheric scatter propagation is a means of beyond the horizon propagation for UHF signals. It uses certain properties of the troposphere, the nearest portion of atmosphere that is 15 kms above the ground.

As shown in Fig. 7.13, two directional antennas are pointed so that their beams intersect midway between them, above the horizon. If one of these is UHF transmitting antenna and the other a UHF receiving one, sufficient radio energy will be directed toward the receiving antenna to make this a useful communication system. Reflection from blobs in atmosphere, similar to the scattering of a searchlight beam by dust particle. The best frequencies most often used, are centred on 900, 2000 and 5000 MHz. The actual proportion of forward scatter to signals incident on the scatter volume is very tiny between –60 dB and –90 dB. High transmitting powers are obviously needed.

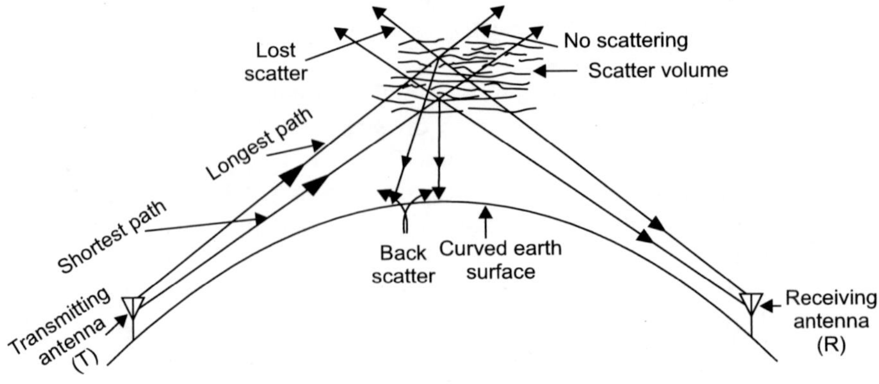

Fig. 7.13: Tropospheric Scatter Communication

It has been found in practice that best results are obtained from troposcatter propagation if antennas are elevated and then directed towards the horizon. This method of propagation is often to provide long distance telephone and other communication links.

(d) *Satellite communication:* A communication satellite is a big microwave repeater.

Theoretically, three satellites properly placed in synchronous orbit could provide 100% coverage of earth. Effective isotropically radiated power (EIRP) is of 20 to 28 dBW.

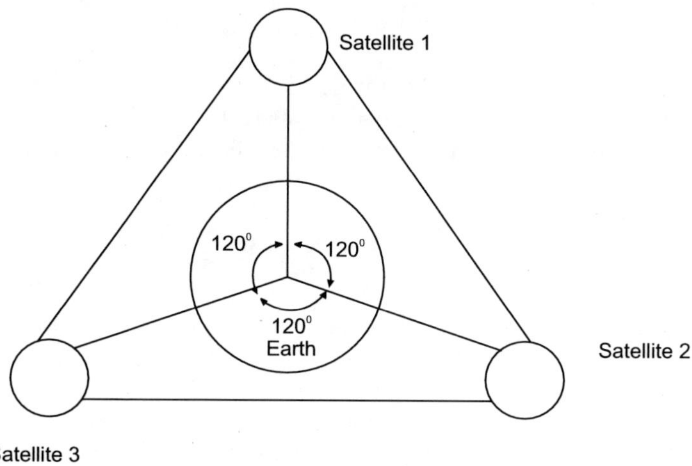

Fig. 7.14: Satellite Communication

7.9.1 Numbering Plan

In numbering plan every subscriber is identified by its own number. Then subscriber connected to a telecommunication network.

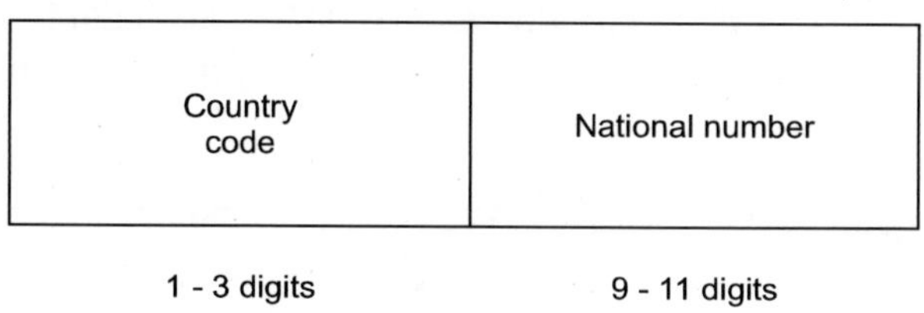

Fig. 7.15: National Telephone Number

As shown in Fig. 7.15, country code is 1-3 digits and national number is 9-11 digits.

7.9.2 International Telephone Number

International telephone number is shown in Fig. 7.16.

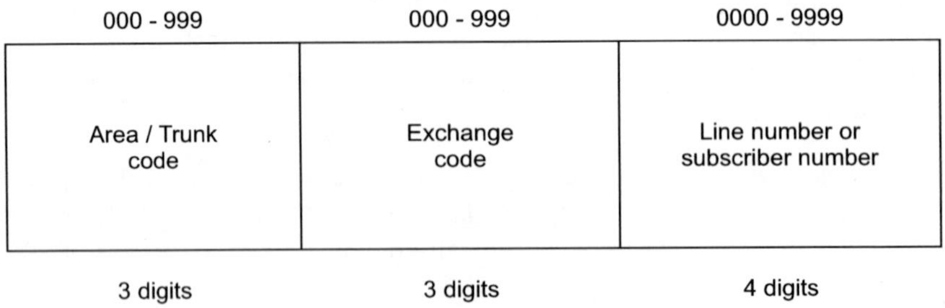

Fig. 7.16: International Telephone Number

(1) Area/trunk code is 3 digits that is 000-999.
(2) Exchange code is again 3 digits that is 000-999.
(3) Line number or subscriber number is 4 digit that is 0000-9999.

7.10 DIALLING

Diallings are classified as:

(1) Pulse dialling, and
(2) Multifrequency (tone) dialling or dual tone multifrequency dialling that is DTMF dialling

(1) **Pulse Dialling:** In pulse dialling, train of pulses is used to represent a digit in the subscriber number. The number of pulses in a train is equal to the digit value it represent except in the case of zero which is represented by 10 pulses. Successive digit in a number are represented by a series of pulse train. Two successive trains are distinguished from by a pause in between them known as interdigit gap. The pulses are generated by alternatively breaking and making the loop circuit between subscriber and exchange.

The pulsing pattern is shown in Fig. 7.17 for digits 3 and 2.

Mark duration = 0.33 T and break duration = 0.67 T and interdigit gap at least = 200 msec. Pulse rate is 10 pulses/sec and interdigit gap minimum as much as 400-500 msecond.

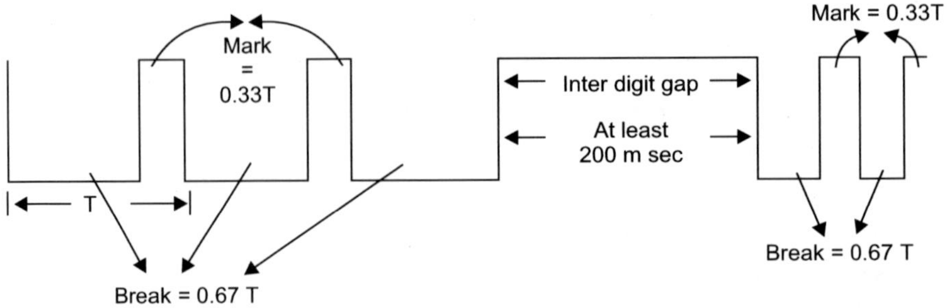

Fig. 7.17: Pulse Dialling

(2) **Drawbacks of pulse dialling:**

(a) Since pulses are produced by make and break of subscriber loop, there is chance of sparking inside the telephone instrument.

(b) Transmitter, receiver and the bell circuit of telephone set may be damaged if dialling pulses are passed through them.

(c) Dialling habits of users vary widely and hence all timing aspects should be independent of user action.

7.11 DUAL TONE MULTI FREQUENCY (DTMF) DIALLING

One more drawback of pulse dialling is that pulse dialling limited to signalling between subscriber and exchange and no signalling is possible between two subscribers. Subscriber to subscriber signalling is in the voice frequency band.

"Touching" a button generates "tone" which is a combination of two frequencies, one from the lower band and other from the upper hand.

For example, pressing the push button 9 transmits 852 Hz and 1477 Hz.

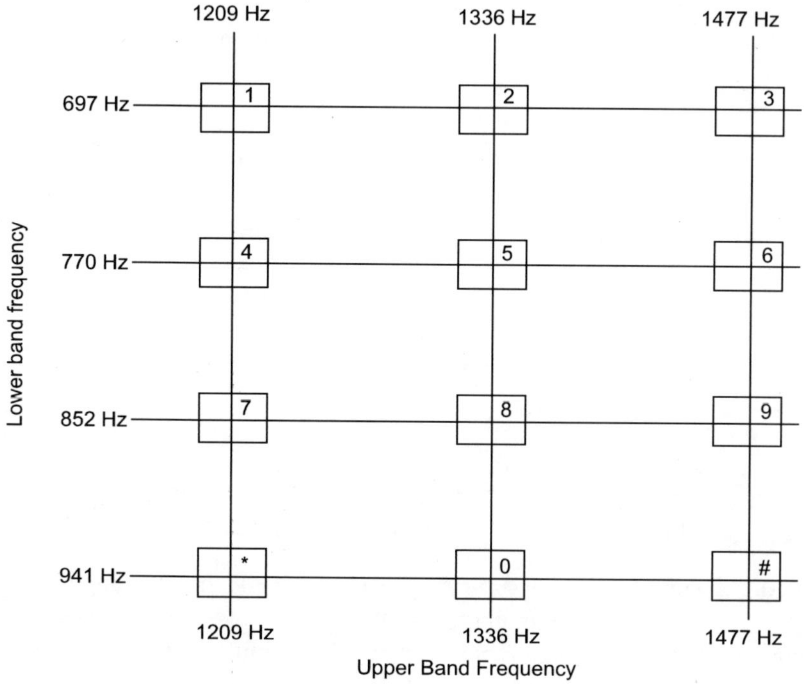

Fig. 7.18: Touch Dial Arrangement

Advantage of tone dialling: The signalling information in case of tone dialling can be transmitted from subscriber to exchange as well as subscriber to subscriber. That is, in short the signalling information can be transmitted at any point in the network since in case of tone dialling voice is transmitted.

7.12 SIGNALLING TONES

For establishing, maintaining and releasing a telephone conversation there is need to perform a number of signalling functions. Above-mentioned functions are performed by operator in manual exchange. In automatic switching system, the verbal signalling of the operator is replaced by a series of distinctive tones.

Five subscriber related signalling functions are performed by operator.

(1) Exchange respond to calling subscriber to obtain the identification of called subscriber. That is, identification of called subscriber done at exchange by responding to calling subscriber.
(2) Exchange inform to calling subscriber that the call is being established.
(3) Ring the bell of called party.
(4) Exchange inform the calling subscriber, if the called party is busy.
(5) Exchange inform the calling subscriber, if the called party line is unobtainable for some reason.

However, most of the modern exchange provide a call-in-progress or routing tone for function.

7.13 TYPES OF TONES

(1) Dial Tone: The signalling function of call identification is understood by sending a "dial tone" to calling subscriber. This tone indicates that the exchange is ready to accept dialled digits from subscriber. The subscriber should start dialling only after hearing the dial tone. Otherwise initial dial pulses may be missed by exchange which may result in call landing on a wrong number. Most often, the dial tone is sent out by the exchange when subscriber lift the handset that is OFF hook. The dial tone is a 33 Hz or 50 Hz or 400 Hz continuous tone as shown in Fig. 7.19(a). The 400 Hz signal is usually modulated with 25 Hz or 50 Hz.

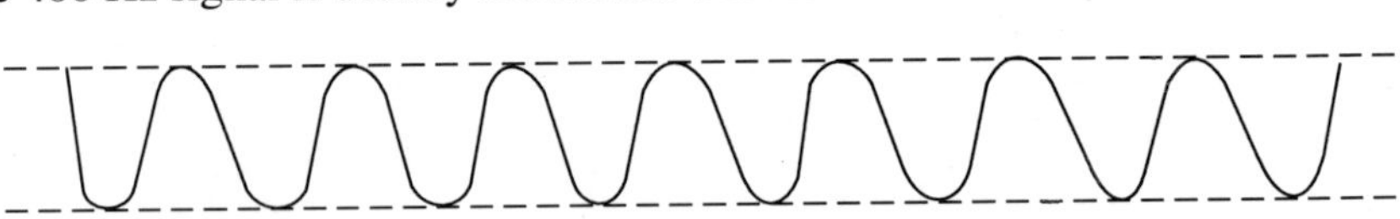

Fig. 7.19(a): 33 or 50 or 400 Hz Continuous Dial Tone

(2) Ringing Tone: When called party line is obtained, the exchange control equipment sends out the ringing current to the telephone set of the called party. This ringing current has the double ring pattern. Simultaneously, the control equipment sends out a "ringing tone" to the called subscriber, which has pattern similar to that of the ringing current shown in Fig. 7.19(b). The two rings in the

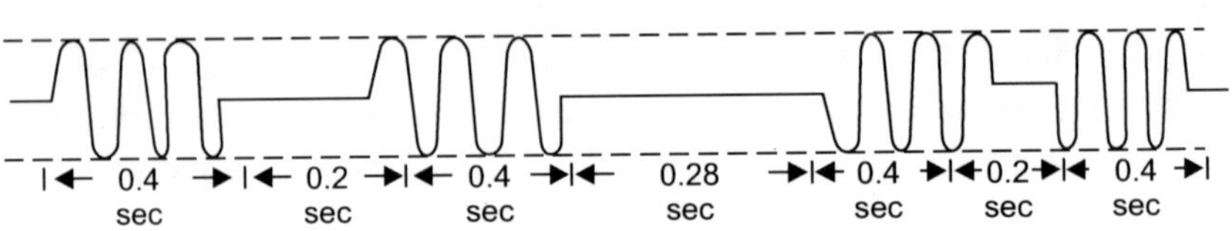

Fig. 7.19(b): Ringing Tone

double ring pattern are separated by a time gap of 0.28 sec and two double ring pattern by a gap of 0.2 sec. The ring burst duration is 0.4 sec. The frequency of the ringing tone is 133 Hz or 400 Hz, sometimes modulated with 25 Hz or 33 Hz.

(3) Busy Tone: It is bursty 400 Hz signal with silence period in between. The brust and silence duration have the same value of 0.75 sec or 0.375 sec. A busy tone is sent to calling subscriber whenever switching equipment is not available to put through the call or the called subscriber line is engaged.

Fig. 7.19(c): Busy Tone

(4) Number unobtainable tone: The number unobtainable tone which is a continuous 400 Hz signal. When called party line is out of order or disconnected, then this tone comes on line.

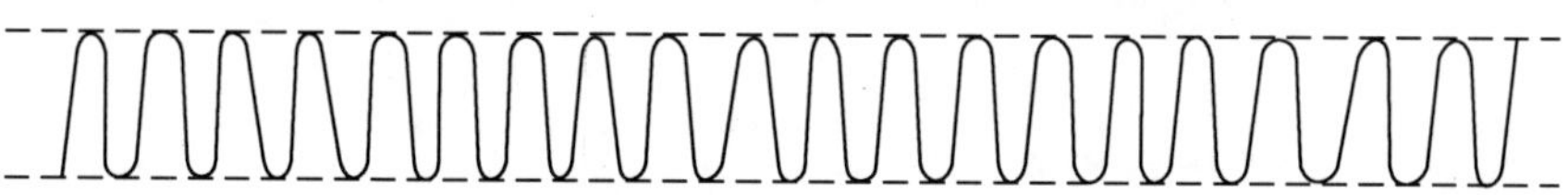

Fig. 7.19(d): Number Unobtainable Tone

(5) Call-in-progress tone: The routing tone or call-in progress tone is a 400 Hz or 800 Hz intermittent pattern. In analog electronics exchanges it is 400 Hz pattern with 0.5 sec off period and 2.5 sec ON period.

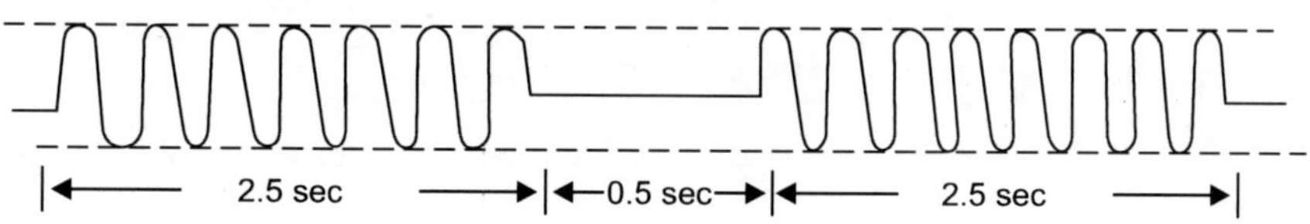

Fig. 7.19(e): Call-in-Progress Tone

7.14 STEP-BY-STEP SWITCHING

Step-by-step switching takes place by using uniselector or two motion selector or combination of both. Wiper contacts of these selector move in direct response to dial pulses or other signal like off hook that lifting of handset; from subscriber telephone. The wiper forward on contact at a time and moves by as many contacts as the number of dial pulses received. Most of the necessary circuit are built-in as an integral part of selector. That is, most of the necessary control circuit are built-in is present within the selector mechanism. Thus, enabling them to receive and respond to user's signalling directly.

7.15 STORED PROGRAM CONTROL

Exchange control functions performed through program control in the memory of computer led to nomenclature stored program control. Hence, led to full-scale automatic exchange. We can introduce a variety of new services to user's due to stored program control and due to automatic switching system. Common channel signalling provided to exchange does centralized maintenance of exchange and it also does automatic fault diagnosis of exchange as well as switching system exchange.

Interactive human machine interface is possible due to stored program control provided to telephone switching. Above-mentioned features that have become possible due to application of S.P.C. to telephone switching. The computer controlling the exchange must be highly tolerant to faults. That is due to stored program control in which part of exchange faults is created is detected by computer automatically. In S.P.C. a program or set of instructions to the computer is stored in its memory and instructions are executed automatically one-by-one by the processor.

7.16 TRAFFIC ENGINEERING

Whenever we have to design and analyse a telecommunication network there is need to understand traffic engineering. Blocking probability is determined due to the non-availability of switching path. That is, P_B exist due to non-availability of switching path.

The traffic patterns on the network varies during the day with heavy traffic at certain times and low traffic at other times.

7.17 NETWORK TRAFFIC LOAD AND PARAMETERS

In a telephone network, the traffic load and typical working day during 24 hours is shown in Fig. 7.20.

The originating call amplitudes are relative and the actual values depends on the area where the statistics is collected. There is little use of network during 0 and 6 hours when most of population is asleep.

There is a large peak around mid-forenoon and mid-afternoon signifying busy office activities. The afternoon peak is slightly smaller. The load during lunch hour period is low that is 12.00 to 14.00.

The period 17.00–18.00 hours is characterized by low traffic signifying that the people are on move from offices to their residence. The peak of domestic calls occur after 18.00 hours when persons reach home and reduced traffic applies.

During holidays and festival days the traffic pattern is different from that shown in Fig. 7.20. Generally, there is a peak of calls around 10.00 hours just

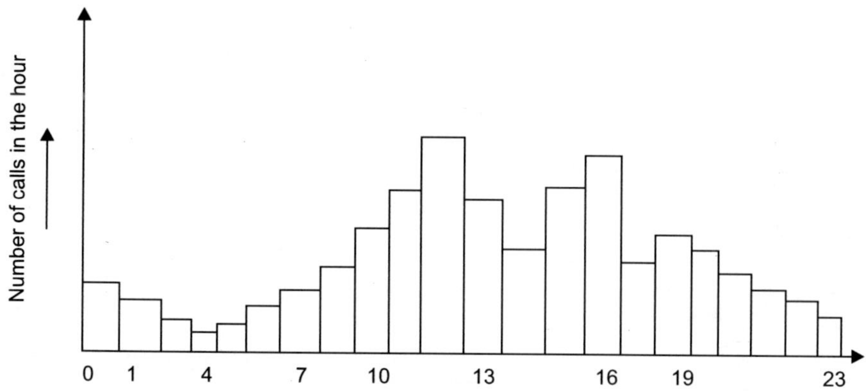

Fig. 7.20: Hours of the Day

before people leave their homes on outgoing and another peak occurs again in the evening.

(1) **Busy hour:** In a day the 60 minutes interval in which the traffic is the highest is called the busy hour. Continuous one-hour period lying wholly in the time interval concerned, for which the traffic volume or the number of call attempts is greatest.

(2) **Peak busy hour:** The busy hour each day usually varies from day to day or over a number of days.

(3) **Time consistent busy hour:** The one-hour period starting at the same time each day for which the average traffic volume or the number of call attempts is greatest over the days under consideration.

(4) **Call completion rate:** Call completion rate (CCR) is defined as the ratio of the number of successful calls to number of call attempts.

$$\text{CCR} = \frac{\text{Number of successful calls}}{\text{Number of call attempts}}$$

(5) **Busy hour call attempt** (BHCA): The number of call attempts in the busy hour is called busy hour call attempts (BHCA) By BHCA parameter we can determine processing capacity common control or SPC system of exchange.

(6) **The busy hour calling rate (BHCR):** BHCR which is definded as the average number of calls originated by subscriber during the busy hour.

Example 7.1: An exchange serves 2000 subscribers. If average BHCA is 10,000 and CCR is 60%, calculate the busy hour calling rate

Solution: (1) Average busy hour call $= \text{BHCA} \times \text{CCR}$

$$= 10{,}000 \times 0.6$$

$$= 6000 \text{ calls}$$

$$\text{(2) Busy hour calling rage} = \frac{\text{Average busy hour calls}}{\text{Total number of subscribers}}$$

$$= \frac{6000}{2000} = \frac{6}{2} = 3$$

All common subsystem of telecommunication network are collectively termed server or trunks. The traffic on the network may then be measured in terms of occupancy of the servers in the network. Such a measure is called as the traffic intensity which is defined as:

$$\text{Traffic intensity} = \frac{\text{Period for which a server is occupied}}{\text{Total number of observation}}$$

Example 7.2: In a group of 10 servers, each is occupied for 30 minutes in an observation interval of two hours. Calculate the traffic carried by group

Solution: (1) Traffic carried per server $= \frac{\text{Occupied duration}}{\text{Total duration}}$

$$= \frac{30}{120} = 0.25\,\text{E}$$

(2) Total traffic carried by group $= 10 \times 2.5$

$$= 2.5 \text{ E}$$

Example 7.3: A group of 20 servers carry a traffic of 10 erlangs. If the average duration of a call is three minutes, calculate the number of calls put through by a single server and the group as a whole in an one hour period.

Solution (1) Traffic per server $= \frac{10}{20} = 0.5 \text{ E}$

That is, a server is busy for 30 minutes in one hour.

(2) Number of calls put through by one server $= \frac{30}{3}$

$= 10$ calls

(3) Total number of calls put through by the group $= 10 \times 20 = 200$ calls.

Example 7.4: A subscriber makes three phone calls of three minutes, four minutes and two minutes duration in a one-hour period. Calculate the subscriber traffic in erlangs, CCS and CM.

Solution: (1) Subscriber traffic in erlangs $= \frac{\text{Busy Period}}{\text{Total Period}}$

$$= \frac{3+4+2}{60} = 0.15 \text{ E}$$

$$\text{(2) Total traffic in CCS} = \frac{(3+4+2)\times 60}{100} = 5.4 \text{ CCS}$$

$$\text{(3) Total traffic in CCN} = 3 + 4 + 2 = 9 \text{ CM}$$

Example 7.5: Over a 20 minutes observation interval, 40 subscribers initiate calls. Total duration of call is 4800 seconds. Calculate the load offered to network by the subscriber and average subscriber traffic.

Solution: (1) Mean arrival rate $= c = \frac{40}{20} = \frac{2 \text{ calls}}{\text{min}}$

$$\text{(2) Mean holding time} = \frac{4800}{40\times 60} = \frac{2 \text{ min}}{\text{call}}$$

$$\text{(3) Offered load} = 2 \times 2 = 4 \text{ E}$$

$$\text{(4) Average subscriber traffic} = \frac{4}{40} = 0.1 \text{ E}$$

7.18 GRADE OF SERVICE AND BLOCKING PROBABILITY

Conventional automatic exchanges behave like loss systems. On the other hand operator oriented manual exchanges can be considered as delay system.

The traffic models used for studying loss system are known as blocking or congestion models and the ones used for studying delay system are called queuing models.

Grade of service (GOS) is definded as the ratio of lost traffic to offered traffic.

$$\text{Grade of service} = \frac{\text{Offered traffic} - \text{Carried traffic}}{\text{Offered traffic}}$$

$$\text{Grade of service} = \frac{\text{Lost traffic}}{\text{Offered traffic}}$$

$$\text{GOS} = \frac{A - A_0}{A}$$

where A = offered traffic

A_0 = carried traffic

$A - A_0$ = lost traffic

Smaller the value of grade of service, better is the service. The recommended value of G.O.S. in India is 0.02 which means that two calls in every 1000 calls may be lost or one call in every 500 calls may be lost.

7.18.1 Blocking Probability

The blocking probability P_B is defined as the probability that all trunks in system are busy. When all trunks are busy no further traffic can be carried by the system and the arriving subscriber traffic is blocked. The blocking probability is the same as measure of GOS. The probability that all trunks are busy may well represent the fraction of calls lost, which is what GOS is all about.

The fundamental difference is that the GOS is a measure from subscriber point of view whereas the blocking probability is a measure from the network point of view. GOS is arrived at by observing the number of rejected subscriber calls, whereas the blocking probability is arrived at by observing the busy trunks in switching system. GOS is called congestion or loss probability and the blocking probability is called time congestion.

A new term called quality of service (QOS) is being used and it is considered to be more general than the GOS and includes other factors like quality of speech, error, free transmission capability, etc.

There are two different important performance measure, one obtained by observing the network behaviour. Blocking probability term used from subscriber and network viewpoints.

Subscriber viewpoint,

$$\text{GOS} = \text{call congestion} = \text{loss probability}$$

Network point of view

$$\text{Blocking probability} = \text{time congestion}$$

Example 7.6: A switching system serves 10000 subscribers with a traffic intensity of 0.1 E per subscriber. If there is sudden spurt in traffic, increasing the average traffic by 50%, what is the effect on the arrival rate?

Solution: Number of active subscriber during (a) normal traffic = 1000 (b) increased traffic = 15000

Number of available subscriberrs for generating new traffic:
(a) Normal traffic = 9000 (b) increased traffic = 8500

$$\text{Change in arrival rate } = \frac{500}{9000} \times 100 = 5.6\%$$

Change in arrival rate = 5.6%

Example 7.7: In telephone system, there are 20 serves and 100 subscribers. On an average, there are 10 busy servers at any time. The probability of all the servers being busy is 0.2. Calculate the grade of service (a) erlang traffic and engest traffic.

Solution: For erlang traffic GOS is the same as P_R. Therefore, GOS = 0.2

For engent traffic, GOS is given by:

Carried traffic = number of average busy server = 10
Therefore,

$$\text{Grade of service} = \frac{100-20}{100-10}\times 0.2$$

$\therefore$ G.O.S. = 0.1875

Example 7.8: During a 2-hour busy period, 2400 calls arrive at an exchange. Average holding time per call is two minutes. What is traffic load in (a) erlangs and (b) in CCS.

Solution: (a) Mean arrival rate $= c = \dfrac{2400}{60+60}$

$$= \frac{2400}{120}$$

$$= 20\frac{\text{calls}}{\text{min}}$$

$$\text{Average holding time} = t_h = 2\frac{\text{min}}{\text{call}}$$

$$\text{Offered traffic load} = 20\frac{\text{calls}}{\text{min}}\times 2\frac{\text{min}}{\text{calls}}$$

$$= 40 \text{ E}$$

$$\text{Traffic in C.C.S.} = 40 \times 36$$

$$= 1440 \text{ CCS}$$

Example 7.9: 10,000 subscribers are connected to an exchange. If the exchange is designed to achieve a call completion rate of 0.8 when the busy hour calling rate is 4.8. What is the BHCA that can be supported by the exchange? What should be call processing time for this exchange?

Solution: Given data

$$\text{CCR} = 0.8, \text{BHCR} = 4.8$$

$$\text{BHCR} = \frac{\text{Average busy hour call}}{\text{Total number of subscriber}}$$

$$4.8 = \frac{\text{Average busy hour call}}{10,000}$$

$$\text{Average busy hour call} = 4.8 \times 10,000$$

$$\therefore \text{Average busy hour call} = 4800$$

$$\text{B.H.C.A.} = \frac{\text{Average busy hour call}}{\text{Call completion rate}}$$

$$\text{B.H.C.A.} = \frac{48000}{0.8} = 60{,}000$$

60,000 call attempts made in one hour

$$\therefore \frac{60 \times 60}{60{,}000} = 60 \text{ msec/call}$$

Example 7.10: A call processor in an exchange requires 120 msec to service a complete call. What is BHCA rating for the processor? If exchange is capable of carrying 700 erlangs of traffic, what is the call completion rate? Assume an average call having holding time of two minutes.

Solution: Given data 120 msec. call process time is required to complete call:

$$\text{BHCA} = \frac{(60 \times 60)}{120 \text{ msec}} = \frac{60 \times 60}{120} \times 1000$$

Busy hour call attempt = BHCA = 30,000

$$700 \text{ E} = \frac{\text{call}}{\text{min}} \times \frac{2 \text{ min}}{\text{call}} \text{(holding time)}$$

$$\frac{\text{calls}}{\text{min}} = \frac{700 \text{ E}}{2\frac{\text{min}}{\text{call}}} = 350\frac{\text{calls}}{\text{min}}$$

$$\frac{\text{calls}}{\text{min}} = 350\frac{\text{calls}}{\text{min}}$$

$$\frac{\text{calls}}{\text{hour}} = 350 \times 60$$

$$\frac{\text{calls}}{\text{hour}} = 21{,}000\frac{\text{calls}}{\text{hour}}$$

Average busy hour calls = 21,000 $\frac{\text{calls}}{\text{hour}}$

Average busy hour calls = busy hour call attempts × call completion rate
Average busy hour calls = BHCA × CCR

$$21{,}000 = 30{,}000 \times \text{CCR}$$

$$\text{Call completion rate} = \frac{21{,}000}{30{,}000} = \frac{7}{10} = 0.7$$

Call completion rate = 0.7

Example 7.11: The traffic statistics of a company using PABX indicates that 180 outgoing calls are initiated every hour during working hours. Equal number of calls comes in. Each call lasts for 200 second on the average. If the G.O.S. required is 0.05, determine the number of lines required between the PABX and the main exchange.

Solution: Given data:

180 outgoing calls/hour
180 incoming calls/hour

$$\text{Total calls} = \text{incoming}\ \frac{\text{calls}}{\text{hour}} + \text{outgoing}\ \frac{\text{calls}}{\text{hour}}$$

$$= 180 + 180 = 360\ \frac{\text{calls}}{\text{hour}}$$

$$\text{Offered traffic} = 360\frac{\text{calls}}{\text{hour}} \times \frac{200}{60 \times 60}\text{sec}$$

$$\text{Offered traffic} = \frac{7200}{360} = 20\ \text{E}$$

$$\text{Carried traffic} = ?$$

$$\text{Grade of service} = \frac{\text{Offered traffic} - \text{carried traffic}}{\text{Offered traffic}}$$

$$\text{G.O.S.} = 0.05 = \frac{20 - X}{20}$$

$$0.05 \times 20 = 20 - X$$

$$1 = 20 - X$$

$$+X = \text{carried traffic} = 20 - 1$$

$$\text{Carried traffic} = 19$$

Example 7.12: In an exchange, the call arrived at the rate of 1100 calls/hour with each call holding for a duration of three minutes. If demand is serviced by a trunk group of 50 lines, determine the grade of service.

Solution: Given data

$$\text{Call arrival rate} = 1100\ \text{calls/hour}$$

$$\text{Call holding time} = t_h = 3\ \text{min.}$$

$$\text{Offered traffic} = \frac{1100}{60} \times 3 = 55\ \text{E}$$

Then,

$$\text{Grade of service} = \frac{\text{Offered traffic} - \text{Carried traffic}}{\text{Offered traffic}}$$

$$\text{GOS} = \frac{A - A_0}{A} = \frac{55 - 50}{55}$$

$$\text{GOS} = 0.0909$$

Example 7.13: An exchange is designed to handle 2000 calls during the busy hour. One day, the number of calls during the busy hour is 2200. What is the resulting grade of service?

Solution: Given data,

$$\text{Offered traffic} = 2200$$

$$\text{Carried traffic} = 2000$$

$$\text{Grade of Service} = \frac{\text{Offered traffic} - \text{Carried traffic}}{\text{Offered traffic}}$$

$$\text{GOS} = \frac{A - A_0}{A} = \frac{2200 - 2000}{2200}$$

$$\therefore \quad \text{GOS} = \frac{200}{2200}$$

$$\therefore \quad \text{GOS} = 0.0909$$

QUESTIONS

7.1. Describe three communication systems that send their signal as an EM wave through the atmosphere.

7.2. Discuss the advantages obtained by the use of a system with registers compared with the use of non-register step-by-step system in large area with a high telephone density.

7.3. Explain briefly the meaning of following terms:
Concentration, expansion, multiple, supervision, class of service.

7.4. Explain briefly the meanings of the following terms: progressive control, common control, stored-program control, register, marker, link system, conditional selection.

7.5. Draw the pulse dialling waveform for the number 41.

7.6. A long distance dialler hears four different types of call-in-progress signals while establishing a call. What can he conclude?

7.7. A busy tone does not imply that the called party is actually engaged in a conversation. Explain.

7.8. How can a suppressor can be designed such that the interdigit gap precedes the dialling pulses?

7.9. What are the basic approaches to the design of subscriber access to strowger system? Describe them.

7.10. What is meant by the Erlang B distribution?

7.11. What is called the Erlang C distribution?

7.12. What is call congestion?

7.13. What is relationship between grade of service and blocking probability?

7.14. How to connect 1000 users to one another without directly connecting every user to every other one?

7.15. Explain constant holding time and general distribution.

PROBLEMS

7.1. On average, during the busy hour, a company makes 120 outgoing calls of average duration 2 minutes. It receives 200 incoming calls of average duration 3 minutes. Find:
(1) the outgoing traffic
(2) incoming traffic
(3) the total traffic.

7.2. During the busy hour, on average a customer with a single telephone line makes three calls and receives three calls. The average call duration is 2 minutes. What is the probability that a caller will find the line engaged?

7.3. During the busy hour, 1200 calls were offered to a group of trunks and six calls were lost. The average call duration was 3 minutes. Find:
1. The traffic offered
2. The traffic carried
3. The traffic lost
4. Grade of service
5. The total duration of the periods of congestion.

7.4. Observations were made of the busy lines in a group of junctions at intervals of 5 minutes during the busy hours. The results obtained were 11, 13, 8, 10, 14, 12, 7, 9, 15, 17, 16, 12.

7.5. On average, one call arrives every 5 seconds. During a period of 10 seconds, what is the probability that
1. No call arrives?
2. One call arrives?
3. Two call arrives?
4. More than two call arrives.

CHAPTER 8

Digital Modulation and Detection Systems

8.1 DIGITAL MODULATION SYSTEM

In analog modulation unmodulated carrier modulated by sine wave which is known as baseband signal. On the other hand, in digital modulation system, instead of analog signal stream of bits are used for modulating unmodulated carrier; here unmodulated carrier is high frequency sine wave which contain no information, i.e. unmodulated carrier modulated by a digital data. The function of carrier is to shift the digital data from the audio frequency region into a pass band region of the frequency spectrum passband region of frequencies are very useful in radio transmission as well as frequency multiplexing on a telephone lines. Digital modulation systems are broadly classified as amplitude shift keying (on-off keying), frequency shift keying and phase shift keying. It is easier to implement than analog modulation system. The waveforms of ASK (OOK), PSK and FSK shown in Fig. 8.1 (a), (b) and (c).

From above waveforms of Fig. 8.1(a), it is observed that in amplitude shift keying (on-off keying) for bit 1 of digital modulating data, frequency and phase of unmodulated carrier remains constant, while amplitude of carrier will become

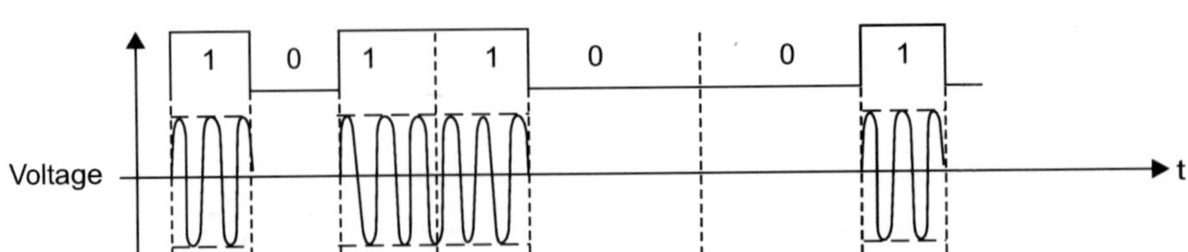

Fig. 8.1(a): ASK (OOK)

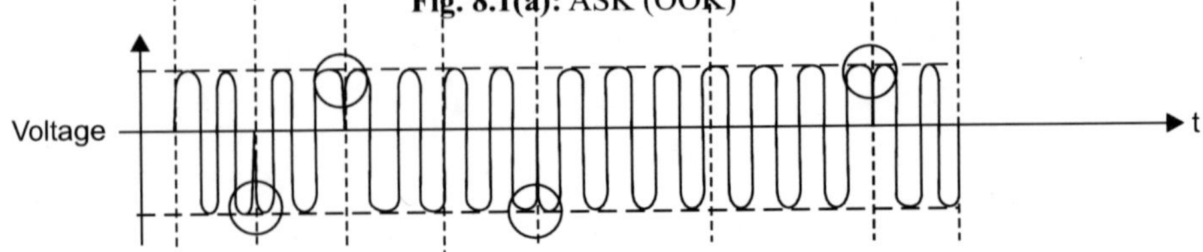

Fig. 8.1(b): Phase Shift Keying

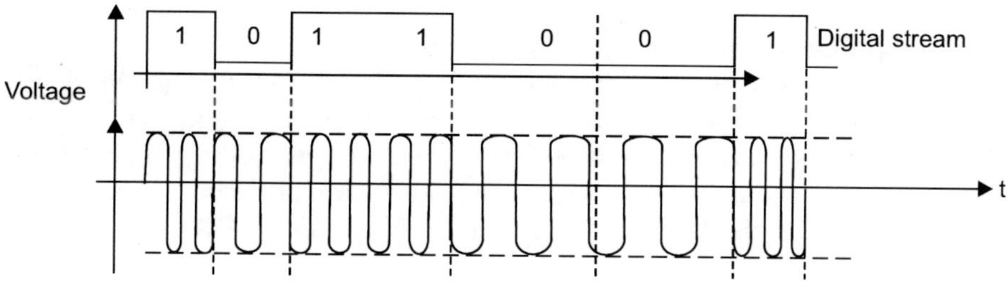

Fig. 8.1(c): Frequency Shift Keying

equal to V_c sin $\omega_c t$ for a load resistance of 1 Ω. On the other hand, for bit 0 of digital modulating bit, frequency and phase of unmodulated carrier remains constant, while amplitude of unmodulated carrier becomes 0 due to digital modulation technique.

From above waveforms of Fig. 8.1(b) it is seen that, in phase shift keying, amplitude of unmodulated carrier remains constant but there will be change in phase of unmodulated carrier at rising and falling edge of modulating data stream, i.e. at rising and falling edge of data stream there is phase reversal, due to phase shift keying technique.

Lastly from waveforms of Fig. 8.1(c) it is seen that for bit 0 of data stream which is known as modulating data unmodulated carrier get frequency shift keying modulated, hence unmodulated carrier frequency get decreased and becomes less than the frequency of unmodulated carrier. For logic 1; of data stream unmodulated carrier frequency get increased due to frequency shift keying; hence modulated carrier frequency is greater than unmodulated carrier frequency due to frequency shift keying.

A very important parameter in digital system is the E_b/N_0 ratio, where E_b is the average bit energy and N_0 the thermal noise power density. In this case thermal noise is internal noise generated due to passive components. In the case of digital modulation technique shown, assuming a carrier of r.m.s. value V_c and a load resistance of 1 Ω, the average received power for the binary phase shift keying and binary frequency shift keying waveforms is simply V_c^2. For OOK waveforms the average received power is $V_c^2/2$, because it is off half the time on average. This is important when we have to compare binary or digital modulation technique. If the average received power is the same in every case, the bit energy will also be same; but this requires that during the on periods for OOK the transmitted power must be doubled or the power amplifier operated at the peak voltage level that is $\sqrt{2}$ times that for continuous carrier system. If comparison is made on the basis of the same value of received voltage, then the average power, and hence the bit energy, of OOK modulation is half the other modulation methods.

8.1.1 Amplitude Shift Keying (ASK) System

With amplitude modulation the data stream is used to switch the carrier signal between amplitude levels, therefore, it is referred to as amplitude shift keying (ASK). ASK binary modulation shown in Fig. 8.2, in this case the modulating waveform consists of unipolar pulses, since in this particular case the carrier is switched on and off, the method is known as on-off keying (OOK).

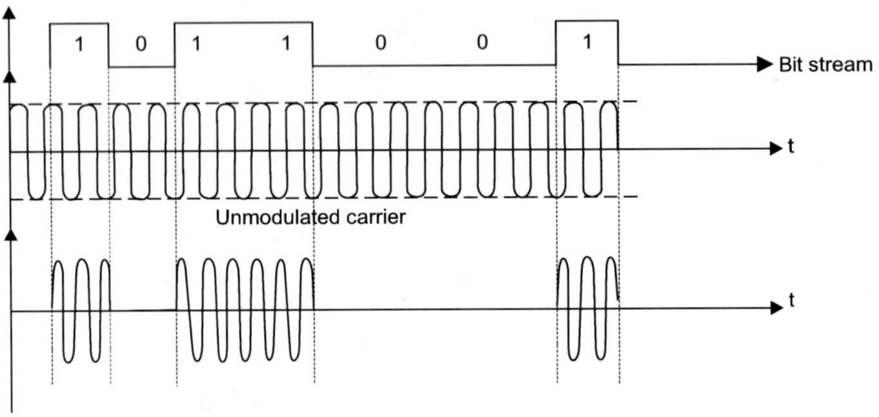

Fig. 8.2: Amplitude Shift Keying by Unipolar Rectangular Pulse.

ASK practically simple to implement. It is less efficient than angle-modulated methods and is not employed in practice. It is useful in radio transmitter and fibre optic communication.

Block diagram given below (Fig. 8.3) shows on-off keying radio transmitter.

(1) Crystal Oscillator: It generates carrier frequency.

(2) Buffer Amplifier: It keeps isolation between crystal oscillator and frequency multiplier amplifier stage. Since buffer amplifier has high input resistance and low output resistance as well as it has unity gain. Hence, loading effect is

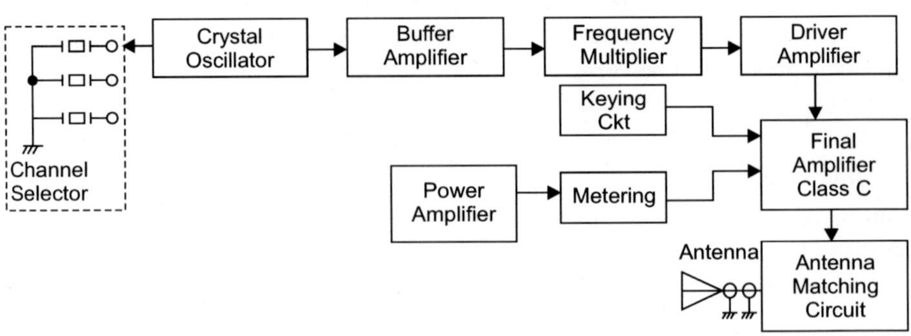

Fig. 8.3: O.O.K. Radio Telegraph Tx.

avoided. Due to this current delivered by crystal oscillator handles stage known as frequency multiplier amplifier. It also maintains good frequency stability.

(3) Frequency multiplier amplifier: To maintain oscillator frequency lower than the required carrier frequency there is need of one or more frequency multiplier.

(4) Driver Amplifier: It is power amplifier that provides drive for final RF amplifier which is operated in class c mode.

(5) Keying circuit, class C final amplifier: Keying circuit used to simply interrupt the current in the final amplifier by means of a "mark and break" keying could take place at a lower power amplifier stage in the transmitter, this is not usually done since the class C bias on the final amplifier is derived from the drive signal, and removal of this could result in excessive current rise in final amplifier.

Where only low power stages are included, a product modulator may be used as shown in Fig. 8.4. Denoting unmodulated carrier by

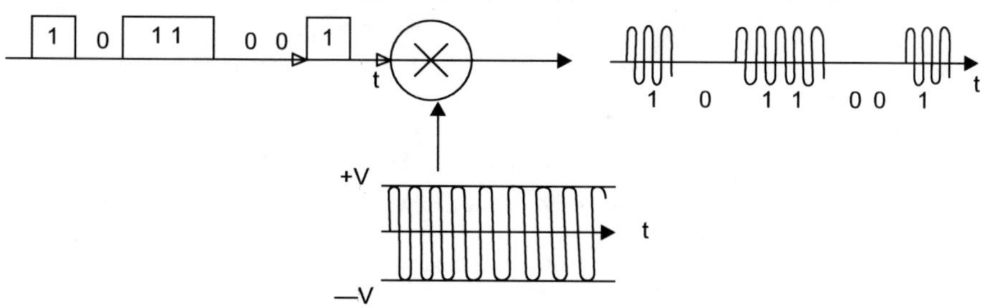

Fig. 8.4: Use of Multiplier Circuit

$$V_c(t) = V_{c\,max} \cos (2\pi f_c t + \phi_c)$$

and binary modulation waveform as $v_m(t)$, then the modulated waveform $v(t) = kv_m(t) \cos (2\pi f_c t + \phi_c)$

Synchronous demodulation of an OOK wave

Demodulation of the OOK waveform may take place using a simple envelope detector. Synchronous detection is more efficient method.

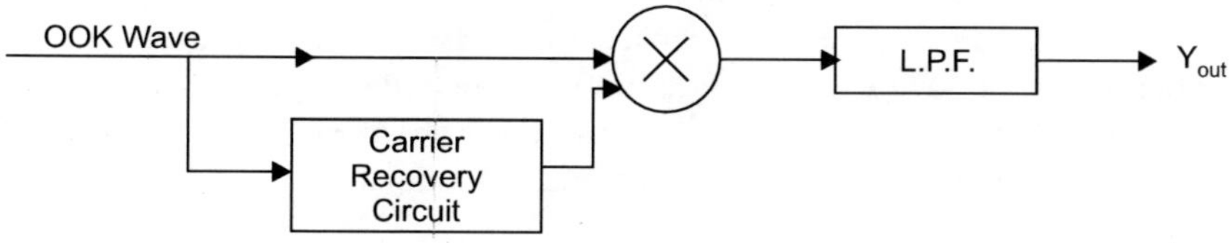

Fig. 8.5: Synchronous Demodulation of an OOK Wave

(1) Carrier Recovery Circuit: Synchronous detection requires a carrier recovery ckt; which is used to generate a local carrier exactly synchronized to the transmitted carrier. As shown spectrum contains a component at the carrier frequency that can be used to phase lock the VCO in a PLL. Applying locally generated carrier and the received signal to multiplier results in an output

$$V_{out}(t) = AV(t) \cos (2\pi f_c t + \phi_c) = AV(t) [\cos (2\pi f_c t + \phi_c)]^2$$

where A is an amplitude constant. Expanding the cosine squared term,

$$V_{out}(t) = AV_m(t)\left[\frac{1}{2} + \frac{1}{2}\cos\left(4\pi f_c t + 2\phi_c\right)\right]$$

(2) LPF: The second harmonic carrier term is easily eliminated by LPF, leaving as the output $V_{out}(t) = \frac{A}{2} V_m(t)$. Thus, baseband signal is recovered. The constant term $\frac{A}{2}$ is used to adjust gain. The synchronous detection referred to as coherent detection. The demodulation of OOK wave can also be carried out by envelope detector (this is known as non-coherent or non-synchronous detection).

The coherent detector is more complicated than envelope detector; but it produces a lower probability of error for a given $\left(\frac{S}{N}\right)$ input. The analysis is not given here, but the results are that the coherent OOK detection has a probability of error identical to that for the baseband system, which for optimum detection is given by $P_{be} = \frac{1}{2}\text{erfc}\sqrt{E_b / N_0}$. The optimum non-coherent detector requires that $E_b/N_o >> 1$ and for this condition

$$P_{be} \cong \frac{1}{2} e^{-\frac{E_b}{2N_0}}$$

8.1.2 Frequency Shift Keying System

In continuous frequency modulation carrier modulated by audio signal. On the other hand, in digital frequency shift keying modulation technique instead of audio signal bit stream is taken as modulating signal and carrier is high frequency sinusoidal (cosine) wave which contains no information, acts as unmodulated carrier. Unmodulated carrier modulated by bit stream of 1 and 0, i.e. digital data. In frequency shift keying (FSK), the carrier frequency is shifted in levels or steps corresponding to the levels of the digital modulating signal. Hence, in case of a binary signal two carrier frequencies are used, one corresponding to the binary 0 and the other to a binary 1. In general, for binary modulation the carriers can be represented by

Binary 0: $V_0(t) = A_0 \cos (2\pi f_0 t + \alpha_0)$ (8.1.2)

Binary 1: $V_1(t) = A_1 \cos (2\pi f_1 t + \alpha_1)$ (8.1.3)

Block diagram for two methods shown in Fig. 8.6 and 8.7 respectively.

(1) Oscillator: The two carrier generated from separate oscillators, independent of one another. These are indicated by separate symbols for the amplitudes and fixed phase angle.

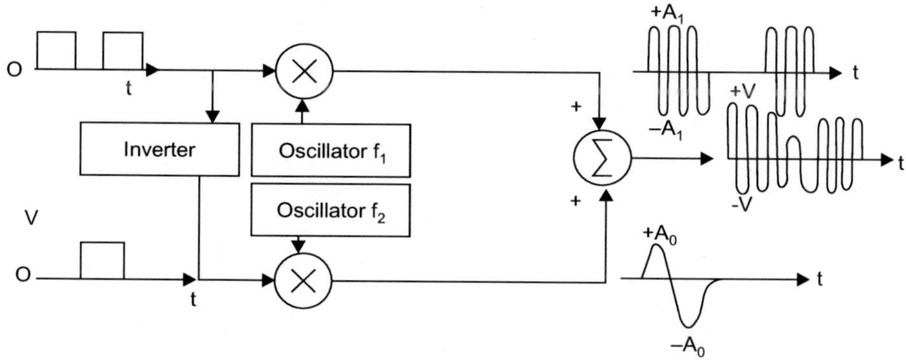

Fig. 8.6: FSK using separate oscillator method

(2) Summer: Whatever ASK modulated signal by product modulators get added in summer and hence output of summer is resulting frequency shift keying output. The combined signal can therefore have discontinuities in amplitude and phase which are not desirable. Alternatively, the modulation can be achieved by frequency modulating a common carrier, which prevents the discontinuties from occurring.

(3) Product modulator: Mean carrier (unmodulated carrier) represented by frequency f_c. When one input to product of modulator is a binary 1 and another input to product modulator is carrier having frequency f_c. After modulation by product modulator resulting ASK signal has frequency $f_1 = f_c + \delta f$ and for binary 0, resulting ASK modulated signal has frequency $f_0 = f_c - \delta f$, where $2\delta f$ is the difference between the two frequencies. The modulated signal is given by:

$$\text{Binary 0: } V_0(t) = A\cos(2\pi f_0 t) \qquad (1)$$

$$\text{Binary 1: } V_1(t) = A\cos(2\pi f_1 t) \qquad (2)$$

The fixed phase angle has been set equal to zero for each signal.

Continuous phase frequency shift keying: CPFSK block diagram shown in Fig. 8.7.

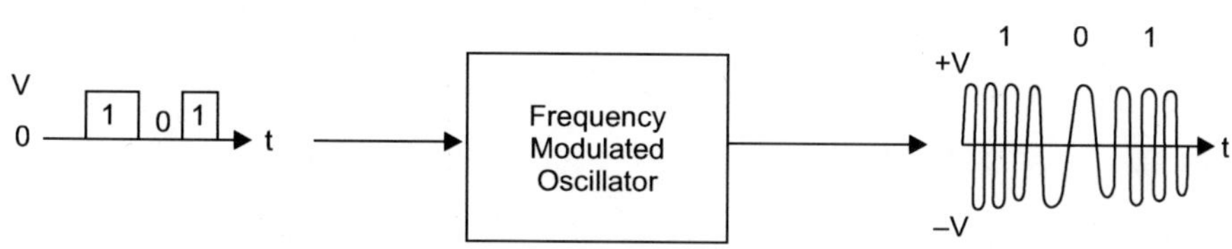

Fig. 8.7: CPFSK

When single oscillator frequency modulated by the digital data stream as modulating signal, the method is known as continuous phase frequency shift keying (CPFSK). Figure 8.7 shows the waveform for the CPFSK wave where for convenience, an integer number of carrier cycles per bit period is shown.

By treating the CPFSK wave as two OOK wave as shown in figure, the spectrum of the OOK wave shown in figure can be used for each, and the resultant

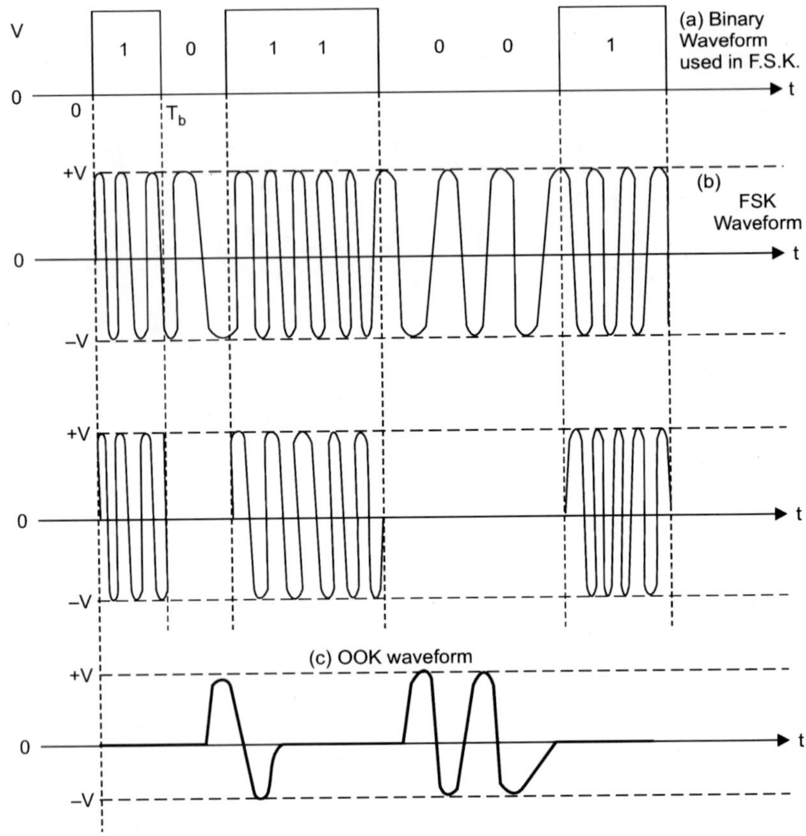

Fig. 8.8: FSK, OOK, resoluted OOK Wave

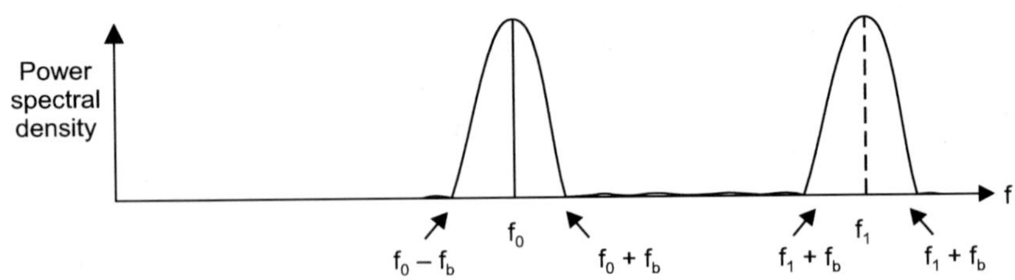

Fig. 8.9a: Power Spectra for CPFSK with Widely Spaced Signalling Frequency

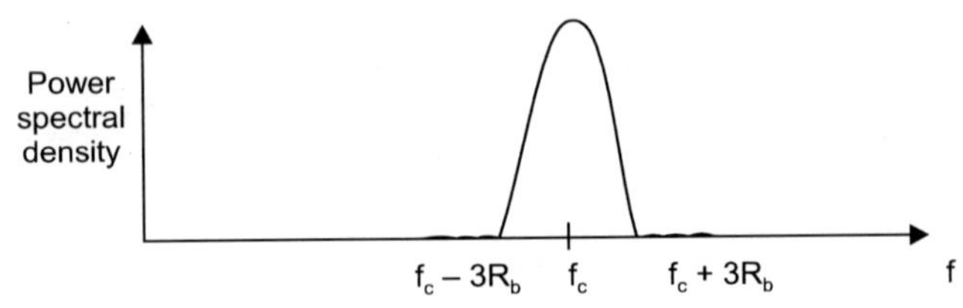

Fig. 8.9b: MSK

spectrum sketched in Fig. 8.9 can be used for each, and the resultant spectrum sketched in Fig. 8.9a and 8.9b.

A special and important case of CPFSK known as minimum shift keying (MSK) occurs when $\delta_f = \frac{R_b}{4}$. This is the minimum separation for which correlation between the two signalling waveforms is zero. It can be shown that for any closer spacing the correlation between the waveform results in an increased probability of bit error.

For MSK, it is assumed that the spectrum beyond the first nulls can be ignored, the overall spectrum bandwidth is given by $B_T = \frac{3}{2} R_b$.

In practice, it is found that the rate of decay of spectrum outside a bandwidth given by $B_T = \frac{R_b}{2}$ is very rapid, with the result that the bps/Hz figure can be improved to unity.

Because FSK appears two OOK waves, the coherent receiver can be constructed by using two separate OOK coherent detector shown in Fig. 8.12.

8.2 DIGITAL DETECTION SYSTEMS

8.2.1 Coherent and Non-coherent Detection for CPFSK System

Because FSK appears as two OOK waves, the coherent receiver can be constructed by using two separate OOK coherent detector as shown in Fig. 8.10 and 8.11.

The outputs are combined to form a polar binary signal, which for optimum detection is then passed through matched filter. Correlation between the two signalling frequencies results in general in an increased probability of bit error, but

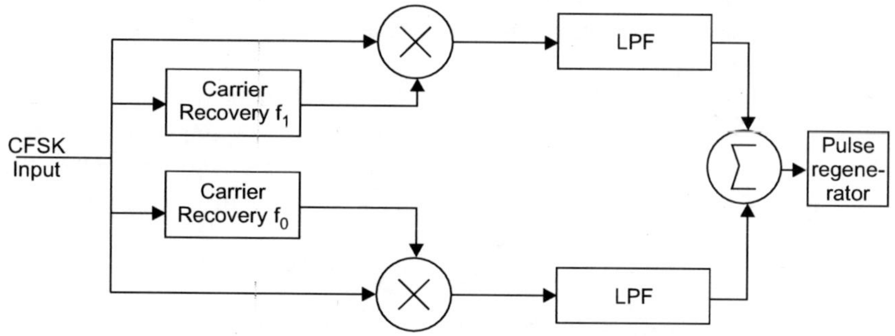

Fig. 8.10: Coherent Detection for CPFSK

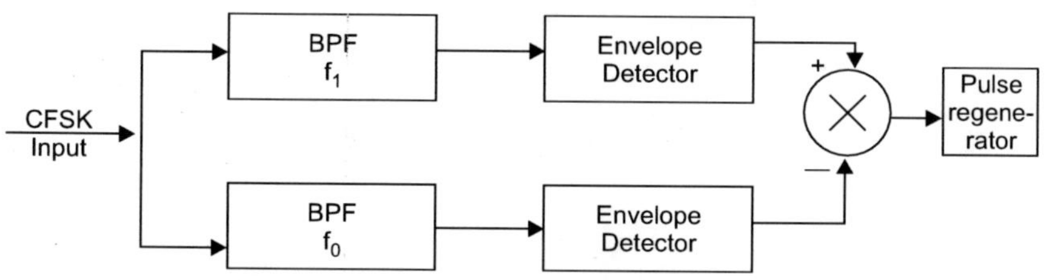

Fig. 8.11: Non-coherent Detection for CPFSK System

with MSK the correlation is zero and expression for bit-error-probability is the same as that for OOK which is reproduced here for convenience:

$$P_{be} = \frac{1}{2}\text{erfc}\sqrt{\frac{E_b}{2N_0}}$$

When comparing OOK with CPFSK, it should be kept in mind that the bit energy with CPFSK is double that of OOK for the same carrier voltage levels at the receiver.

Noncoherent detection can also be used with FSK signals. Again, because FSK appears as two OOK waves, the non-coherent receiver need consists only two separate paths with bandpass filters tuned to the individual frequencies as shown in Fig. 8.11. Each filter is followed by an envelope detector. The outputs are combined to form a polar waveform, which is then passed as input to the pulse regenerator operating at zero voltage threshold. When properly adjusted to optimum performance, the probability for bit error for the non-coherent FSK detector is given by

$$P_{be} \cong \frac{1}{2}e^{-E_b/2N_0}$$

The non-coherent receiver is much simpler to build than the coherent receiver, and for many applications the degradation in bit error probability is acceptable.

8.2.2 Phase Shift Keying and Coherent, Polar NRZ Detector Systems

In case of analog phase modulation, carrier modulated by analog signal. On the other hand, in case of phase shift keying modulation technique analog signal replaced by digital data stream.

By digital data stream when carrier modulated then such digital modulation technique is known as phase shift keying. In phase shift keying (PSK), the binary signal is used to switch the phase of carrier between 0° and 180°. It is also known as phase reversal keying (PRK). The modulated carrier is described by:

$$V(t) = \begin{cases} V_{C\max}\cos(2\pi f_c t + \phi_c)\ldots & \text{binary 1} \\ V_{C\max}\cos(2\pi f_c t + \phi_c + 180°),\ldots & \text{binary 0} \end{cases}$$

The circuit for implementing binary phase shift keying is a balanced modulator, as shown in Fig. 8.12.

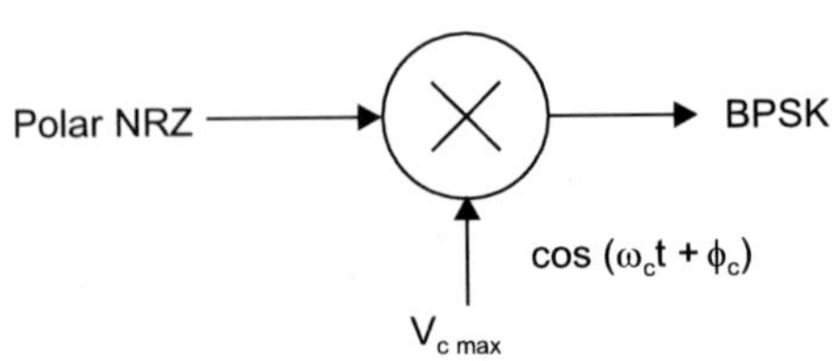

Fig. 8.12: Balanced Modulator Producing BPSK

Binary signal applied to balanced modulator is known as modulating signal. The modulating signal is polar NRZ and another signal applied to balanced modulator is carrier signal. Carrier signal in this case is $V_{C\ max} \cos(w_c t + \phi_c)$. Output of balanced modulator is modulated signal and this modulated signal is binary phase shift keying (BPSK).

The modulating signal is polar NRZ, and when this has value +1, the modulated output is $+1 \times V_{c\ max} \cos(2\pi f_c t + \phi_c) = V_{c\ max} \cos(2\pi f_c t + \phi_c)$ and when it is –1; the modulated output is $-1 \times V_{c\ max} \cos(2\pi f_c t + \phi_c) = V_{c\ max} \cos(2\pi f_c t + \phi_c + 180)$.

Coherent Detection System: Coherent detection must be used with BPSK since the envelope does not contain the modulating information. The coherent BPSK receiver is shown in Fig. 8.13.

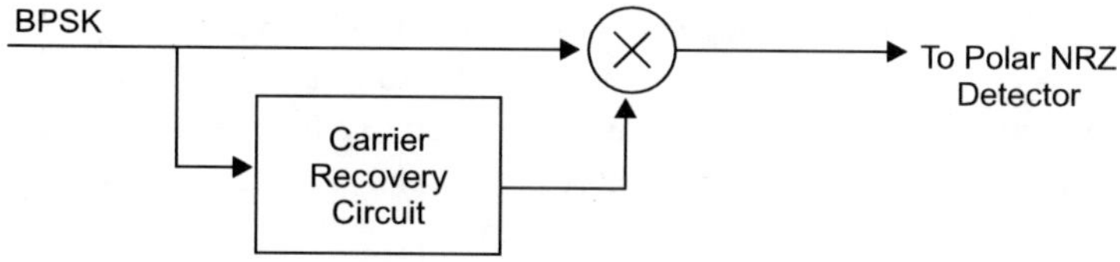

Fig. 8.13: Detection of BPSK

The BPSK modulator is similar to the OOK modulator, the difference being that no d.c. component is present in the modulating waveform and therefore no carrier component transmitted. The spectrum is shown in figure of OOK, but with no spike at the carrier frequency.

The BPSK wave has in effect a DSBSC spectrum shown in Fig. 8.14. With raised-cosine filtering on the baseband signal, the bps/Hz figure of merit also given by eq $\alpha = \frac{1}{1+\zeta}$ as $\frac{1}{1+\zeta}$. Coherent detection of BPSK followed by matched filter detection results in a bit-error probability given by $P_{be} = \frac{1}{2}\text{erfc}\sqrt{\frac{E_b}{N_o}}$

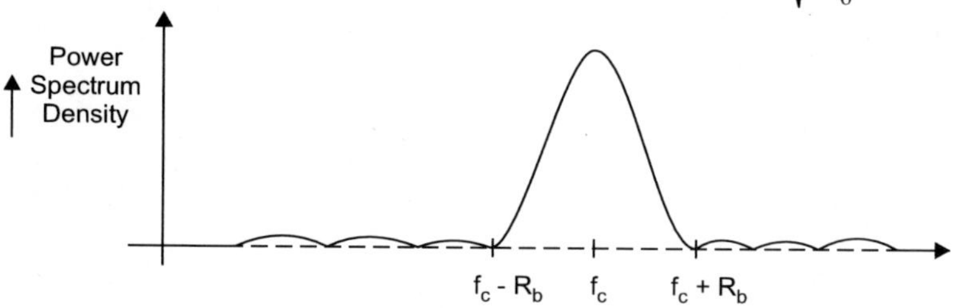

Fig. 8.14: Spectrum for the BPSK Wave

Polar NRZ detector system:

Figure 8.15 shows Polar NRZ detector system. The received signal is shown as V(t), the noise voltage referred to the receiver input as n(t) and the noisy signal is

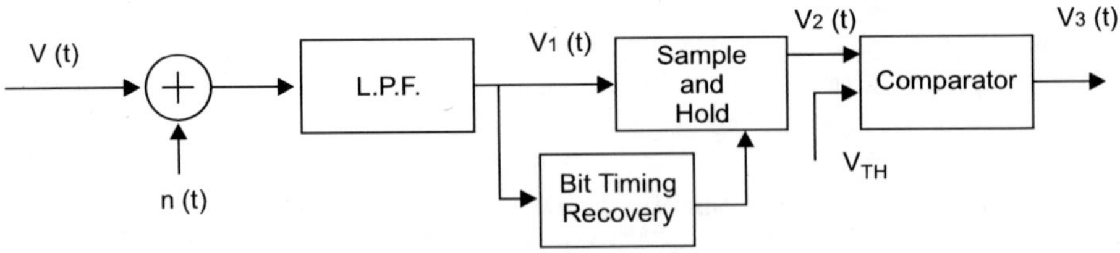

Fig. 8.15: Polar NRZ Detector

the sum of these. The noisy signal passed through a low-pass filter to limit the noise. The overall filtering needed to eliminate intersymbol interference. The received signal after filtering is shown as $V_1(t)$. This is sampled at the bit rate by means of the sample-and-hold circuit. This is at the bit rate by means of the sample-and-hold circuit. The timing of the sampling circuit is critical.

To analyze the process, a polar signal is shown where the peak values ± A represents binary 1 and 0 respectively. The sample-and-hold circuit produces output level, shown as $V_2(t)$ and these in turn are compared to a threshold level. Received levels above the threshold result in a + A output pulse and below threshold a – A output pulse, shown as $V_3(t)$.

8.2.3 Quadrature Phase Shift Keying (QPSK) Modulator and Detection System

Binary baseband signals can be reformatted as M-ary signals with a consequent reduction in transmitted bandwidth, so M-ary level modulation can be used to similar effect.

Quadrature Phase Shift Keying (QPSK) utilizes four distinct levels of phase shift and is a widely used form of multilevel modulation. In this method a serial-to-parallel converter is used to convert the binary digital signal P(t) into two separate binary signals in which the bit period is doubled, as shown in Figs. 8.16(a), (b) and (c).

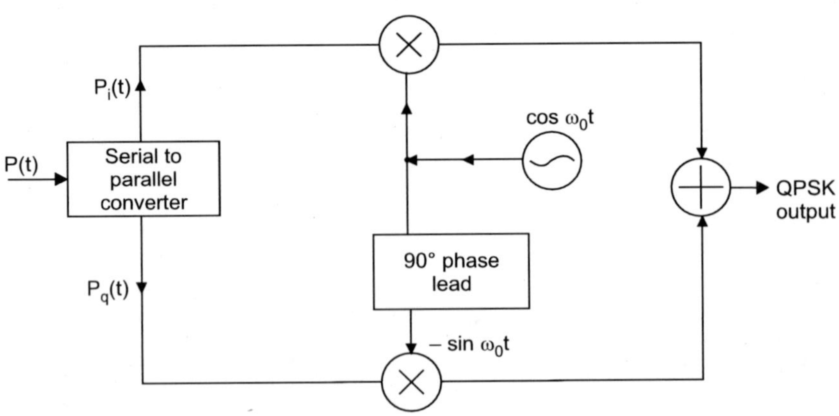

Fig. 8.16(a): QPSK Modulator

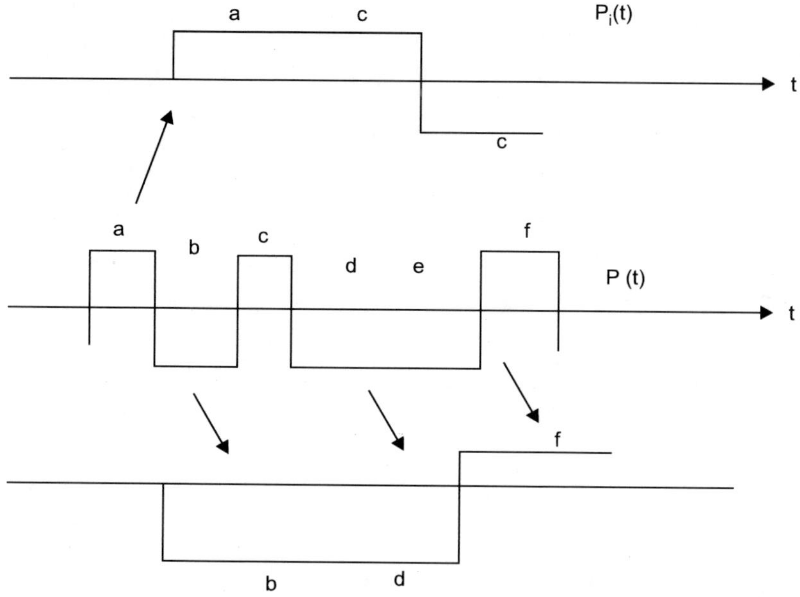

Fig. 8.16(b): Waveforms for (a) and (c)

$P_i(t)$	$P_q(t)$	QPSK
1	1	$\cos \omega_0 t - \sin \omega_0 t = \sqrt{2}\ \cos(\omega_0 t + 45°)$
1	–1	$\cos \omega_0 t + \sin \omega_0 t = \sqrt{2}\ \cos(\omega_0 t - 45°)$
–1	1	$\cos \omega_0 t - \sin \omega_0 t = \sqrt{2}\ \cos(\omega_0 t + 135°)$
–1	–1	$\cos \omega_0 t + \sin \omega_0 t = \sqrt{2}\ \cos(\omega_0 t - 135°)$

Fig. 8.16(c): QPSK Modulator States

These two binary signals has given the name as $P_i(t)$ for in-phase and $P_q(t)$ for quadrature-phase components respectively.

The in-phase component modulates a carrier to produce a BPSK signal, while the quadrature component modulates a carrier component shifted by 90° (hence name quadrature), also to produce a BPSK signal. This two BPSK signals are added to produce the QPSK signal, the modulator states being as shown in Fig. 8.16(c).

Thus, the QPSK signal is equivalent to two BPSK signals, but with the carriers 90° out of phase with one another. Each BPSK waveform has a DSBSC spectrum as described previously, and these spectra overlap. However, they do not interfere with one another, because of the phase difference between the carriers. Thus, QPSK signalling requires one-half the bandwidth of BPSK signal for the same input bit rate in both cases and the bps/Hz figure of merit is $2/(1 + \rho)$ where raised cosine filtering is used.

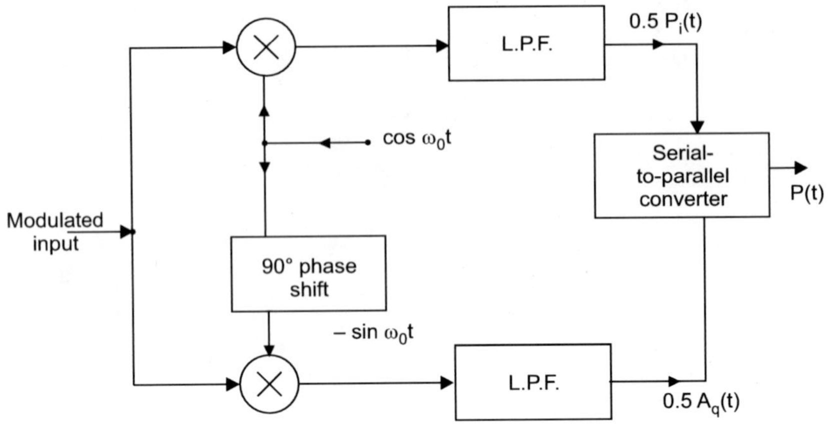

Fig. 8.16(d): QPSK demodulator circuit

QPSK Detection System: Detection of QPSK is similar to that for BPSK with difference that the recovered carriers must also have the 90° phase difference. A demodulator circuit is shown in Fig. 8.16(d).

Assuming that the demodulated output is followed by a matched filter detector, bit error probability for QPSK is the same as that for BPSK as given by equation

$$P_{be} = \frac{1}{2}\text{erfc}\sqrt{\frac{E_b}{N_0}}$$

At this point it may be useful to summarize some of results obtained for ideal systems, that is where $\rho = 0$ for the raised cosine spectrum and the transmission channel is distortionless so that only a constant attenuation and constant delay introduced. Using these conditions the comparisons shown in Table 8.1 may be made, assuming that coherent demodulation for the modulated signal and optimum detection for the pulse.

Table 8.1

Bit error Probability	Signal Type	BPS/Hz
$P_{be} = \frac{1}{2}\text{erfc}\sqrt{\frac{E_b}{N_0}}$	Polar NRZ BPSK QPSK	2 1 2
$P_{be} = \frac{1}{2}\text{erfc}\sqrt{\frac{E_b}{2N_0}}$	Unipolar NRZ OOK MSK	2 1 1

It will be recalled that the average bit energy is obtained by multiplying the average transmitted power by the bit period and that in the case of unipolar (including OOK) transmission, the voltage levels have to be increased by a factor of $\sqrt{2}$ compared to continuous wave transmissions.

8.2.4 Differential Phase Shift Keying (DPSK) Modulator and Detection System

Differential encoding of baseband signals is already discussed in previous chapter, where a binary message sequence d_k is shown differentially encoded as sequence V_k. For case of comparison, this same sequence and (c) is used here and is shown in Fig. 8.17, (a), (b), (c) the V_k phase modulates a carrier cos $\omega_c t$ as shown.

The output from the modulator is a BPSK signal, but because it is modulated by a differentially encoded signal, it will be referred to as a differentially phase shift keying or DPSK signal.

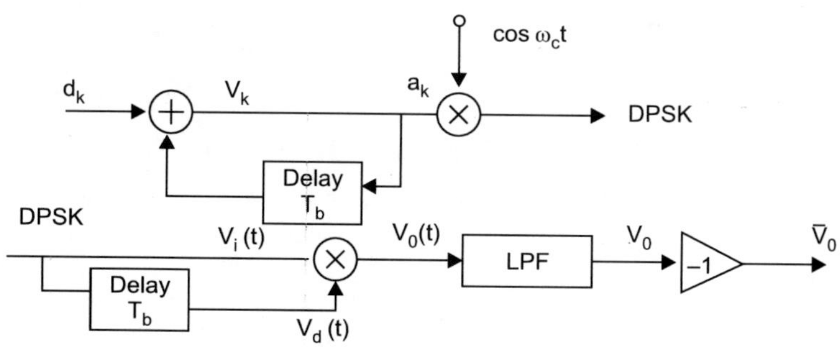

(a): Differential PSK System

d_k (Logic Levels)	1	1	0	1	0	0	1
V_k (Logic Levels)	1	0	0	1	1	1	0
d_k (Volts)	+1	–1	–1	+1	+1	+1	–1

(b): DPSK Logic Levels

a_k (volts)	+1	–1	–1	+1	+1	+1	–1
a_{k-1} (volts)	–1*	+1	–1	–1	+1	+1	+1
	*initial state						
V_0 (volts)	–1	–1	+1	–1	+1	+1	–1
V_0 (Logic Levels)	0	0	1	0	1	1	0
$\overline{V}_0$ (Logic Levels)	1	1	0	1	0	0	1

(c): Logic Level After Detection

Fig. 8.17(a–c): Differential Phase-shift Keying

By assuming $a_k = +1$ V to a binary 1 and $a_k = -1$ V to a binary 0, the DPSK signal can be represented by

$$\text{DPSK} = a_k \cos \omega_c t \tag{8.2.1}$$

It will be seen that a binary 1 corresponds to no phase shift and a binary 0 to a 180° phase shift in the carrier. At the receiving end, the decoder consists of a multiplier, the inputs to which are the DPSK signal and a 1-bit delayed version of this. Ignoring the attenuation and delay produced in the transmission path (since these are common to both inputs), the output from the multiplier is

$$\begin{aligned} V_0(t) &= KV_i(t)\,V_d(t) \\ &= K(a_k \cos \omega_c t) \times (a_{k-1} \cos \omega_c (t - T_b)) \end{aligned} \tag{8.2.2}$$

where K is multiplier constant. By making $f_c = \frac{N}{T_b}$; where N is an integer (and usually $N \geq 1$), $\cos \omega_c(t - T_b) \equiv \cos \omega_c t$ and

$$V_0(t) = K\, a_K\, a_{K-1} \cos^2 \omega_c t$$

$$V_0(t) = K\, a_K\, a_{K-1} \left(\frac{1}{2} + \frac{1}{2} \cos 2\omega_c t\right) \tag{8.2.3}$$

The low-pass filter at the output of the multiplier removes the second harmonic term of carrier, leaving only the baseband component. Denoting the constants by A, the low-pass output is

$$\begin{aligned} V_0 &= A\, a_K\, a_{K-1} \\ V_0 &= +A, \text{ for } a_K = a_{K-1} \\ V_0 &= -A, \text{ for } a_K \neq a_{K-1} \end{aligned} \tag{8.2.4}$$

The logic levels corresponding to V_0 are shown in the lower table of Fig. 8.2.6. These are the message levels inverted, and an inverter is required to restore the message sequence at the output. Comparison of the inverted output $\overline{V}_0$ shown in Fig. 8.2.6 with the binary input d_k shows these to be some. More generally, it can be shown that the multiplication followed by inversion is equivalent to the exclusive –OR decoder shown in Fig. 8.18 for the differentially encoded baseband signal.

A disadvantage with DPSK is that bit errors tend to occur in pairs, because the polarity of a given bit depends on the polarity of the preceding bit. The average bit-error probability for DPSK can be shown to be given by

$$P_{be} = \frac{1}{2} e^{-(E_b/N_0)} \tag{8.2.5}$$

The process of decoding where the received bit is compared with a threshold level and a binary decision made is referred to as hard-decision decoding. In other words, the output is either right or wrong. Furthermore, where encoding for the purpose of error correction is employed. The bit stream is partitioned into

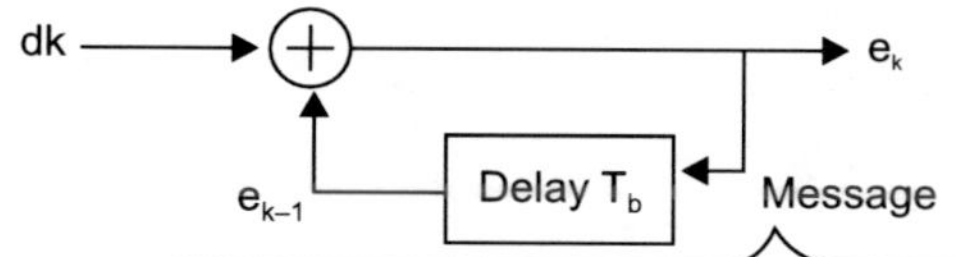

d_k (reference bit)	1	1	0	1	0	0	1
e_{k-1}	Initial state 0	1	0	0	1	1	1
$e_k = d_k \otimes e_{k-1}$	1	0	0	1	1	1	0

Fig. 8.18: Differential Cncoding of a Binary Message.

codewords. Again with hard decision decoding, a firm right or wrong decision is made for each received codeword. An alternative to hard decision decoding is soft decision decoding. In soft decision decoding, a received codeword is compared with a (possibly large) number of alternatives, and the one that shows the closest, but not necessarily exact, correlation is assumed to be correct. Soft decision decoding results in about a 2-dB improvement in performance over hard decision decoding, meaning that the same bit-error probability can be achieved with a 2-dB reduction in the received signal to noise ratio. However, the cost and complexity of the additional equipment needed means that, it is only used where absolutely necessary.

QUESTIONS

8.1. Describe in short block diagram of digital communication system.
8.2. What is meant by a matched filter?
8.3. What is meant by CFSK and MSK?
8.4. Explain forward error correction.

PROBLEMS

8.1. A binary unipolar signal has an average power of 6 pw, and pulse duration of 0.02 μs. The equivalent noise temp. at the Rx, input is 550 K. Determine bit-error probability.
8.2. Compare the bit-error probabilities for polar and unipolar transmission for which the received $\left(\frac{S}{N}\right)$ ratio is 9 dB in each case.

CHAPTER 9

Digital Modulation Technique, Error Probability and Line Coding

Digital modulation technique is broadly classified as:
(1) Amplitude shift keying
(2) Frequency shift keying, and
(3) Phase shift keying.

9.1 AMPLITUDE SHIFT KEYING OR ON-OFF KEYING

When modulating signal is m(t) that is data bits 10110001 and unmodulated carrier is cos ω_ct. By this digital data, when unmodulated carrier (cos ω_ct) amplitude get modulated by keeping frequency and phase constant, then resulting digital modulation is known as amplitude shift keying or on-off keying. For logic 1 output of amplitude shift keying is modulated signal m(t) × cos ω_ct because modulation is multiplicative process and for logic 0, resulting output of amplitude shift keying is zero level. Figure 9.1 shows waveforms of amplitude shift keying or on-off keying.

9.2 FREQUENCY SHIFT KEYING (FSK)

When modulation signal is m(t) that is digital data 10110001 and unmodulated carrier is cosω_ct. By digital data when unmodulated carrier (cos ω_ct) get digital modulated by keeping amplitude and phase constant then resulting digital modulation is known as frequency shift keying (FSK). For logic 1 resulting frequency shift keying digital modulated signal has frequency to be ω_{c_1} or $2\pi f_{c_1}$. On the other hand, for logic 0 resulting output of frequency shift keying digital modulated signal has frequency to be ω_{c_0} or $2\pi f_{c_0}$ but amplitude and phase remains constant.

Note that in frequency shift keying digital modulated output waveform ω_{c_1} or $2\pi f_1 > \omega_{c_0}$ or $2\pi f_0$. Then simply we can say $f_1 > f_0$. By how much amount it is greater that depends upon amplitude of digital data pulse magnitude of its amplitude.

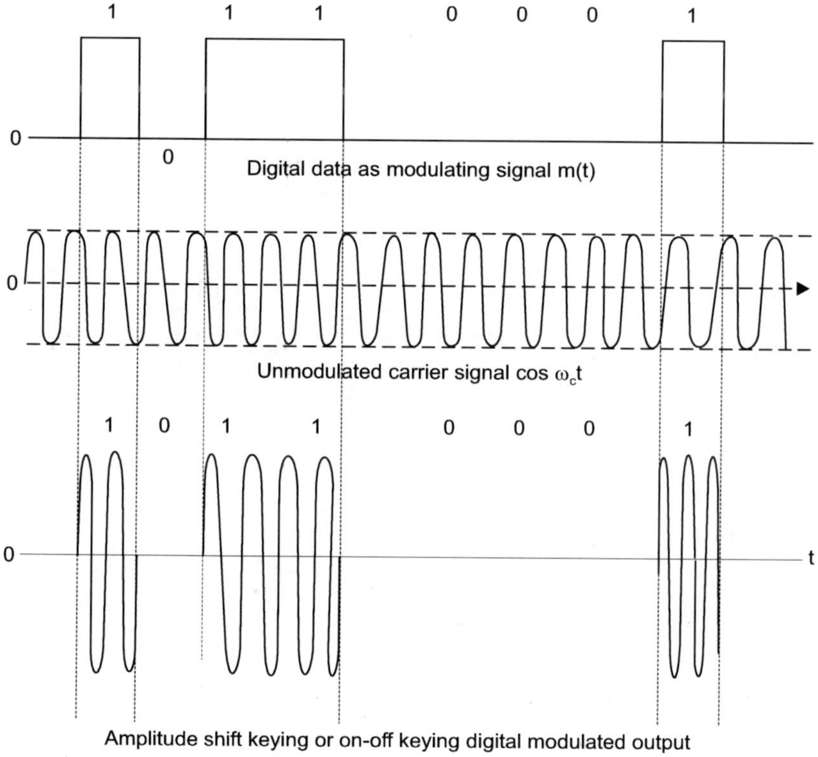

Fig. 9.1: ASK or OOK Output Waveform

9.3 PHASE SHIFT KEYING (PSK)

When modulating signal is m(t) that is digital data 10110001 and unmodulated carrier is cos ω_ct. In this phase shift keying frequency and amplitude is kept constant, but by phase shift keying (PSK) at leading or rising edge there is 180° phase change. At falling edge or trailing edge of digital modulating data phase changes from cos ω_ct to cos (ω_0t – 180°). On the other hand, at rising edge or leading edge phase changes from cos ω_ct to cos (ω_ct + 180°) Figure 9.3 given shows phase shift keying waveforms.

9.4 DIFFERENTIAL PHASE SHIFT KEYING (DPSK)

The input sequence of binary bits is modified such that the next bit depends upon the previous bit. Therefore, in the decoder the previous received bit are used to detect the present bit. The digital information content of the binary data is encoded in terms of signal transitions. For symbol 'O' may be used to represent transition in a given binary sequence with respect to the previous encoded bit and symbol '1' to indicate no transition. The new digital signalling technique which combines differential encoding with phase shift keying (PSK) is known as differential phase

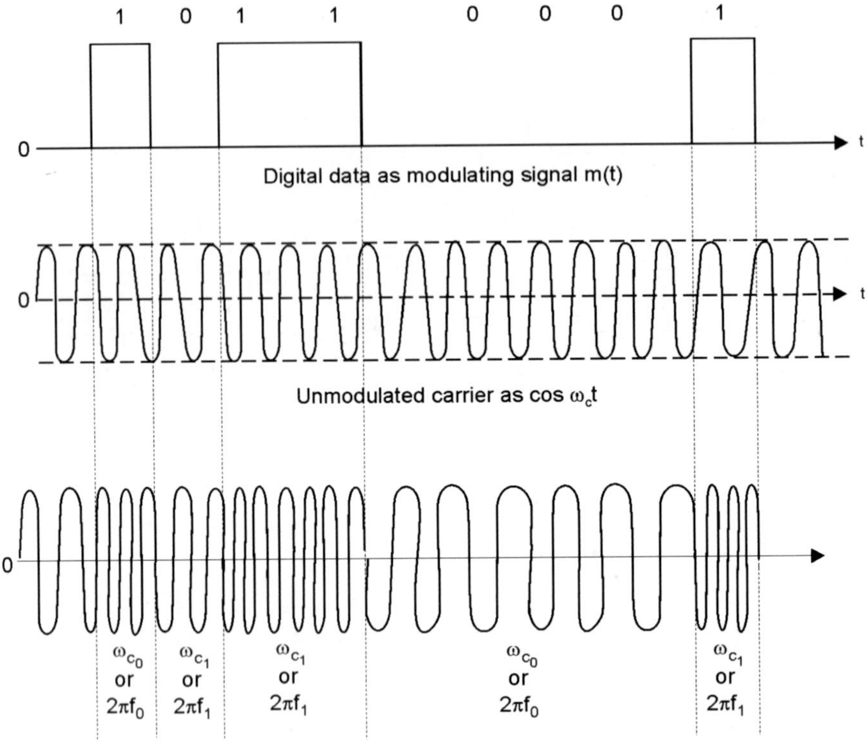

Fig. 9.2: Frequency Shift Keying Digital Modulated Output Waveform

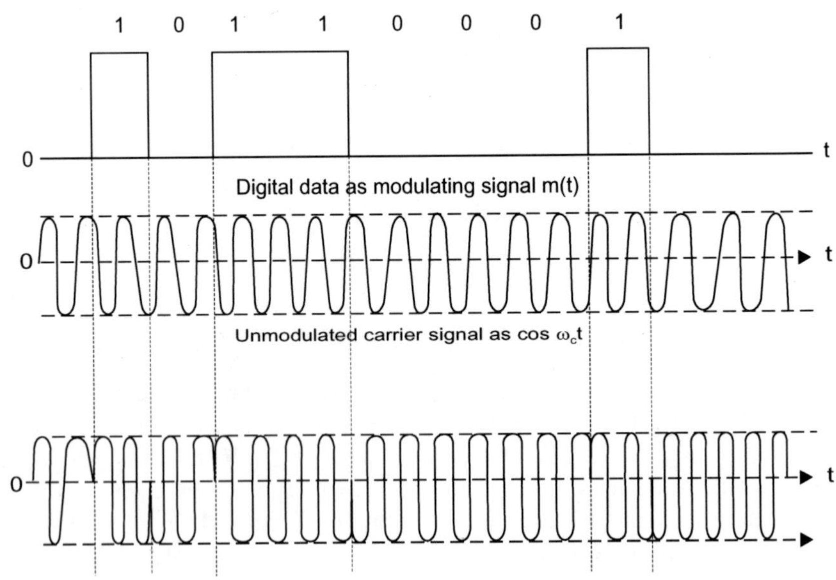

Fig. 9.3: Phase Shift Keying Output Waveform

shift keying (DPSK). For data sequence 00100100111. For this data stream the encoding has been done in such a way that transition in the given binary stream with respect to the previous encoded bit is represented by a digit 0 and no transition for by bit '1'. An extra bit 1 has been added as an initial bit. The phase of generated DPSK is shown below.

Binary data b_i		0	0	1	0	0	1	0	0	1	1	1
Differentially encoded data d_i	1*	0	1	1	0	1	1	0	1	1	1	1
Phase of D.P.S.K.	0	π	0	0	π	0	0	π	0	0	0	0

9.5 QUADRATURE PHASE SHIFT KEYING (QPSK)

In quadrature phase shift keying (QPSK) two successive bits in the data sequence are grouped together. This reduces the bit rate or signalling rate (f_b) and thus reduces the bandwidth of channel. In QPSK, two successive bits are combined, this combination of two bits forms distinct digits. When the symbol changed to next symbol, then the phase of carrier is changed by 45° or $\pi/4$ radians. Table shows these symbols and their phase shifts

Symbols and corresponding phase shifts in QPSK

S. No.	Input successive bits		Symbol	Phase shift carrier
i = 1	1(1 V)	0(– 1 V)	S_1	$\pi/4$
i = 2	0(– 1 V)	0(– 1 V)	S_2	$3\pi/4$
i = 3	0(– 1 V)	1(1 V)	S_3	$5\pi/4$
i = 4	1(1 V)	1(1 V)	S_4	$7\pi/4$

There are four symbols and the phase is shifted by $\pi/4$ for each symbol.

9.5.1 Waveforms of QPSK

Binary stream of data is first converted to bipolar not return to zero type signal. The signal denoted by b_i. It represents binary 1 by + 1 V and binary 0 by –1 V as shown in Fig. 9.4(a). The demultiplexer divides b_i into two separate streams of the odd numbered and even numbered digits. Here b_c represents even numbered sequence and b_0 represents odd numbered sequence. The symbol duration of both these odd and even numbered sequence is $2T_b$. Hence, each symbol consists of two bits. Figure 9.4(b) and (c) shows waveforms of b_e and b_0.

Even bit occurs after the first odd bits. Hence, even bit sequence b_e starts with the delay of one bit due to first odd bit. Thus, symbol of b_e is delayed by one bit period T_b with respect to first symbol of $b_0(t)$. This delay of T_b is known as offset.

Also bit b_e modulates carrier cos $(2\pi f_c t)$ and b_0 modulates sin $(2\pi f_c t)$. This carrier is known as quadrature carrier.

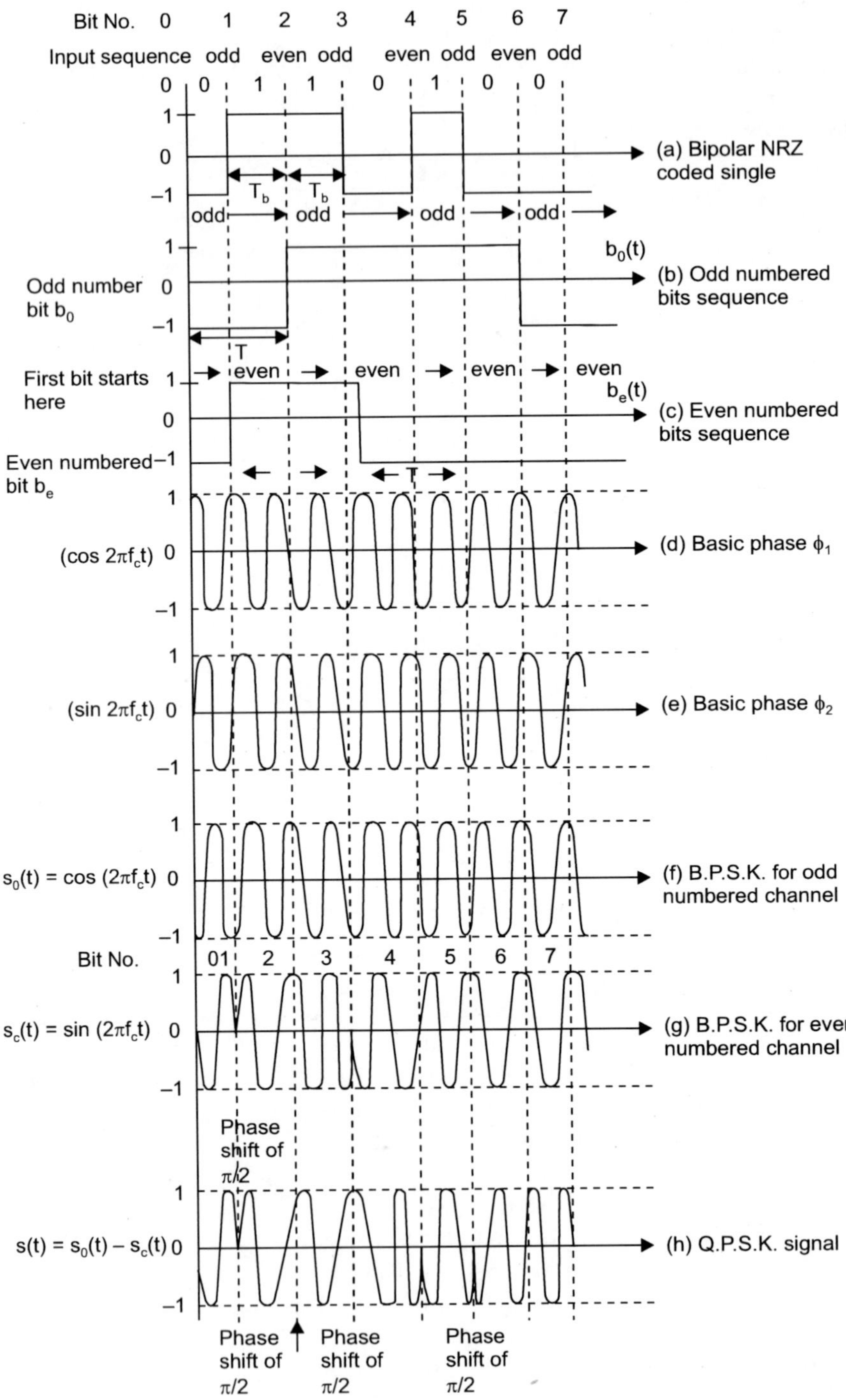
Bit No. 0 1 2 3 4 5 6 7
Input sequence odd even odd even odd even odd
0 0 1 1 0 1 0 0
(a) Bipolar NRZ coded single
T_b
odd
(b) Odd numbered bits sequence
b_0(t)
Odd number bit b_0
T
First bit starts here
even
b_e(t)
(c) Even numbered bits sequence
Even numbered bit b_e
(cos 2πf_ct)
(d) Basic phase φ_1
(sin 2πf_ct)
(e) Basic phase φ_2
s_0(t) = cos (2πf_ct)
(f) B.P.S.K. for odd numbered channel
Bit No. 01 2 3 4 5 6 7
s_c(t) = sin (2πf_ct)
(g) B.P.S.K. for even numbered channel
Phase shift of π/2
s(t) = s_0(t) – s_c(t)
(h) Q.P.S.K. signal
Phase shift of π/2
Phase shift of π/2
Phase shift of π/2

Fig. 9.4

The adder adds two signals b_e and b_0. The output of QPSK signal is s(t)

$$= s_0(t) + s_e(t)$$

$$= \cos (2\pi f_c t) + \sin (2\pi f_c t)$$

In QPSK signal of Fig. 9.4(b) phase change occurs at minimum duration of T_b. This is because s_e and s_0 have an offset of 'T_b'. Due to this offset, the phase shift in QPSK signal is $\pi/2$.

Figure 9.5 given below shows phasor diagram of QPSK signal.

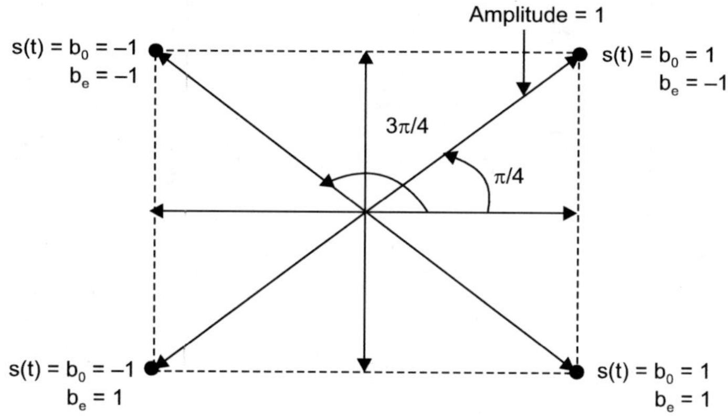

Fig. 9.5: The Phasor Diagram of QPSK

9.6 MINIMUM SHIFT KEYING AND ITS WAVEFORMS

In M.S.K.; the output waveforms is continuous in phase hence there are no abrupt changes in amplitude. The sidelobe of MSK are very small hence bandpass filtering is not needed to avoid interchannel interference. Fig. 9.6 shows waveform of MSK. The binary bit sequence is shown at the top. Fig. 9.6(a) shows corresponding NRZ wave for b. From b two waveforms are generated for odd and even bits b_1 representing odd bits and b_e represents even bits. Figures 9.6(a) and (b) show the waveforms of b_0 and b_e. As shown in those waveforms b_1, b_3, b_5 are representing odd waveforms b_0.

The duration of each bit in b_0 and b_e is $2T_b$, whereas it is T_b in b(t), i.e. $T_s = 2T_b$. The waveforms b_0 and b_e have an offset of T_b. This offset is essential in MSK. Two waveforms $\sin\left(\frac{2\pi t}{4T_b}\right)$ passes through zero at the end of symbol time in b_0. Hence, one symbol duration of b_e contains half cycle of $\cos 2\pi\left(\frac{t}{4T_b}\right)$. This means that similarity one symbol duration of b_e contains half cycle of $\sin\ 2\pi\left(\frac{t}{T_b}\right)$. Thus, there is a phase shift of T_b in sine and cosine waveforms be multiplied by $\sin\ 2\pi\left(\frac{t}{4T_b}\right)$ and b_0 is multiplied by $\cos\ 2\pi\left(\frac{t}{4T_b}\right)$. These

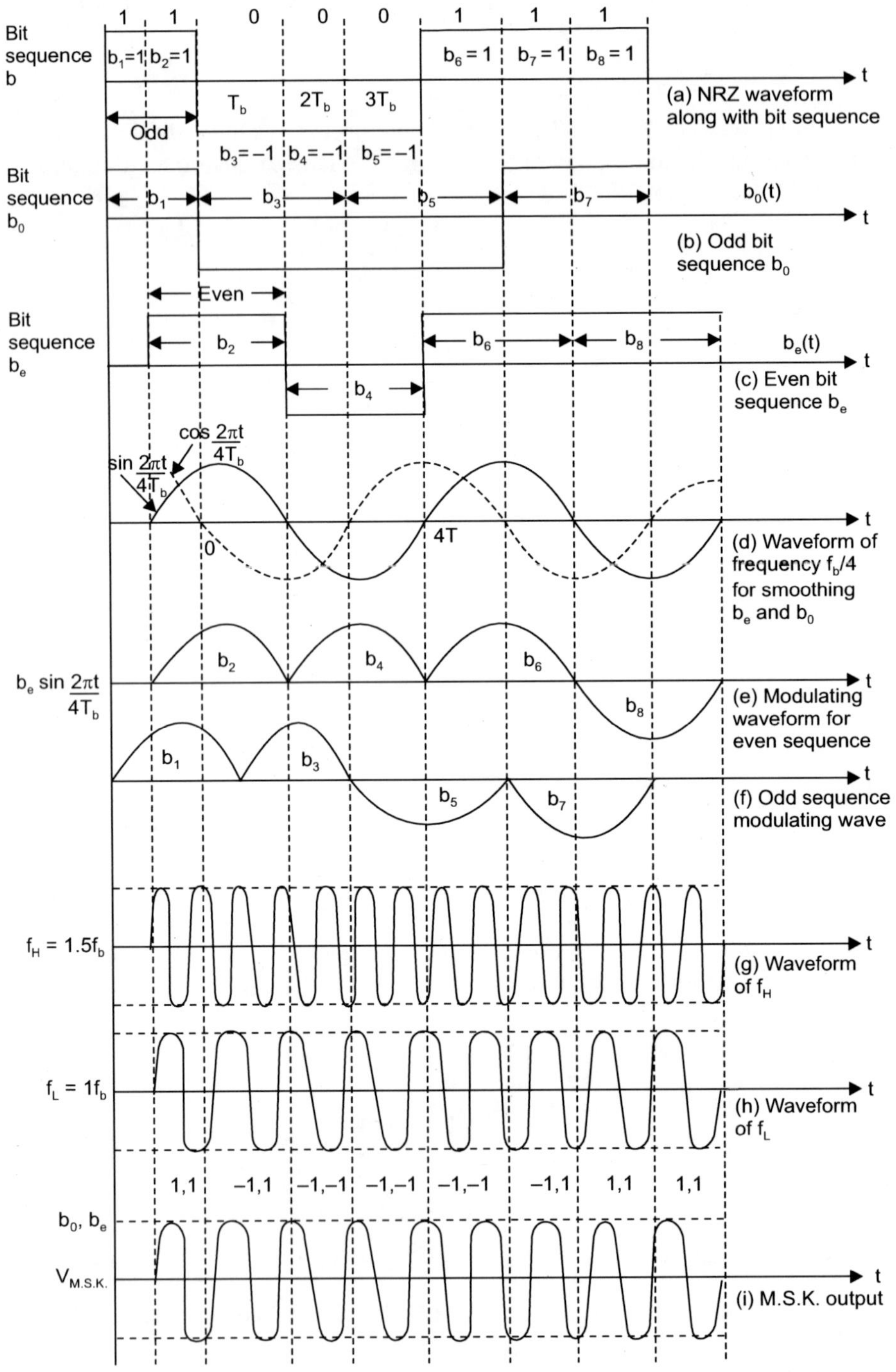

1 1 0 0 0 1 1 1
Bit sequence b
b1=1 b2=1
b6 = 1 b7 = 1 b8 = 1
Tb 2Tb 3Tb
Odd
b3=–1 b4=–1 b5=–1
(a) NRZ waveform along with bit sequence
Bit sequence b0
b1 b3 b5 b7
b0(t)
(b) Odd bit sequence b0
Even
Bit sequence be
b2 b6 b8
be(t)
b4
(c) Even bit sequence be
cos 2πt/4Tb
sin 2πt/4Tb
0
4T
(d) Waveform of frequency fb/4 for smoothing be and b0
be sin 2πt/4Tb
b2 b4 b6 b8
(e) Modulating waveform for even sequence
b1 b3 b5 b7
(f) Odd sequence modulating wave
fH = 1.5fb
(g) Waveform of fH
fL = 1fb
(h) Waveform of fL
1,1 –1,1 –1,–1 –1,–1 –1,–1 –1,1 1,1 1,1
b0, be
VM.S.K.
(i) M.S.K. output
t

Fig. 9.6

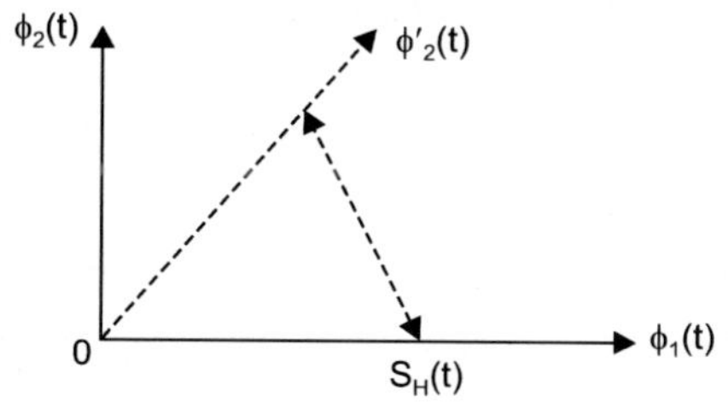

Fig. 9.7: Geometrical Representation of Non-orthogonal BPSK Signal

waveforms are shown in Fig. 9.6(e) and (f). The transmitted M.S.K. Signal is represented by as follows:

$$s(t) = 1\left[b_e \sin\left(\frac{2\pi t}{4T_b}\right)\right]\cos(2\pi f_c t) + 1\left[b_0 \cos\left(\frac{2\pi t}{4T_b}\right)\right]\sin\times(2\pi f_c t)$$

9.7 QUADRATURE AMPLITUDE MODULATION (QAM)

The bandwidth efficiency of PAM S.S.B. can also be obtained by simultaneously impressing two separate k-bit symbols from information sequence $\{a_n\}$ one two quadrature carriers cos $2\pi f_c t$ and sin $2\pi f_c t$. The resulting modulation technique is called quadrature PAM or QAM and corresponding signal waveforms may be expressed as:

$$S_m(t) = \text{Re}\ [(A_{me} + jA_{ms})]\ g(t)\ e^{j2\pi f_c t}];\ m = 0, 1, 2, \ldots M,\ 0 \le t \le T$$

$$= A_{mc}g(t)\cos 2\pi f_c t - A_{ms}g(t)\sin 2\pi f_c t \qquad (9.1)$$

where A_{mc} and A_{ms} are information bearing signal amplitudes of quadrature carriers and g(t) is the signal pulse.

Alternatively, the QAM signal waveform may be expressed as:

$$s_m(t) = \text{Re}\ [V_m\ e^{j\theta m}\ {}_{g(t)}e^{j2\pi fct}]$$

$$\therefore \qquad s_m(t) = V_m g(t)\cos[(2\pi f_c t + \theta_m)] \qquad (9.2)$$

where $V_m = \sqrt{A_{mc}^2 + A_{ms}^2}$ and $\theta = \tan^{-1}\left(\frac{A_{ms}}{A_{mc}}\right)$. From this expression, it is apparent that the QAM signal waveforms may be viewed as combined amplitude and phase modulation.

In fact, we may select any combination of M_1-level PAM and M_2-phase PSK to construct an $M = M_1 M_2$ combined PAM-PSK signal constellation. If $M_1 = 2^n$ and $M_2 = 2^m$, the combined PAM–PSK signal constellation results in symbol rate $\frac{R}{(m+n)}$. Examples of signal space diagram for combined PAM–PSK are shown in Fig. 9.8 for m = 8 and m = 16.

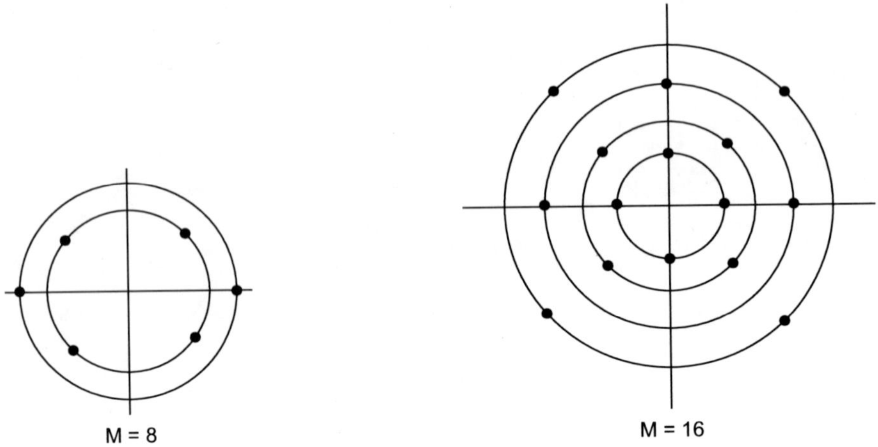

Fig. 9.8: Examples of Combined PAM–PSK Signal Space Diagram

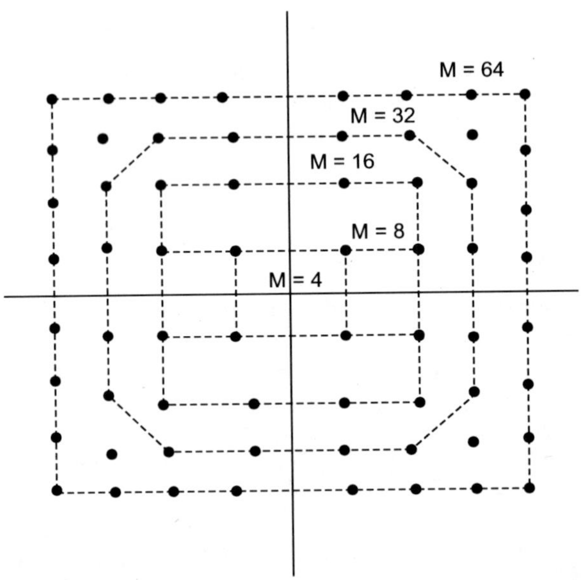

Fig. 9.9: Several Signal Space Diagram for Rectangular QAM

As in the case of PSK signal the QAM signal waveform may be represented as a linear combination of two orthogonal signal waveform $f_1(t)$ and $f_2(t)$, i.e.

$$s_m(t) = s_{m_1} f_1(t) + s_{m_2} f_2(t) \tag{9.3}$$

where
$$f_1(t) = \sqrt{\frac{2}{\varepsilon_g}} g(t) \cos 2\pi f_c t$$

$$f_2(t) = \sqrt{\frac{2}{\varepsilon_g}} g(t) \sin 2\pi f_c t \tag{9.4}$$

and $$s_m = [s_{m_1} s_{m_2}] = \left[A_{me} \sqrt{\frac{1}{2}\varepsilon_g} A_{ms} \sqrt{\frac{1}{2}\varepsilon_g} \right] \tag{9.5}$$

ε_g is the energy of signal pulse g(t)(9.5).

The Euclidean distance between any pair of signal vector is:

$$d_{run}^{(e)} = |s_m - s_n| = \sqrt{\frac{1}{2}\varepsilon_g (A_{mc} - A_{nc})^2 + (A_{ms} - A_{ns})^2}$$

In the special case where signal amplitudes takes the set of discrete values {(2m – 1 – M)d, m = 1, 2, …, m)}, the signal space diagram is rectangular as shown in Fig. 9.7. In this case, the Euclidean distance between adjacent points, i.e. the minimum distance, is:

$$d_{min}^{(e)} = d\sqrt{2\varepsilon_g} \tag{9.6}$$

which is the same as result as for PAM.

9.8 PROBABILITY OF ERROR, P_e OF BPSK SIGNAL

The general equation derived using union bound approximation gives probability of error as under

$$P(e) = \frac{1}{2} \sum_{\substack{k=1 \\ k \neq i}}^{M} \text{erfc}\left(\frac{d_{ik}}{2\sqrt{N_0}} \right) \text{ for all i.} \tag{9.7}$$

Here we have only two points in BPSK. Let $S_H(t)$ be $s_1(t)$ and $S_H(t)$ be $s_2(t)$. Then distance between these two points can be written as

$$d = d_{12} = \sqrt{2E_b}$$

Since there are only two points (i.e., M = 2) take i = 2 and k = 1, then (9.7) becomes:

$$P(e) = \frac{1}{2} \text{erfc}\left(\frac{d_{12}}{2\sqrt{N_0}} \right)$$

or $$P(e) = \frac{1}{2} \text{erfc}\left[\left(\frac{\sqrt{2E_b}}{2\sqrt{N_0}} \right) \right]$$

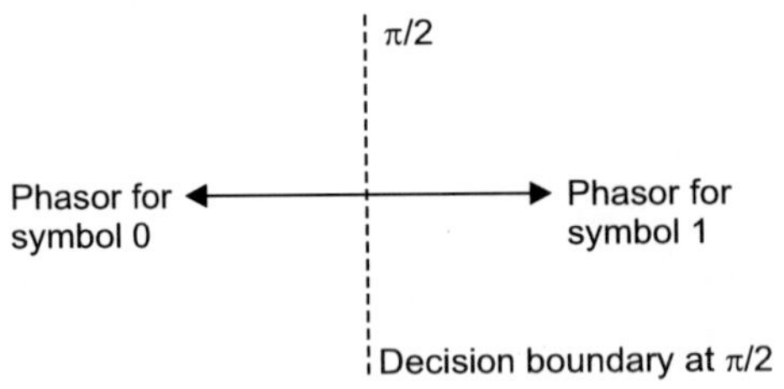

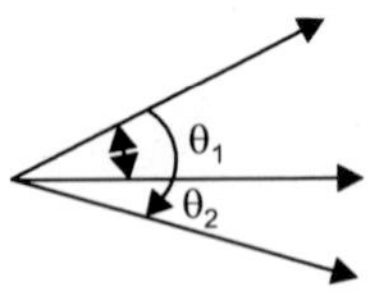

Fig. 9.10: (a) D.P.S.K. Phasor in Absence of Noise **(b)** Three Consecutive Bits

$$P(e) = \frac{1}{2}\text{erfc}\sqrt{\frac{E_b}{2N_0}}$$

This equation shows that error probability of BFSK is higher compared to that of BPSK.

9.9 ERROR PROBABILITY OF DPSK

Figure 9.10 shows the phasor diagram of DPSK signal when no noise is present. This means that in the absence of noise and transmission delay, the phase shift of the DPSK signal is either 0 or π. Therefore, a decision boundary is draws at π/2 as shown in Fig. 9.10(a). Therefore, we consider that the transmitted symbol 1 if phase difference between two consecutive bit differs by less than π/2. If phase difference between two consecutive bit differs by more than π/2, then decision is taken in favour of zero.

Figure 9.10(b) shows three consecutive bits. The first bit signal contains no noise hence its phasor is along the horizontal line. Therefore, symbol transmitted in first bit is assumed to be '1'. Because of noise, there is phase difference of 'θ_1' between first and second bit. Since $\theta_1 < \pi/2$ second bit is also taken as symbol '1'. The phase difference between second and third bit is θ_2.

From Fig. 9.10(b) it is clear that $\theta_2 > \pi/2$, hence third bit is taken as symbol zero. Since phase difference are not exact between two successive bits, some error is introduced in the detection. Therefore, DPSK system is called sub-optimum' in nature. If the synchronization is used to make phase differences exact, then it becomes PSK system.

Because of the sub-optimum nature of DPSK, the error probability is higher than that of BPSK. The average probability of error of non-coherent receiver is given as:

Here E is the energy per symbol and $\frac{N}{2}$ is spectral density of white Gaussian noise. We know that symbol duration $T = 2T_b$. Hence, energy of the symbol will be $E = 2\,E_b$. Hence, expression for P(e) becomes: $P(e) = \frac{1}{2}e^{-\frac{E_b}{N_0}}$. This is called average probability of error or bit error rate (BER) of D.P.S.K. system.

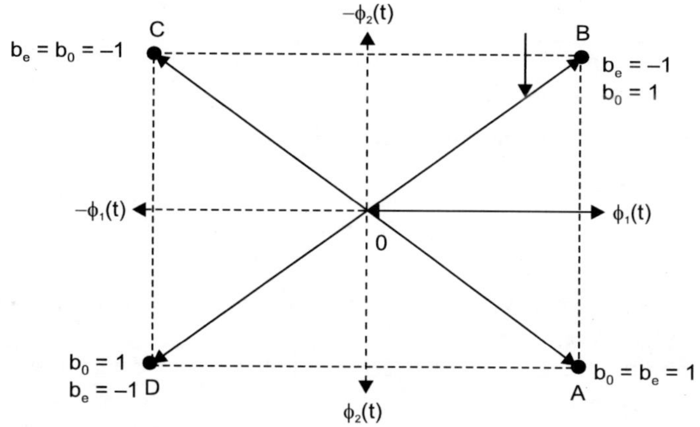

Fig. 9.11: Signal Space Representation for QPSK Signal

9.10 PROBABILITY OF ERROR OF QPSK SYSTEM

Observe Fig. 9.11 carefully. Between signal phasor 'OA' and 'OB' the axis $\phi_1(t)$ can be called decision boundary. Let us say signal vector 'OB' is present but because of imperfect phase synchronization, it is detected as 'OA'. For this to happen, the phase shift must be at least $\pi/4$. The same thing can happen in case of other phasor also.

There are two correlates in the QPSK decoder. One correlator is used to detect even bits and other detected as odd bits. Thus, any correlator can make a mistake if phase shift of $\pi/4$ occurs in the corresponding carrier. Therefore, the probability that correlator 1 or correlator 2 will mistake is given by:

$$P_1(e) = P_2(e) = \frac{1}{2}\text{erfc}\sqrt{\frac{E\cos^2\theta}{N_0}}$$

This equation has been written because each correlator is independent F.S.K. decoder 'E_b' is replaced by E in above equation, since energy of one bit in b_e or b_0 is E or E_s. Substituting phase shift $\theta = \frac{\pi}{4}$ in above equation, we obtain:

$$P_1(e) = P_2(e) = \frac{1}{2}\text{erfc}\sqrt{\frac{E\cos^2\frac{\pi}{4}}{N_0}}$$

or $$P_1(e) = \frac{1}{2}\text{erfc}\sqrt{\frac{E}{2N_0}}$$

We can verify the above relation for BPSK as under. We know that $E = E_s = P_s T_s$ and $T_s = 2T_b$

$\therefore$ $E = P_s\, 2T_b = 2P_sT_b$ and $E_b = P_s\, T_b$, and $E = 2E_b$. By using the above relation, we obtain error probability of B.P.S.K. i.e.

$$P_1(e) = P_2(e) = \frac{1}{2}\text{erfc}\sqrt{\frac{E_b}{N_0}}$$

The above equation gives bit error probability of QPSK. Thus, bit error probability of QPSK and BPSK is same.

The probability P(e) that the QPSK decoder will correctly detect the transmitted signal is equal to product of probabilities that both correlator 1 and correlator 2 will receive their bits correctly. The probability of correct reception of correlator 1 and 2 is:

$$P_1'(e) = 1 - P_1(e) \text{ and } P_2'(e) = 1 - P_2(e)$$

$$P(e) = P_1'(e) * P_2'(e) = [1 - P_1'(e)] \times [1 - P_2'(e)]$$

$$P(e) = 1 - P_1'(e) - P_2'(e) + P_1'(e) \times P_2'(e)$$

$P_1'(e) = P_2'(e)$ we can write above equation as:

$$P(e) = 1 - 2P'(e) + P'^2(e)$$

Here $$P_1'(e) = P_2'(e) = P'^2(e)$$

The term $P'^2(e)$ will be very small and can be neglected.

Hence, $P(e) = 1 - 2P'(e)$

$\therefore$ Probability of error of QPSK system is:

$$P(e) = 1 - P'(e)$$

or $$P(e) = 1 - 1 + 2P'(e)$$

or $$P(e) = 2P'(e)$$

Substituting value of $P'(e) = \frac{1}{2}\text{erfc}\sqrt{\frac{E}{2N_0}}$, we get

$$P(e) = \text{erfc}\sqrt{\frac{E}{2N_0}} = \text{erfc}\sqrt{\frac{E_b}{N_0}} (\text{since } E = 2E_b)$$

The individual probabilities $P_1'(e)$ and $P_2'(e)$ correlators are sometimes called as bit error probabilities or Bit Error Rate (BER). Thus, bit error rate of QPSK is given by:

$$\text{B.E.R.} = \frac{1}{2}\text{erfc}\sqrt{\frac{E_b}{N_0}}$$

9.11 MATCHED FILTER

For this filter, we have considered generalized Gaussian noise. When this noise is white Gaussian noise, then optimum filter is known as matched filter. For the white Gaussian noise, the power spectral density is given as:

$$S_{ni}(f) = \frac{N_0}{2}$$

9.11.1 Calculation of Impulse Response of Matched Filter

The transfer function of the optimum filter is expressed as:

$$H(f) = k \times \frac{X^*(f)}{S_{n_i}(f)} e^{-j2\pi fT}$$

If we substitute $S_{n_i(f)} = \frac{N_0}{2}$, i.e. PSD of white noise, then H(f) becomes transfer function of matched filter. This means that transfer function of matched filter

$$H(f) = k\frac{X^*(f)}{\frac{N_0}{2}} e^{-j2\pi fT}$$

or
$$H(f) = \frac{2k}{N_0} X^*(f) e^{-j2\pi fT} \qquad \text{(X)}$$

From property of Fourier transform we know that X*(f) = x(–f). Using this property, we can write eqn (X) as under

$$H(f) = \frac{2k}{N_0} X(-f) e^{-j2\pi fT}$$

The impulse response of matched filter can be evaluated by taking inverse Fourier transform of above equation.

Thus, we have $h(t) = \text{IFT}\,|\,H(f)\,| = \text{IFT}\left[\frac{2k}{N_0} X(-f) e^{-j2\pi fT}\right]$

The inverse Fourier transform of X (– f) is x (–t) and $e^{-j2\pi fT}$ represents time shift of T seconds. Hence, we have.

$$FT[x(-t)] = x(-f)$$

and $$FT[x(T-t)] = X(-f)e^{-j2\pi fT}.$$

With all these properties of Fourier transform, the eq. given below takes the form

$$h(t) = \frac{2k}{N_0} x(T-t) \qquad \text{(Y)}$$

Note that we have considered $x(t) = x_1(t) - x_2(t)$ therefore equation given below will become:

$$h(t) = \frac{2k}{N_0}[x_1(T-t) - x_2(T-t)] \qquad \text{(Z)}$$

The two equations (Y) and (Z) give the required impulse response of the matched filter.

9.12 COMPANDING

Companding is a non-uniform quantization. It is required to be implemented to improve the signal to quantization noise ratio of weak signal. We know that the quantization noise is given by $N_q = \frac{\Delta^2}{12}$. In uniform quantization, once step size is fixed, the quantization noise power remains constant. However, the signal power is not constant. It is proportional to the square of amplitude signal. Therefore, the signal to quantization ratio for weak signal is very poor. This will affect the quality of signal. Companding is a term derived from two words, i.e. compression and expansion as under.

In practice, it is difficult to implement the non-uniform quantization because it is known as in advance about the changes in the signal level. Therefore, a particular method is used. The weak signals are amplified and strong signal are attenuated before applying them to a uniform quantizer. This process is called as compression and the block that provides it is called as a compressor.

At the receiver exactly opposite is followed which is called expansion. The circuit used for providing expansion is called as an expander. The compression of signal at the transmitter and expansion at the receiver is called companding. The process of companding has been shown in the form of block diagram as Fig. 9.12 given below.

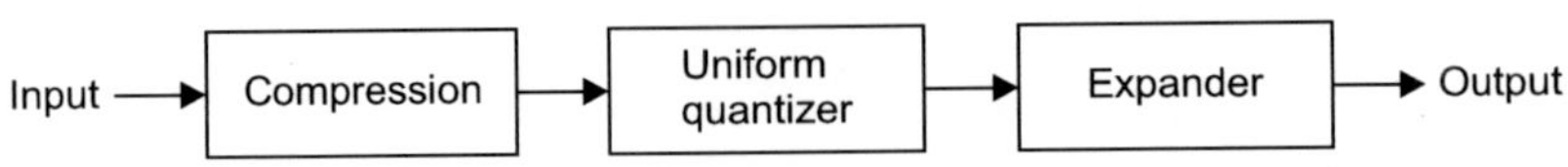

Fig. 9.12: A Companding Model

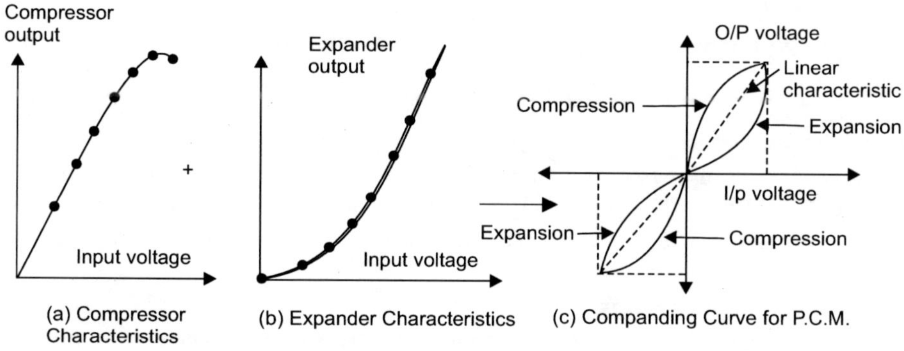

Fig. 9.13

9.12.1 Compression Characteristics

Figure 9.13(a) shows the compressor characteristics. As shown in Fig. 9.13(c), the compressor provide higher gain to the weak signal and smaller gain to strong signal input signal. Thus, weak signals are artificially boosted to improve the signal to quantization noise ratio. It may be noted that this compressor characteristics has been shown only for the positive input signal but we can draw it even for negative input signals using the same principle. In fact, the compressor is included at P.C.M. transmitter.

9.12.2 Expander Characteristics

Figure 9.13(b) shows the expander characteristics. This characteristic is exactly the inverse of the compressor characteristics. This ensures that all the artificially boosted signal by the compressor are brought back to their original amplitudes at the receiver end.

9.13 DIFFERENT TYPES OF COMPRESSOR CHARACTERISTICS

Ideally we need a linear compressor characteristics for small amplitudes of the input signal and a logarithmic characteristics elsewhere. In practice, this is achieved by using (1) μ-law companding and (ii) A-law companding.

9.13.1 *μ*-law Companding

In μ-law companding, the compressor characteristics is continuous. It is approximately linear for smaller values of input level and logarithmic of high input signal. The μ-law compressor characteristics is mathematically expressed as under:

$$z(x) = \frac{(\text{sgn } x)\ln|1+\mu|x|/X_{max})}{\ln(1+\mu)} \text{ where } 0 \leq \frac{|x|}{x_{max}} \leq 1.$$

Here z(x) represents the output and x is the input to the compressor. Also, $\frac{|x|}{X_{max}}$ represents the normalized value of input with respect to the maximum value x_{max}. Further, (sgn x) term represents ± 1.

The μ-law compressor characteristics for different values of μ have been shown in Fig. 9.4(a). The practically used value of μ is 255. For μ = 0 there is linear quantization. The μ-law companding used for speech and music signal.

9.13.2 A-law Companding

In A-law companding, the compressor characteristics is piecewise, made up of linear segment for low level amplitudes and a logarithmic for high level inputs. Figure 9.14(b) shows the A-law compressor characteristics for different values of A. For A = 1 it is seen that characteristics is linear which means uniform quantization. The practically used value of A is 87.56.

It can be mathematically expressed as:

$$\frac{Z(x)}{x_{max}} = \begin{cases} \dfrac{\dfrac{A|x|}{x_{max}}}{1+\log_e A} & \text{for } 0 \le \dfrac{|x|}{x_{max}} \le 1 \\ \dfrac{1+\log_e[A|x|/x_{max}]}{1+\log_e A} & \text{for } \dfrac{1}{A} \le \dfrac{|x|}{x_{max}} \le 1 \end{cases}$$

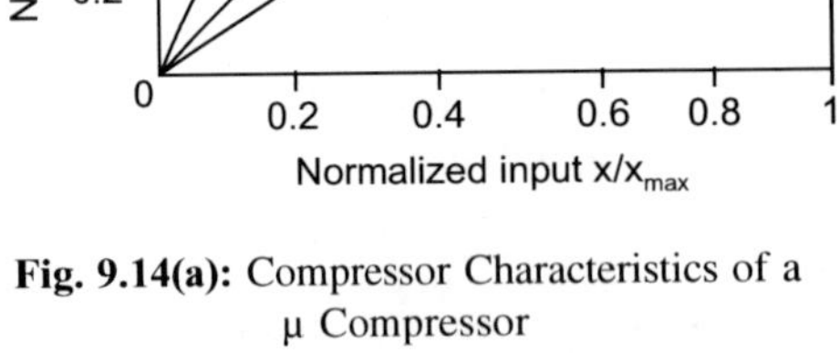

Fig. 9.14(a): Compressor Characteristics of a μ Compressor

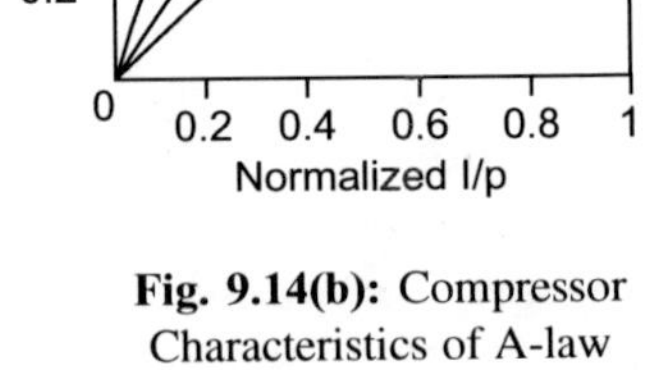

Fig. 9.14(b): Compressor Characteristics of A-law

9.14 INTER SYMBOL INTERFERENCE (ISI)

When data being transmitted in a form of pulses, i.e. bits the output produced at the receiver due to other bits interferes with output produced by the desired bit. This is known as intersymbol interference (ISI). The ISI will introduce error in detected signal. Figure 9.15 given below shows baseband binary data transmission by binary P.A.M. system. The input signal consists of a binary data sequence b with bit duration T_b.

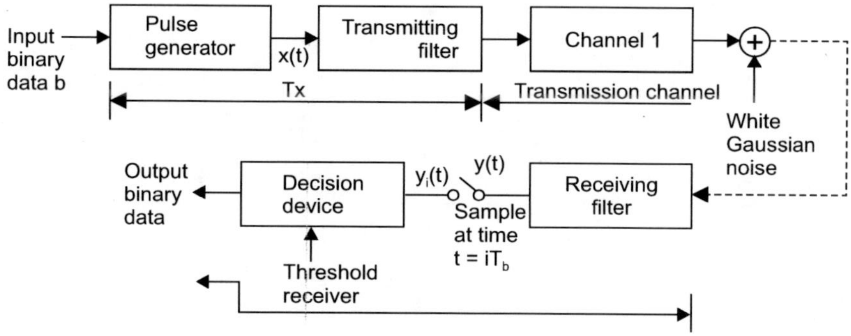

Fig. 9.15

The P.A.M. signal x(t) is then passed through a transmitting filter. The output of transmitting filter is then transmitted over transmission channel. A random noise added to transmitted signal when it travels over the transmission channel. Thus, the signal received at the receiving end is contaminated with noise.

The channel output is passed through a receiving filter. This filter is sampled synchronizly with transmitter. The sampling instants are determined by a clock or timing signal which is extracted from the receiving filter output.

The sequence of sample obtained at the output of receiving filter is used to reconstruct the original data with help of a decision-making device. Each sample is compared to a threshold level in the decision-making device. If the amplitude of sample is higher than the threshold level, then it is decided that bit 1 is received. On the other hand, if the signal has an amplitude lower than the threshold, then the decision is that a 0 is received.

I.S.I. arises due to imperfection in the overall frequency response of system. The presence of ISI will introduce error in the decision at the receiver output.

9.15 LINE CODING

9.15.1 Unipolar Return to Zero and Non-Return to Zero Code

In unipolar format, the waveform has a single polarity. This waveform can have + 5 V when high. The waveform is simple on-off. In the unipolar RZ form, the waveform has zero value when bit 0 is transmitted and waveform has 'A' volts when 1 is transmitted. In return to zero form the A volts is present for $\frac{T_b}{2}$ period if digit 1 is transmitted and for remaining $\frac{T_b}{2}$, waveform returns to zero value, i.e. for unipolar RZ form, we have.

If digit '1' is transmitted, then we have

$$x(t) = \begin{cases} A \text{ for } 0 \le t \le \frac{T_b}{2} \text{(half interval)} \\ 0 \text{ for } \frac{T_b}{2} \le t < T_b \text{(half interval)} \end{cases}$$

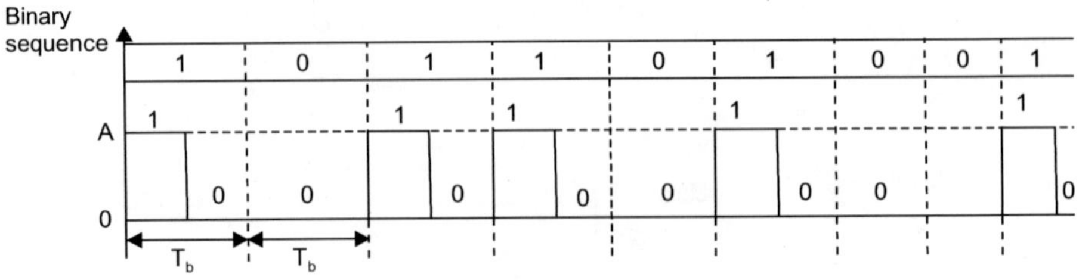

Fig. 9.16: Unipolar RZ Code

and if bit '0' is transmitted, then

$$x(t) = 0 \text{ for } 0 \leq t \leq T_b \text{ (complete interval)}$$

Hence, in unipolar RZ code, each pulse returns to a zero value. Fig. 9.16 shows this signal code. A unipolar NRZ code shown in Fig. 9.17. When bit 1 is to be transmitted, the signal has A volts for full duration. When bit 0 is to be transmitted, the signal has zero volts for complete bit duration.

Thus, for unipolar NRZ code,

If bit 1 is transmitted, we have

$$x(t) = A \text{ for } 0 \leq t \leq T_b \text{ (complete interval)}$$

If bit 0 is transmitted, we have

$$x(t) = 0 \text{ for } 0 \leq t \leq T_b \text{ (complete interval)}$$

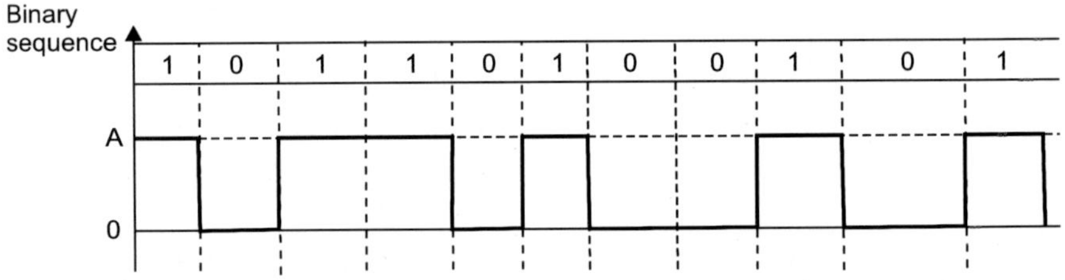

Fig. 9.17: Unipolar NRZ Code

Internal computer waveform usually of unipolar NRZ types.

9.15.2 Polar RZ and NRZ Code

In polar RZ code, bit 1 is represented by positive voltage polarity whereas bit 0 is represented by negative voltage polarity. Because this is RZ code, the pulse is transmitted only for half duration. Thus, for polar RZ, if digit '1' is transmitted then

$$x(t) = \begin{cases} +\dfrac{A}{2} \text{ for } 0 \leq t < \dfrac{T_b}{2} \\ 0 \text{ for } \dfrac{T_b}{2} \leq t < T_b \end{cases}$$

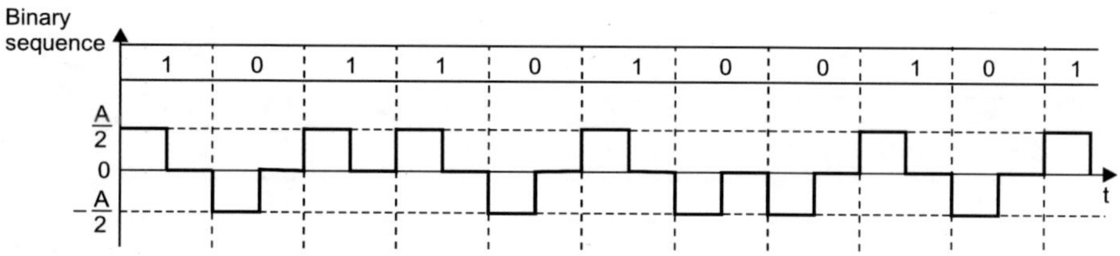

Fig. 9.18: Polar RZ Code

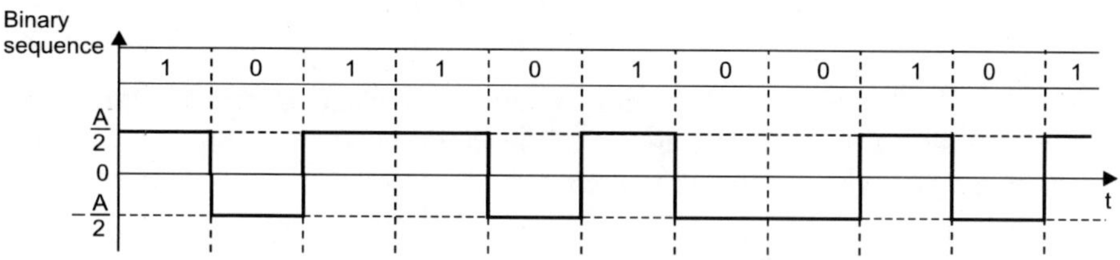

Fig. 9.19: Polar NRZ Code

and if digit 0 is transmitted, then

$$x(t) = \begin{cases} -\dfrac{A}{2} \text{ for } 0 \le t < \dfrac{T_b}{2} \\ 0 \text{ for } \dfrac{T_b}{2} \le t < T_b \end{cases}$$

Polar RZ waveform has been shown in Fig. 9.18. The polar NRZ is shown in Fig. 9.19. In polar NRZ code, bit 1 is represented by positive polarity whereas bit 0 is represented by negative polarity. The polarities are maintained over the complete pulse duration, i.e. for polar NRZ, we have

If bit '1' is transmitted, then

$$x(t) = +\frac{A}{2} \text{ for } 0 \le t < T_b$$

and if digit '0' is transmitted, then

$$x(t) = -\frac{A}{2} \text{ for } 0 \le t < T_b$$

Since polar RZ and NRZ code are bipolar, therefore, the average D.C. values is minimum in these waveforms. If probabilities of occurrence of bits 1 and 0 are same then D.C. components of waveform would be zero.

9.15.3 Bipolar NRZ [Alternate Mark Inversion] (AMI)

In this code, the successive 1's are represented by pulses with alternate polarity and '0' are represented by no pulses. From Fig. 9.20 shows the Bipolar NRZ or

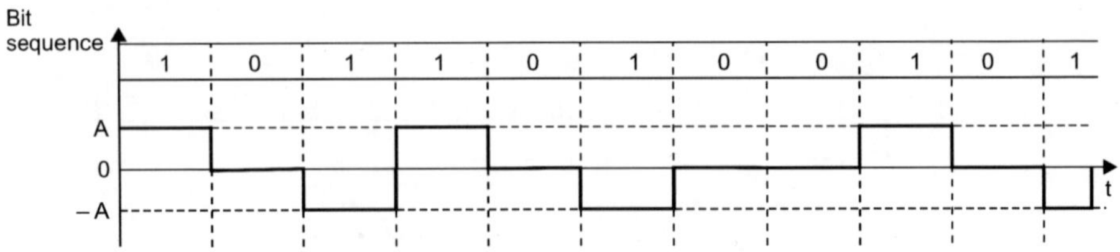

Fig. 9.20: Bipolar NRZ Code (AMI Code)

AMI waveform. If there are even number of 1's, the DC components of the waveform would be zero. The advantage of this code is that the ambiguities due to transmission sign inversion are eliminated.

9.15.4 Split Phase Manchester Code

This type of code waveform shown in Fig. 9.21. If bit 1 is to be transmitted, then a positive half interval of pulse is followed by a negative half interval of pulse. If bit 0 is to be transmitted, then a negative half interval pulse is followed by a positive half interval pulse. Hence, for any digit the pulse takes positive as well as negative value.

If bit '1' is to be transmitted, then

$$x(t) = \begin{cases} \dfrac{A}{2} & \text{for } 0 \le t < \dfrac{T_b}{2} \\ -\dfrac{A}{2} & \text{for } \dfrac{T_b}{2} \le t < T_b \end{cases}$$

and if bit 0 is to be transmitted then

$$x(t) = \begin{cases} -\dfrac{A}{2} & \text{for } 0 \le t < \dfrac{T_b}{2} \\ \dfrac{A}{2} & \text{for } \dfrac{T_b}{2} \le t < T_b \end{cases}$$

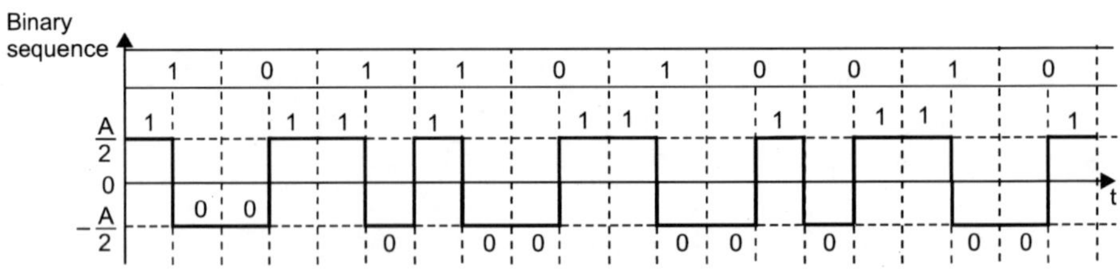

Fig. 9.21: Split Phase Manchester Code

The primary advantage of this code is that irrespective of probability of occurrence of digit 1 and 0 the waveform has zero average value. Therefore by this mode the power saving is quite more. However, the drawback of this code is that it requires absolute sense of polarity at the receiver end.

QUESTIONS

9.1. What is meant by synchronization in relation to digital transmission?
9.2. Explain matched filter.
9.3. Explain an integrate-and-dump detector used for the detection of unipolar pulse in noise.
9.4. With help of waveforms explain two level code and multilevel code.
9.5. What is meant by compander and expander?
9.6. Explain what is meant by ASK.
9.7. Explain with help of waveform FSK.
9.8. With help of waveform explain PSK.
9.9. With help of waveform explain QPSK.
9.10. Draw block diagram of digital modulation system and explain each block in detail.
9.11. Explain inter symbol interference and draw block diagram.
9.12. With the help of diagram explain μ-law and A-law of companding.
9.13. What is meant by companding?
9.14. Explain what is meant by optimum filter?
9.15. Explain the difference between coherent and noncoherent detection of OOK signals.
9.16. Explain coherent and optimum noncoherent detection.
9.17. With the help of waveform, explain AMI waveform.
9.18. Draw the waveform and the corresponding eye diagram.
9.19. Explain what is meant by coding gain.
9.20. Explain what is meant by carrier recovery.
9.21. Show how carrier may be recovered from a QPSK signal by the use of a frequency quadrupler circuit.
9.22. What is meant by phase ambiguity and how this may be corrected?
9.23. Draw the block diagram of QPSK demodulator.

PROBLEMS

9.1. For a binary waveform, the received signal to noise voltage ratio is $\frac{V_s}{V_n} = 4$.

Determine values of (a) erf $\left(\frac{4}{\sqrt{2}}\right)$, (b) erfc $\left(\frac{4}{\sqrt{2}}\right)$ and (c) the bit-error probability.

9.2. For a binary unipolar signal, the received signal voltage has a maximum value of 4 mV. The r.m.s. noise voltage is 0.5 mv. Determine the bit-error probabilities.

9.3. Compare the bit-error probabilities for polar and unipolar transmissions for which the received $\left(\frac{S}{N}\right)$ ratio is 9 dB in each case.

9.4. A binary unipolar signal has an average power of 6 pw, and the pulse duration of 0.02 μs. The equivalent noise temperature at the receiver input is 550 kHz. Determine the bit-error probability.

9.5. Calculate the bit-error probability for OOK using (a) synchronous carrier demodulation and (b) nonsynchronous carrier demodulation when the bit energy to noise density ratio is 10 dB.

9.6. A digital signal utilizes 8-bit codewords. Calculate the probability of a received codeword containing three errors, given that the bit-error probability in transmission is 0.01.

9.7. For a binary transmission link, the uncoded $\frac{E_b}{N_0}$ ratio is 9 dB, and matched filtering with polar transmission is used. Calculate bit-error rate.

CHAPTER 10

Random Process

10.1 PROBABILITY RANDOM VARIABLE

A function which can take on any value from the sample space and its range is some set of real numbers is called a random variable of an experiment. Random variables are denoted by upper case letters such as X, Y, etc. and values taken by them are denoted by lower case letters with subscripts such as x_1, x_2, y_1, y_2, etc.

Random variables may be classified as:

(1) Discrete Random Variable

(2) Continuous Random Variable

(1) **Discrete random variable:** A discrete random variable may be defined as the random variable which can take on only finite number of values in a finite observation interval. This means that the discrete random variable has countable number of distinct values.

(2) **Continuous random variable:** A random variable that takes infinite number of values is called a continuous random variable. Continuous random variable may be used to define the output of such system.

The noise voltage generated by an electronic amplifier has continuous amplitude. This means that sample space S of noise voltage amplitude is continuous. Therefore, in this case, the random variable X has a continuous range of values.

10.2 PROBABILITY DISTRIBUTION OF DISCRETE RANDOM VARIABLE

Let X be discrete random variable and also let $x_1, x_2, x_3, \ldots$ be the values that x can take. Then $P(x = x_j) = f(x_j)$ where $j = 1, 2, 3, \ldots$ will be probability of x_j.

This $f(x_j)$ or simply f(x) is called the probability function or probability distribution of the discrete random variable.

10.3 PROBABILITY DENSITY FUNCTION (PDF)

Probability density function (PDF) is generally denoted by $f_x(x)$. Mathematically, PDF may be expressed as: PDF: $f_x(x) = \frac{dF_x}{dx}(x)$ where x is dummy variable.

Probability Density Function (PDF) is the more convenient representation for continuous random variable.

Property (1): Probability density function (PDF) is always non-zero for all values of x. Mathematically $f_x(x) \geq 0$ for all values of x.

Property (2): The area under the PDF curve is always equal to unity. Mathematically

$$\int_{-\infty}^{+\infty} f_x(x) \cdot dx = 1$$

Property (3): The cumulative distribution function (CDF) may be obtained by integrating probability density function (PDF). Mathematically,

$$F_x(x) = \int_{-\infty}^{+\infty} f_x(x) \cdot dx$$

Property (4): Probability of event $(x_1 < x \leq x_2)$ is simply given by the area under the probability density function (PDF) curve in the range $x_1 < x \leq x_2$. Mathematically,

$$P(x_1 < x \leq x_2) = \int_{x_1}^{x_2} f_x(x) \cdot dx$$

10.4 MEAN OR AVERAGE

The mean or average of any random variable is expressed by the summation of the values of random variable X weighted by their probabilities. When value of a random variable is denoted by m_x.

Mean value is also known as expected value of random variable X.

$$m_x = E(X)$$

where E[ ] represents expectation operator. Therefore, both m_x and E(X) have the same meaning.

Also, in general term, mean or average is given by:

$$\text{Mean or average of } X = \frac{\text{Arithmetic sum of all values of } X}{\text{Total number of values of } X}$$

10.5 MOMENTS AND VARIANCE

The *n*th moment of any random variable X may be defined as the mean value of X^n, i.e. $g(x) = X^n$ then the mean is written as: mean = $\overline{g(x)}$

$$\overline{x^n} = \int_{-\infty}^{+\infty} x^n f_x(x) \cdot dx = E(X^n)$$

$$\overline{x^n} = \int_{-\infty}^{+\infty} x^n f_x(x) \cdot dx$$

Substituting n = 1 the above equation becomes

$$\bar{x} = E(x) = \int_{-\infty}^{+\infty} x^n f_x(x) \cdot dx$$

or

$$\bar{x} = m_x$$

Hence, the first moment of random variable x will be same as its mean value.

$$\bar{x}^2 = E(x^2) = \int_{-\infty}^{\infty} x^2 f_x(x).dx$$

where $\bar{x}^2$ is known as mean square value of random variable x.

Similarly, the central moments are the moments of the difference between random variable x and its mean m_x. Therefore, the *n*th central moments may be given as:

$$E[(x - m_x)^n] = \int_{-\infty}^{+\infty} (x - m_x)^n f_x(x) \cdot dx$$

The second central moment for n = 2 is known as variance of random variable x i.e.

$$\text{variance } [x] = E\ (x - m_x)^2\ f_x(x) \cdot dx$$

The variance provides an indication about randomness of random variable. Variance is generally represented by σ_x^2, i.e.

$$\sigma_x^2 = \text{variance } (x)$$

$$\sigma_x^2 = \text{var}(x) = E(x - m_x)^2 = \int_{-\infty}^{+\infty} (x - m_x)^2 f_x(x) \cdot dx \qquad (1)$$

Simplifying eqn. (1), we get:

$$\sigma_x^2 = E[x^2 - 2m_x + m_x^2] \because (x - m_x)^2 = x^2 + m_x^2 - 2m_x x$$

The E [ ] operator represents mean or average value, therefore, it is linear. Thus,

$$\sigma_x^2 = E[x^2] - 2m_x E(x) + m_x^2$$

$$\sigma_x^2 = E[x^2] - 2m_x m_x + m_x^2$$

$$\sigma_x^2 = E[x^2] - m_x^2$$

σ_x^2 = mean square value – square mean value

The above equation may also be written as:

$$\text{variance} = \sigma_x^2 = \overline{x}^2 = m_x^2$$

10.6 GAUSSIAN OR NORMAL DISTRIBUTION

Gaussian distribution is also known as normal distribution. It is defined for a continuous random variable. The PDF for a Gaussian random variable is expressed as:

$$f_x(x) = \frac{1}{\sigma\sqrt{2\pi}} e^{-(x-m)^2/2\sigma^2}$$

In this equation, m = mean value of random variable, σ^2 = variable of random variable. Figure 10.1 shows the plot of Gaussian PDF $f_x(x)$ with respect to x.

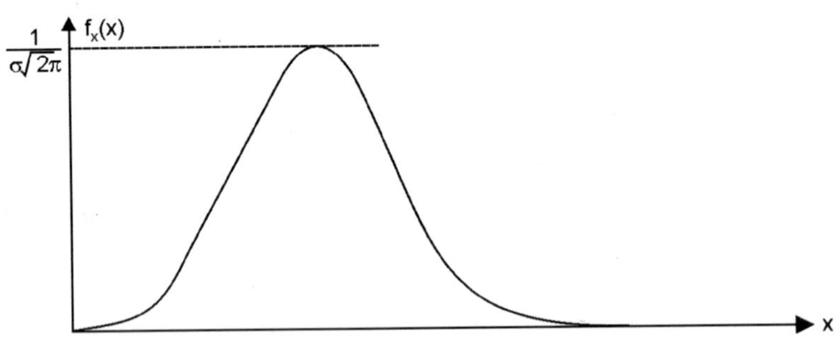

Fig. 10.1: Plot of Gaussian PDF

There are several properties of Gaussian PDF.

Property (1): The peak value occurs at x = m. Mathematically,

$$f_x(x) = \frac{1}{\sigma\sqrt{2\pi}} \text{ at } x = m$$

Property (2): Gaussian PDF exhibit even symmetry around mean value. Mathematically,

$$f_x(m - \sigma) = f_x(m + \sigma)$$

Property (3): The area under PDF curve is $\frac{1}{2}$ for all values of x below mean value and $\frac{1}{2}$ for all values of x above mean value. Mathematically,

$$P(x \leq m) = P(x > m) = \frac{1}{2}$$

Property (4): As $\sigma \to 0$ the Gaussian function approaches to impulse function located at x = m.

10.7 STATIONARY AND NON-STATIONARY PROCESS

A random process x(t) is called stationary if its statistics are not affected by any phase shift in the time origin. We may define stationary process in terms of ensembles averages as:

(1) Ensemble mean is independent of time. Mathematically, $m_x(t) = m_x(t_1) = m_x(t_2) = m_x(t_3)$ = constant at all time instants.

(2) Auto correlation function $R_\tau(t_1, t_2)$ depends only upon time difference $t_2 - t_1$.

$$R_\tau(t_1, t_2) = R_\tau(t_1 + t, t_2 + t)$$

Therefore, autocorrelation function of any stationary process is a function of time difference and is given as: $R_\tau(t_1, t_2) = R_\tau(t_2 - t_1) = R_\tau(\rho)$ where $\rho = t_2 - t_1$.

Thus, system register producing thermal noise voltages represents a stationary process due to the fact that the noise generated by resistor does not depend on time.

10.8 ERGODIC PROCESS

A random process is known as ergodic process if the time average are equal to ensemble averages. Hence, for an ergodic process, we have

$$m_x = \langle m_x \rangle,\ R_\tau(t_1, t_2) = \langle R_\tau(\varepsilon) \rangle$$

For any ergodic process, all ensemble averages will be equal to corresponding time averages for any particular sample function. The time averages are not a function of time. When time and ensembles averages are the same, it implies that ensembles averages also are not function of time. Thus, an ergodic process is always stationary but converse is not true.

Ergodicity of process may be definded in terms of some statistical averages like mean and autocorrelation.

The random process is ergodic in the mean if $m_x = \langle m_x \rangle$ and variance of $\langle m_x \rangle \rightarrow 0$ as $T \rightarrow \infty$. Similarly, the random process is ergodic in the autocorrelation if $R_\tau(t_1, t_2) = \langle R_\tau(\varepsilon) \rangle$ and variance of $\langle R_\tau(\varepsilon) \rangle \rightarrow 0$ as $T \rightarrow \infty$.

Hence, the ergodicity of any random process may be determined by evaluating statistical averages of single sample function. This means that a single sample function represents entire random process.

10.9 AUTOCORRELATION AND CROSS-CORRELATION FUNCTION

Autocorrelation function may be defined as a measure of similarity between a signal or process and its replica by a variable amount.

The autocorrelation function of stationary process x(t) may be defined as:

$$R_x(t_j - t_i) = E[x(t_j) \times (t_i)] \text{ for any } t_j \text{ and } t_i \tag{2}$$

where $x(t_j)$ and $x(t_i)$ are the random variables obtained by observing the process x(t) at time t_j and t_i respectively. Also, the autocorrelation function depends only on the time difference $(t_j - t_i)$. Using $\tau = t_j - t_i$ in eqn. (2), we get $R_x(\tau) = E[x(t)\cdot x(t-\tau]$. The expression for x(t) and $x(t-\tau)$ are viewed as random variable. The variable τ is known as time lag or time delay parameter. The autocorrelation function for stationary random process is independent of a phase shift of time origin.

10.10 CROSS-CORRELATION FUNCTIONS

An autocorrelation function is determined for single random process but cross-correlation function is determined for two random processes. Let us consider two random processes x(t) and x(t) with autocorrelation functions $R_x(t, u)$ and $R_y(t, u)$ respectively. There will be two cross-correlation functions between two random process x(t) and y(t). These cross-correlation function of two random processes x(t) and y(t). These cross-correlation may be defined as:

$$R_{xy}(t, u) = E[X(t)\, Y(u)] \text{ and } R_{yx}(t, u) = E[Y(t) \cdot X(u)].$$

where t and u are the two values for time at which the process are observed. These cross-correlation function of two random processes X(t) and Y(t) may be expressed conveniently in matrix form as:

$$R(t, u) = \begin{bmatrix} R_X(t, u) & R_{XY}(t, u) \\ R_{YX}(t, u) & R_Y(t, u) \end{bmatrix} \tag{a}$$

R(t, u) is called the correlation matrix of random processes x(t) and Y(t).

Equation $s_X(\omega) = \int_{-\infty}^{+\infty} R_X(\tau)\cdot e^{-j\omega\tau}\cdot d\tau$ may also be written as

$$R(t, u) = \begin{bmatrix} R_X(t-u) & R_{XY}(t-u) \\ R_{YX}(t-u) & R_Y(t-u) \end{bmatrix} \tag{b}$$

If both random process x(t) and y(t) are wide sensy stationary processes, then equation (a) can be written as eqn. (b). In addition, both random processes x(t) and y(t) are said to be joint wide sense stationary.

It may be noted that in general the cross-correlation function is not an even function of τ like an auto-correlation function. It is having following symmetry relation:

$$R_{xY}(\tau) = R_{Yx}(-\tau)$$

Moreover, cross-correlation does not have a maximum at the origin like the autocorrelation function. The two random processes x(t) and y(t) are said to be incoherent or orthogonal if cross-correlation function of x(t) and y(t) is zero.

$$R_{xY}(\tau) = 0$$

The two random processes are said to be non-correlated, if their cross-correlation functions $R_{xY}(\tau)$ are equal to multiplication of mean values.

$$R_{xY}(\tau) = E[x(t)] \cdot E[y(t)]$$

The incoherent or orthogonal processes are non-correlated process with E[x(t)] and/or E[Y(t)] = 0.

10.11 POWER SPECTRAL DENSITY (PSD)

Power spectral density $s_x(\omega)$ of a wide-sense stationary (WSS) random process x(t) may be defined as:

$S_x(\omega)$ = Fourier transform $\{R_x(\tau)\}$

$$S_x(\omega) = \int_{-\infty}^{+\infty} R_x(\tau) \cdot e^{-j\omega\tau} \cdot d\tau \qquad (*)$$

where $R_x(\tau)$ = Autocorrelation function of random process x(t).

Thus, we can say that the Fourier transform of autocorrelation function $R_x(\tau)$ of a random process x(t) is called power spectral density (PSD) of random process. Equation (*) was initially developed by Albert Einstein. Later it was developed by N. Weiner and Khenchine. Hence, this equation is known as Einstein-Weiner Khenchine (EWK) equation. Power spectral density (PSD) is used to measure the energy of signal in frequency band i.e. $[\omega, \omega + d\omega]$.

10.12 RESPONSE OF LINEAR SYSTEMS TO RANDOM SIGNAL

We shall study transmission of random process through linear system or liner filters.

Suppose that a random process x(t) is applied to the input of a linear time invariant (LTI) system having an impulse response h(t). Let the input random process x(t) produces a random process Y(t) at the system output. The block diagram of this LTI system has been shown in Fig. 10.2.

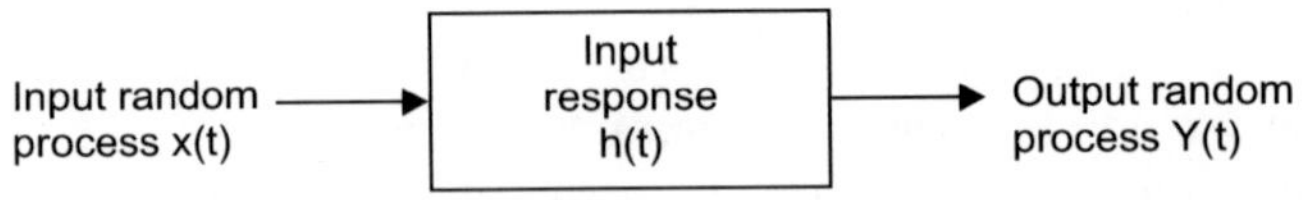

Fig. 10.2: Response of LTI System to Random Signal

The mean output of random process is expressed as

$$m_Y(t) = E[Y(t)] = E\int_{-\infty}^{+\infty} [h(T)X(t-\tau)\cdot d\tau]$$

or
$$m_Y(t) = \int_{-\infty}^{+\infty} h(\tau)E[x(t-\tau)]\cdot d\tau = \int_{-\infty}^{+\infty} h(\tau)m_x(t-\tau)\cdot d\tau \quad (1)$$

Random process x(t) is stationary then mean $m_x(t)$ is a constant m_x. Hence, equation (1) is written as:

$$m_Y = \int_{-\infty}^{+\infty} h(\tau)m_x d\tau = m_x \int_{-\infty}^{+\infty} h(\tau)\cdot d\tau = m_x H(0)$$

Here H(0) is zero frequency response of the system.

We can say that the mean of the output random process of a stable linear time invariant system is equal to the mean of input random process multiplied by the zero frequency response of the system.

10.13 CENTRAL LIMIT THEOREM

We can conclude this chapter with an extremely useful theorem concerning the cdf of a sum of random variable in the limit as the number of terms in the sum approaches infinity. We shall explain the theorem for the case in which the random variable x_i, i = 1, 2, ..., n being summed are statistically independent and identically distributed, each having a finite mean m_x and a finite variance σ_x^2. For convenience we define the normalised random variable.

$$U_i = \frac{x_i - m_x}{\sigma_x}, \; i = 1, 2, \ldots, n$$

Thus, U_i has zero mean and unit variance; it follows that normalized random variable Y has zero mean and unit variance.

We wish to determine the cdf of Y in the limit as $n \to \infty$.

The characteristics function Y is

$$4_Y(jv) = E(e^{jvY}) = E\left[\exp.\frac{\left(jv\sum_{i=1}^{n} u_i\right)}{\sqrt{n}}\right]$$

whose solution by talor's series and simplification and manipulation comes to be

$$\ln \psi_Y(jv) = n\left[-\frac{v^2}{2n} + \frac{R(V, n)}{n} - \frac{1}{2}\left(-\frac{v^2}{2n} + \frac{R(V,n)^2}{n} + \ldots\right)\right] \quad (2)$$

Finally, when we take the limit as $n \to \infty$ eq. (2) reduces to $\lim_{n\to\infty} \ln \psi_Y(jv) = -\frac{1}{2}V^2$

or equivalently,

$$\lim_{n\to\infty} \psi_Y(jv) = e^{-\frac{v^2}{2}}$$

But, this is just the characteristics function of a gaussian variable with zero mean and unit variance. Thus, we have important result that the sum of statistically independent and identically distributed random variable with finite mean and variance approaches a gaussian cdf as $n \to \infty$. This result is known as the central limit theorem.

QUESTIONS

10.1. What is meant by random variable, probability distribution?
10.2. Define term mean, moments and variance.
10.3. Define term CDF, PDF and PSD.
10.4. Explain stationary, nonstationary and ergodic process.
10.5. Define cross-correlation and autocorrelation function.
10.5. Define central limit theorem.

PROBLEMS

10.1. The PDF of a random variable x is P(x). A random variable Y is defined as $Y = ax + b$ where $a < 0$. Determine PDF of Y in terms of the PDF of x.

10.2. Suppose that x is a Guassian random variable with zero mean and unit variance, let

$$Y = ax^3 + b \quad a > 0$$

Determine and plot the pdf of Y.

10.3. The random variable Y is defined as

$$Y = \sum_{i=1}^{n} x_i$$

where $x_i = 1, 2, \ldots, n$ are statistically independent random variables with

$$x_i = \begin{cases} 1 \text{ with probability } p \\ 0 \text{ with probability } 1 - p \end{cases}$$

(a) Determine characteristics function of Y.

(b) From characteristics function, determine the moments E(Y) and $E(Y^2)$.

10.4. Let x(t) be a stationary real normal process with zero mean. Let a new process Y(t) be defined by

$$Y(t) = x^2(t)$$

Determine the autocorrelation function of Y(t) in terms of autocorrelation function of x(t).

CHAPTER 11

Fundamental of Data Communication

11.1 A COMMUNICATION MODEL

We begin our study with simple model of communication shown by block diagram shown in Fig. 11.1. The main purpose of a communication system is to exchange of data between two parties. Figure 11.2 presents one particular example, which is the communication between a workstation and a server. Workstation to server connected by public telephone lines. Another example is the exchange of voice signals between two telephones over the same network.

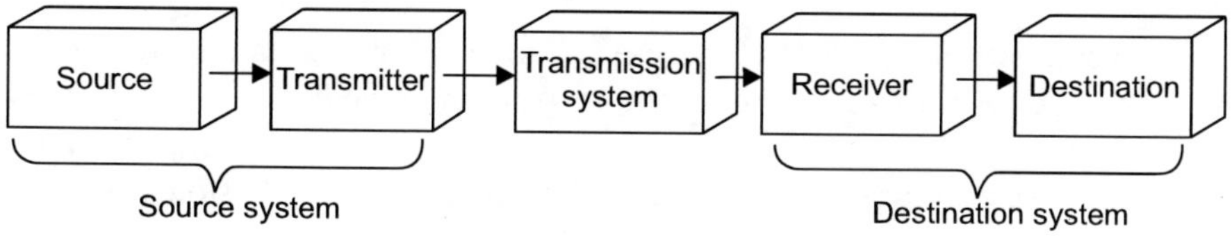

Fig. 11.1: General Block Diagram

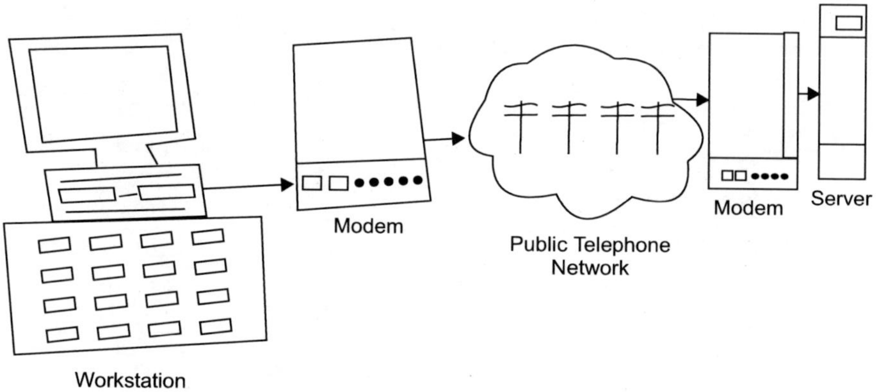

Fig. 11.2: Simplified Communication Model

The key elements of model are:

(1) Source: The device generates the data to be transmitted. For example, telephones and personal computers.

(2) Transmitter (Modem): For example, a modem takes a digital bit streams from an attached device such as personal computer and transforms that bit stream into an analog signal; that can be handled by the telephone network.

(3) Transmission System: This can be a single transmission line or complex network connecting source and destination.

(4) Receiver (Modem): The receiver accepts the signal from the transmission system and converts it into a form that can be handled by the destination device. For example, a modem will accept an analog signal coming from a network or transmission line and converts it into a digital bit stream.

(5) Destination: Takes incoming data from the receiver.

11.2 DATA COMMUNICATION

The most fundamental aspect of the communication function, focusing on the transmission of signal in a reliable and efficient manner.

Simplified data communication model is shown in Fig. 11.3(a). Let us understand details of Fig. 11.3(b) using electronics mail as an example.

Consider that the input device and transmitter are components of a personal computer.

(1) Workstation (modem): The user of the personal computer wishes to send a message to another user: For example, "meeting scheduled for March 25 is cancelled" (m). The user activates the electronics mail package on the personal computer and enter message via keyboard. The character string is stored in main memory. Character stored observed as bit streams.

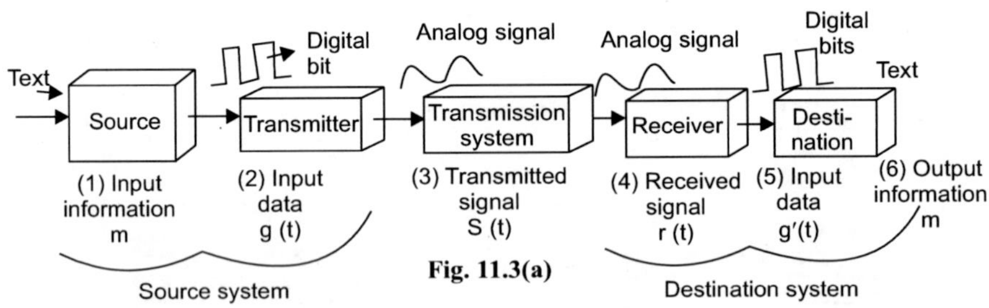

Fig. 11.3(a)

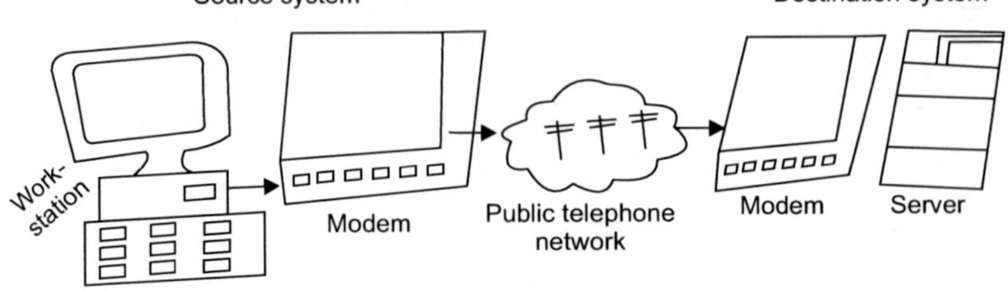

Fig. 11.3(b): Simplified Data Communication Model

(2) Transmission medium: The personal computer is connected to some transmission medium such as a local network or telephone line by an input device that is transmitter. Such as local network transceiver or a modem. The input data are transferred to the transmitter as a sequence of voltage shift g(t) representing bits on same communication bus or cable. The transmitter is connected directly to the medium and converts the incoming stream g(t) into a signal s(t) suitable for transmission.

(3) Receiver (modem): The received signal r(t) may differ to some degree from s(t). The receiver will try to recover the original signal s(t) based on r(t) and its knowledge of the medium, producing a sequence of bits g′(t).

(4) Destination: These bits are sent to the output personal computer where they are stored in memory as a block of bits (g).

11.3 CIRCUIT SWITCHING

In a circuit switched network, a communication path is established between two suitable stations through the nodes of network. That path is connected sequence of physical links between nodes. On each link a logical channel is dedicated to connection.

For example, to send message "meeting scheduled for March 25 is cancelled" (m) along the dedicated path as rapidly as possible. At each node, incoming data are routed or switched to the appropriate outgoing channel without delay. The most common example of circuit switching is the telephone network.

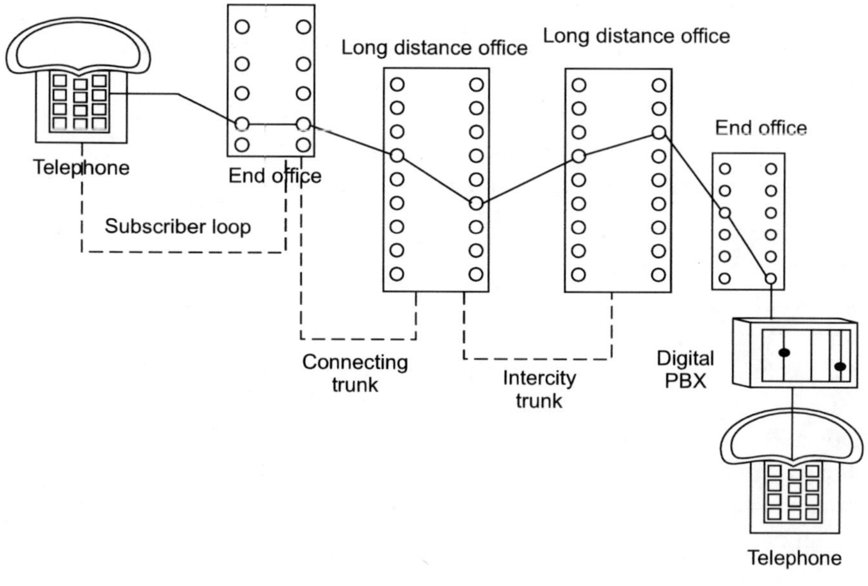

Fig. 11.4: Public Circuit Switching Network

(1) Subscriber: The device that attached to the network. Most subscriber devices to public telecommunication networks are telephone.

(2) Local loop: The link between the subscriber and the network; also referred to as the subscriber loop. Almost all local loop used twisted pair of wires.

(3) Exchange: A switching centre that directly supports subscriber is known as end office. Typically, an end office will support many thousands of subscribers in a localized area.

(4) Trunks: The branch between exchange. Trunks carry multiple voice frequency circuits using either frequency division multiplexed or synchronous time division multiplexed.

Following are the tasks, that must be performed in data communication system. These tasks must be performed at different levels.

Also following facts must be remembered, while transmitting data over a communication system or transmission system.

There is no difference between data processing (computers) and data communication (transmission and switching equipment). There are no differences among data, voice and video communications. Single processor computer, multiprocessor computer, local network, metropolitan network and long-haul network technology are now easily available.

(1) Transmission System Utilization: Transmission system, communicating devices must be reliably and efficiently utilized. For efficient utilization, multiplexing techniques are employed.

(2) Interface: To have reliable communication, communication devices must be properly interface with communication system, that is channel. Form of propagation over a communication channel may be wireless that is electromagnetic waves or wire communication.

(3) Signal generation: Once an interface is established, for efficient communication signal should be generated from source.

Form and intensity of generated signal should be such that:

Signal should propagate over a transmission channel. Data should be derived from receiver.

(4) Synchronization: There should be some form of synchronization between transmitter and receiver.

(5) Exchange management: Once the nature and timing of signal is decided, there are variety of requirement for communication between two parties that might be known as Exchange management.

For example, for two parties to engage in a telephone conversation, only party must dial the number of the other, causing signals to be generated that result in the ringing of called party phone. The called party completes a connection by lifting the receiver.

(6) Error detection and correction: In communication system transmitted signal distorted while transmitted over a channel and where error can not be tolerated there is need of error detection and correlation.

(7) Flow control: Data transfer should not be greater than generation and processing of data. Above-mentioned fact is possible by flow control.

(8) Addressing and routing: Whenever communication system communicate more than two devices then we should know the address of each device. Also specific route through this network must be chosen.

(9) Recovery: It is related with fault in network. When fault in network is corrected it is known as recovery.

(10) Message formatting: Source and destination should use same binary code for character.

(11) Security: In data communication system, it is important to provide security to the communication network.

So, for proper arrangement of network for communication data there should be network management.

11.4 PACKET SWITCHING

In packet switching network switching methodology is absolutely different than circuit switching. In this case how many signals are transmitted through the path of network are not important. That is network capacity is not important but type of switching is important. Packet switching is also known as software switch. In this case, data are sent out in sequence of small chunks called packets. Each packet is passed through the network from node to node along some path leading from source to destination. At each node, entire packet is received, stored and then transmitted to next node. Packet switched networks are commonly used for computer to computer communication.

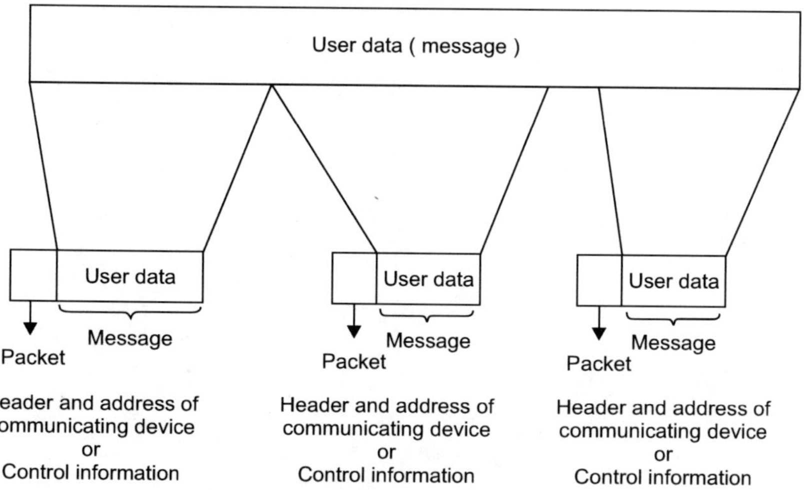

Fig. 11.5: Header and Used Data Within Packet

Figure 11.5 shows header and address of communicating device and also user's data known as message. Data are transmitted in short packet. If source has a longer message to send, the message is broken up into a series of packet.

Each packet contains a portion of the user's data or message plus some control information. The control information contains the address of communication device that is packet header as well as control information. Depending upon address of communication device route of network is decided.

11.5 DATA COMMUNICATION NETWORKING

When two communicating devices are connected by transmission medium then communication takes place between two communicating devices. It is impractical to connect two communicating devices by transmission medium because of following reasons:

The devices are very far apart. To install thousands of miles apart bus becomes expensive. To connect many devices or n devices there is need to connect many buses or n buses it is also impractical. Solution to this problem are as follows:

Connect source system to destination system through communication network. Figure 11.6 shows source system connected to destination system through communication network.

Source system connected to destination system through a communication network which is known as local area network or LAN. Wide area network, that is communication network is LAN or WAN.

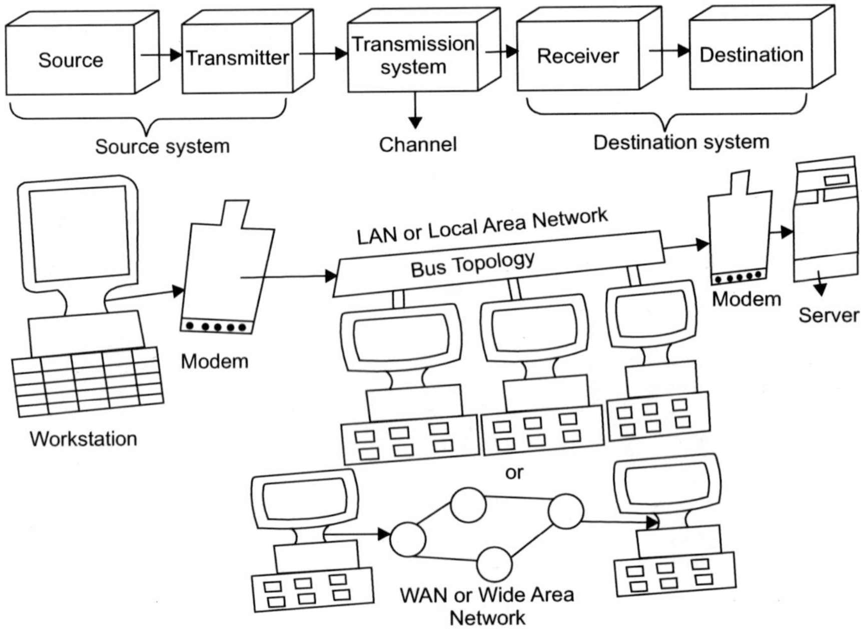

Fig. 11.6: Simplified Network Models

Local area network (LAN): Computer network within single organization is known as local area network. It is practically not possible to provide a wire between two computers without use of very few devices. For example, devices like server.

Wide Area Network: WAN covers a large geographical area. WAN consists of a number of interconnected switching nodes. Suppose you have to make transmission from source system to destination system and source system connected to destination system through a network known as wide area netowrk. Then message is routed through internal nodes present within WAN. Thus, transmission takes place between source system to destination system. These nodes provide switching facility but nodes are not related with data. Thus, data passes from source system to destination through a network known as WAN, in which switching facility provided by nodes. In WAN circuit switching or packet switching technology is provided.

11.6 PROTOCOL STRUCTURE IN NETWORK

Whatever data communication, data communication networking is seen, its scope is much broader and it is not limited to whatever we have already studied. It is related with computers, terminals and or other data processing devices for exchange of data.

Suppose we have to transfer a file between two computers then data path is established between two computers directly or via a communication network. Given task performed in following steps:

(1) The source system activate the direct data communication path or information given communication network of the identity of desired destination system.

(2) The source must understand that the destination system is ready to receive data.

(3) The file transfer application on the source system must understand that the file management program on the destination system is ready to accept and store the file for this particular user.

(4) If file formats used on the two systems are not compatible one or other system must perform a format translation function.

There must be high degree of cooperation between two computer systems. The exchange of information between computers for the purpose of cooperative action is generally referred to as computer communication. When two or more computers are interconnected through a communication network, such a arrangement is known as a computer network.

While discussing computer communication and computer networks, two concepts are important:

- protocols
- computer communication architecture or protocol architecture

While you have to communicate between different system's entities then protocols are used. Examples of entities or communicating entities are computer program file transfer packages, database management system, electronics mail facilities.

Examples of system are computers, terminals and remote sensors. In general, an entity is anything capable of sending or receiving information and a system is physically different object that contain one or more information entities. For two entities to communicate successfully, they must "speak the same language" that is binary. What is communicated, how it is communicated and when it is communicated must confirm to some mutually acceptable conventions between the entities involved. The conventions are referred to as protocols. Protocols is defined as a set of rules set up for the exchange of data between two entities. The main elements of protocols are:

- Syntax: It includes the things like data format and signal levels.
- Semantics: It includes control information for coordination and error handling.
- Timing: It includes speed matching and sequencing.

Concepts of protocol architecture can be understood from Fig. 11.7.

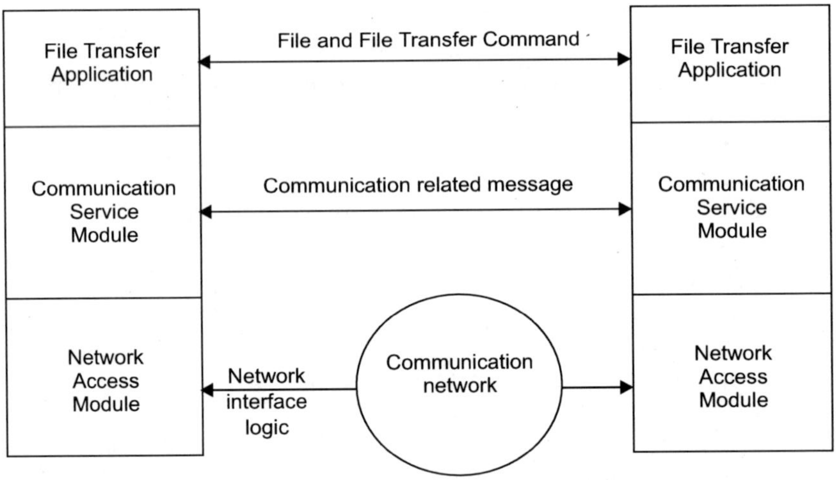

Fig. 11.7: A Simplified Architecture of File Transfer

To understand concepts of protocol, there should be very good cooperation between two computers. Instead of implementing the logic for this as a single module, the task is broken into sub-tasks and each task is implemented separately.

As an example given in Fig. 11.7 gives information how a file transfer facility could be implemented.

(1) File transfer module: When file transfer module transmit file record and file commands, password reliably between computer 1 and computer 2. Then file exchanged between computer 1 and computer 2 reliably and efficiently.

(2) Communication service module: When communication service module activate computer 1 and computer 2, then data exchanged between two computers reliably and efficiently. Data in this case being document transfer or file transfer and electronics mail is also example of data.

(3) Network access module: Network access module exchange data reliably between computer 1 and communication complex network. Also data exchanged between complex communication network and computer 2. Instead of single module, suppose data exchanged between computer 1 and computer 2 by file transfer module, communication service module and network access module, then such arrangement is known as protocol architecture.

11.7 PROTOCOL STRUCTURE IN ACTUAL NETWORK

When data exchanged from computer 1 to computer 2 through a communication network.

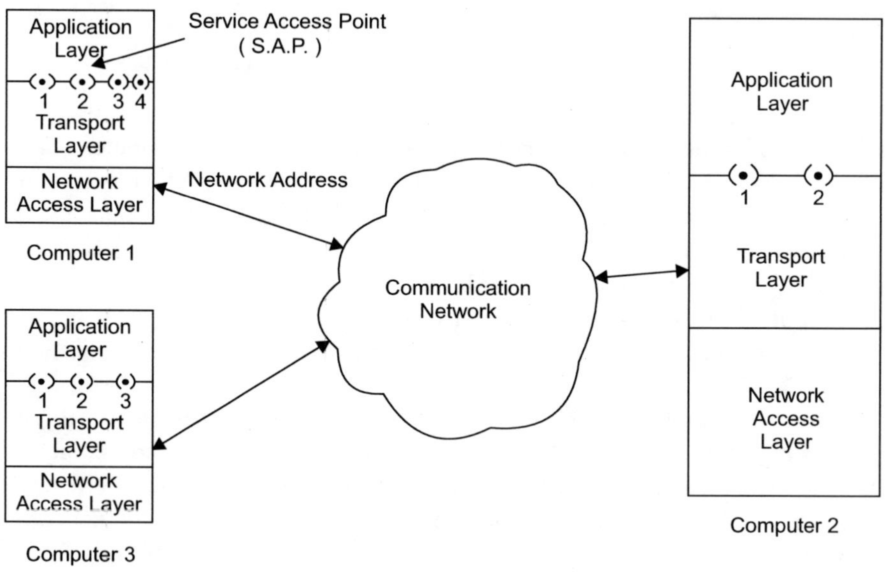

Fig. 11.8: Protocol Architecture and Networks

While performing above-mentioned task, the system goes through three independent layers:

- Network access layer
- Transport layer
- Application layer

(1) Network access layer: Network access layer related with exchange of data between computer and communication network. Sending computer knows the address of destination computer. Hence, sending computer exchange data to destination computer through a communication network. The specific software is used for this layer and this software depends upon type of network. For circuit switching, packet switching and LAN different standards have been developed.

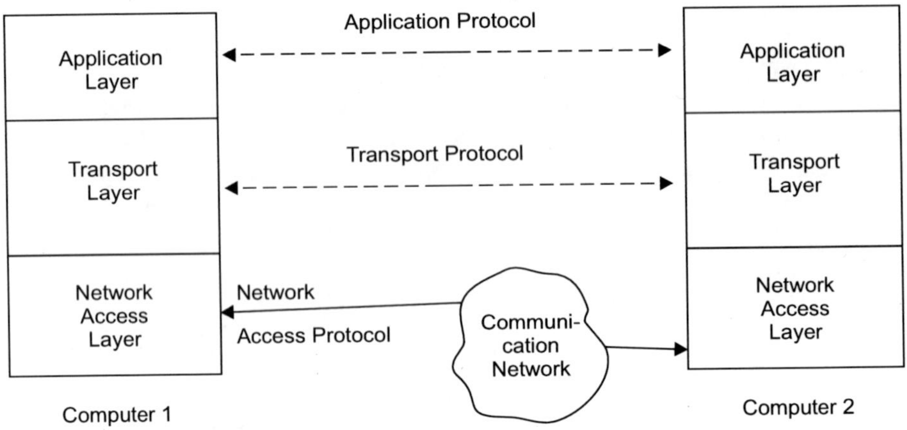

Fig. 11.9: Protocols in a Simplified Architecture

(2) Transport Layer: Whatever may be type of application that is file transfer or electronics mail but data should be reliably exchanged between computer 1 and computer 2. In which order data is sent, in that order data should be received at destination, independent upon type of application. Thus, when data transmitted collected by above-mentioned mechanism by same or common layer shared by all applications, this is known as transport layer.

(3) Application layer: For different type of application, for example, file transfer or electronics mail a separate module is needed and it is known as application layer module.

Figures 11.8 and 11.9 show simple architecture. Figures shows three computers connected to network. For one or more application, each computer contains software at network access layer. For successful communication, every entity in the overall system have unique address. Each computer on the network must have unique address, this allows the network to deliver data to proper computer. Each application on a computer must have an address that is unique within that computer, this allows the transport layer to support multiple applications at each computer.

Figure 11.9 gives information how data exchanged between computer 1 and computer 2 by means of protocol. A protocol is a set of rules or conventions governing the ways in which computer 1 and computer 2 cooperate to exchange data.

(1) Network access protocol: Network access protocol exchange the data between computer 1 and communication network. Also data exchanged between communication network and computer 2.

(2) Transport protocol: When transport protocol activate computer 1 and computer 2, then data exchanged between two computers. Data in this case being document transfer or file transfer and electronics mail. Data transfer becomes reliably and efficiently.

(3) Application protocol: When application protocol transmit file record and file command, password reliably between computer 1 and computer 2. Then file exchanged between computer 1 and computer 2 reliably and efficiently.

11.8 NETWORK INTERFACE X.25 STANDARD

Best known and most widely used protocol standard is X.25. The standard specifies an interface between host system and packet switching system or network. This standard is almost universally used for interfacing to packet switching networks and is employed for packet switching in I.S.D.N.

X.25 standard consists of three layers of functionality.

- Physical layer
- Link layer
- Packet layer

(1) Physical layer: This layer deals with the physical interface between an attached station. For example, computer, terminals and the link that attaches that station to the packet switching node.

(2) Link layer: The link layer provides for the reliable transfer of data across the physical link. Data is transmitted as a sequence of frame. The link layer standard referred to as LAPB that is link access protocol – balanced.

(3) Packet layer: Packet layer provides an external virtual circuit service.

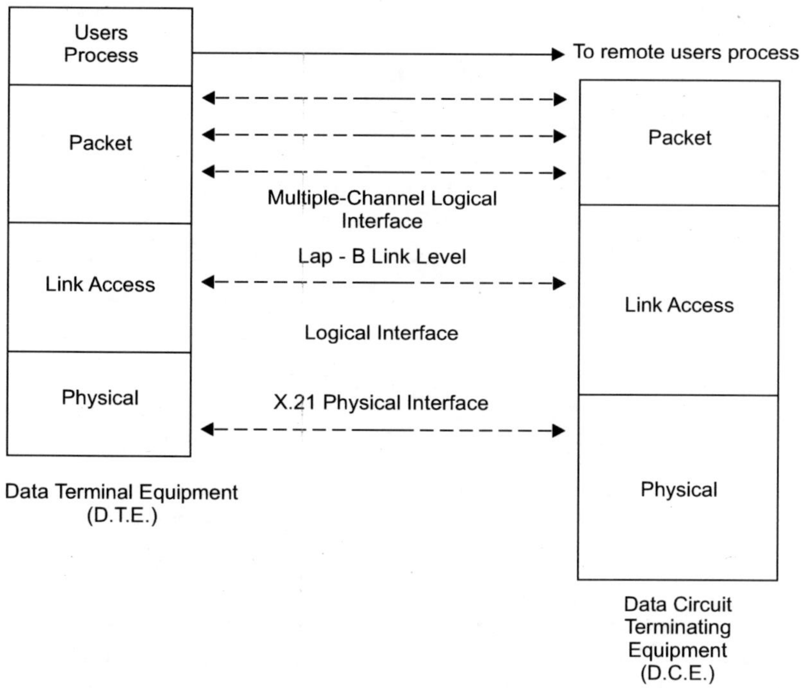

Fig. 11.10: X.25 Protocol Standard

11.9 X.5 PROTOCOL PACKET

(1) LAP-B Header: It contains address of entity and control information and it is known as LAP-B header packet.

(2) Message or data: Remaining part of X.25 protocol packet contains data or message. It is known as message or data packet.

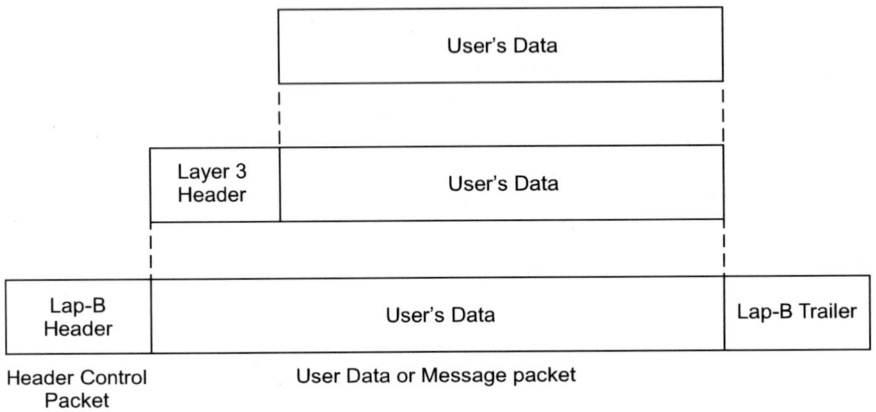

Fig. 11.11: X.5 Protocol Packet

11.10 ROUTING IN PACKET NETWORKS

The primary function of packet switched network is to accept packet from source station and deliver them to destination station.

To achieve this, a path or route through the network must be determined, generally more than one route is possible. Thus, routing function must be performed.

Routing in packet switched networks broadly classified into four categories:

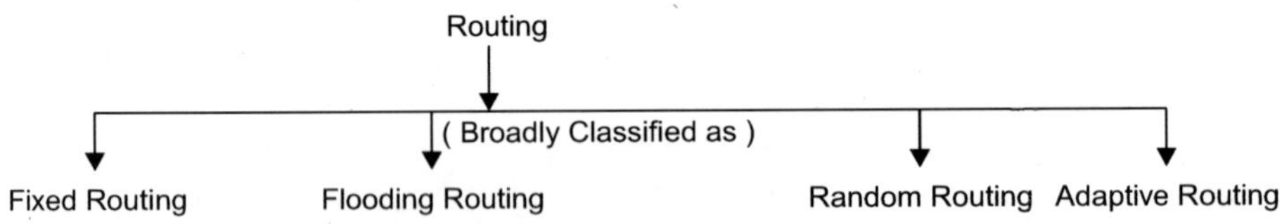

(1) Fixed Routing: For fixed routing, a route is selected from each source node to destination node in the network. The route is fixed, this route will change if network topology is changed. The route also depends on expected traffic or capacity.

(2) Flooding routing: Another simple routing technique is flooding. This technique requires no network information. A packet is sent by a source node to every one of its neighbour. At each node, an incoming packet is retransmitted on all outgoing links except for the link on which it is arrived.

(3) Random routing: Random routing has the simplicity and robustness of flooding with far less traffic load. With random routing, a node selects only one outgoing path for retransmission of an incoming packet. The outgoing link is chosen at random, excluding the link on which the packet arrived. If all links are equally likely to be chosen, then node may utilize outgoing link.

(4) Adaptive routing: In virtually all packet switching networks, adaptive routing technique is used. That is the routing decisions that will change as condition on the network changes. The principal condition that effect routing decisions are:

- *Failure:* When a node or trunk fails, it can no longer be used as part of a route.
- *Congestion:* When a particular portion of the network is heavily congested, it is desirable to route packets around, rather than through, the area of congestion.

The more information that is exchanged and more frequently it is exchanged, better will be routing decision that each node makes. On the other hand, this information is itself load on the network, causing a performance degradation.

11.11 VALUE ADDED NETWORK (ISDN)

Figure 11.12 shows conceptual view of ISDN connection features along with subscriber loop and ISDN channel structures.

When data, voice and image are transmitted by same digital technique, then it is known as integrated system digital networking.

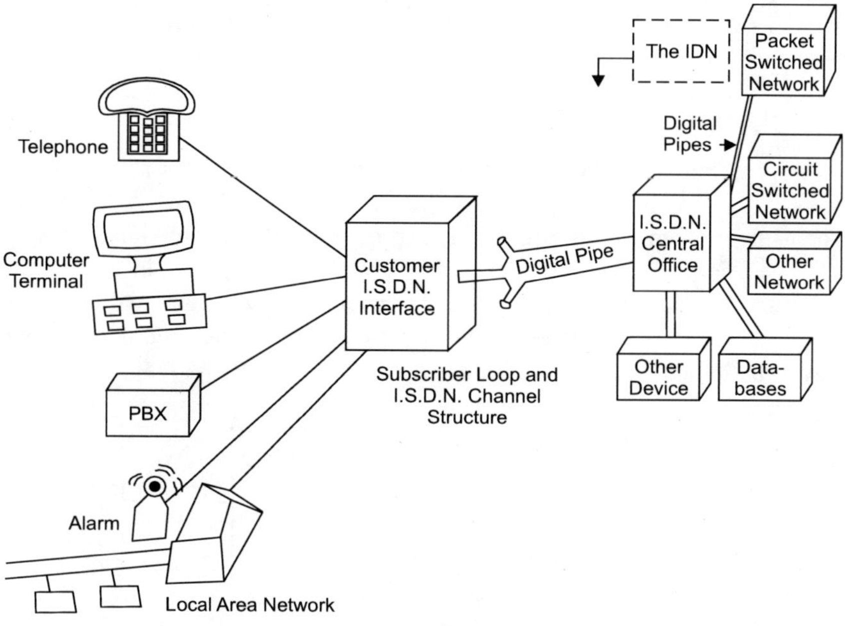

Fig. 11.12: Conceptual View of ISDN Connection Features

Due to evolving technologies, coupled with increasing demands for efficient and timely collection, processing and discrimination of information are leading to the development of integrated system that transmit and process all types of data. The ultimate goal of this evolution is the integrated services digital network (ISDN). ISDN provides value-added processing services.

Figure 11.12 is a conceptual view of the ISDN from a user's point of view. The user has to access to the ISDN by means of a local interface to a "digital pipe" of certain bit rate. Pipes of various sizes are available to satisfy different needs. For example, a residential customer may require only sufficient capacity to handle a telephone and a video text terminal. An office will undoubtedly wish to connect to the ISDN via premise digital PBX. and will require a much higher capacity pipe.

At any given point in time, the pipe to user's premises has a fixed capacity but the traffic on the pipe may be a variable mix up to the capacity limit. Thus, a user may access circuit switched and packet switched services as well as other services in a dynamic mix of signal types and bit rates. To provide these services, the ISDN requires rather complex control signal to instruct it how to sort out the time multiplexed data and provide the required services. These control signals are also multiplexed onto the same digital pipe.

An important aspect of the interface is that the user may at any time employ less than the maximum capacity of the pipe and will be charged according to the capacity used rather than "connect time". This characteristics significally diminishes the value of current user design efforts that are geared to optimise circuit utilization by use of concentrators, multiplexer, packet switches and other line sharing arrangements.

11.12 MODEM

Modem are placed at both ends of the communication circuit as shown in Fig. 11.13.

The name modem is a short form of term modulator and demodulator. As the name implies, both functions are included in a modem. When used in transmitting mode, the modem accepts digital data and converts it to analog signal for use in modulating a carrier signal. At the received end of system, the carrier is demodulated to recover the data.

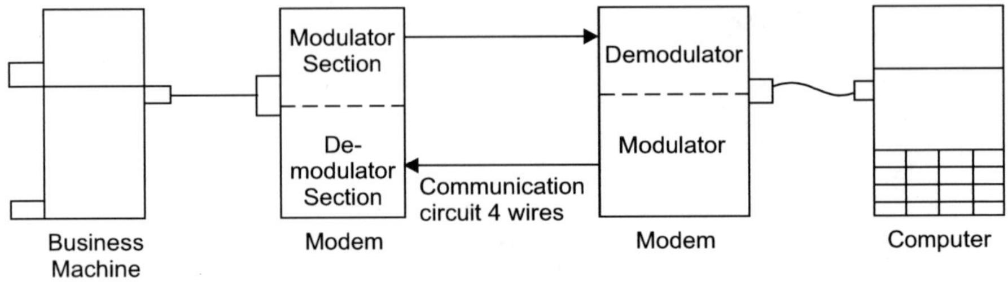

Fig. 11.13: Communication Circuit Using Modem

QUESTIONS

11.1. Explain T.D.M., F.D.M., synchronous T.D.M. statistical T.D.M.

11.2. It was shown that flooding can be used to determine the minimum-hop route. Can it be used to determine the minimum delay route?

11.3. With random routing, only one copy of the packet is in existence in time. Nevertheless, it would be wise to utilize a hop count field. Why?

11.4. Assuming no malfunction in any of the stations or nodes of a network, is it possible for the packet to be delivered to the wrong destination.

11.5. Flow control mechanisms are used at both levels 2 and levels 3 of X.25. Are both necessary or is this redundant? Explain.

11.6. There is no error-detection mechanism (frame check sequence) in X.25. Is not this needed to assure that all of the packets are delivered properly?

11.7. In X.25, why is the virtual circuit number used by one station of two communicating stations of different from the virtual-circuit number used by the other station? After all, it is the same full-duplex virtual circuit.

CHAPTER 12

Multiple Access

12.1 A SATELLITE BASE ALOHA TECHNIQUE OR RANDOM ACCESS TECHNIQUE

Satellite communication using random multiple access technique is known as ALOHA technique or satellite base ALOHA technique or random access technique. Here the packets are not being send in FDMA or TDMA that is the user does not wait his 'turn' as in TDMA nor is he assigned a narrow bandwidth as in FDMA. Each station transmits a packet whenever its buffer has one and the station do not need synchronization. The random multiple access scheme works on four different steps, namely—the transmission, listening, retransmission and time out.

(1) The transmission: In transmission step which is first step, user's transmit at any time they desire encoding their transmission with an error detection code.

(2) Listening: The users listen an acknowledgement (ACK) from the receiver. There is a possibility that the packets send from different terminals may overlapping packets are said to have collied and the users receive a negative acknowledgement (NAK). After having received NAK, the third step of random access technique is followed which is actually the retransmission of collided packets.

(3) Retransmission: In order that these retransmitted packets may not collide again the interval of packet retransmission is randomized in each terminal. In other words, now the user's retransmit after a random delay. In case if after a transmission the user does not receive either an ACK or NAK within specified time the user retransmit the message.

(4) Time out: This is final step of random multiple access and is termed time out.

The analysis of ALOHA system involves message arrival statistics. Suppose that in the satellite system the rate of successful arrival of packet is λ per second. Since there would be some packets lost due to collision so if λ_r is the rate of such lost packets per second, then the total traffic arrival rate λ_t would be

$$\lambda_t = \lambda + \lambda_r \tag{1}$$

where, λ_t = the total traffic arrival rate
λ = rate of the successful arrival of packets per second
λ_r = packet lost due to collision per second.

Let the length of each message or packets be b bits. Then the throughput of the system which is actually the average amount of successful traffic would be expressed as:

$$\rho' = b\lambda \text{ bits per second} \tag{2}$$

where, ρ' = throughput of system
b = length of each packet or message
λ = rate of successful arrival of packets per second.

Similarly, the total traffic G′ on the satellite channel would be

$$G' = b\lambda_t \text{ bits per second} \tag{3}$$

where, G′ = Total traffic on the satellite
b = Message or packet length
λ_t = Total traffic arrival rate per second.

If R bits per second represents the channel capacity (maximum bit rate) then normalised throughput P and the normalized total traffic G could be expressed as:

$$R = \frac{b\lambda}{R} \tag{4}$$

where, P = Normalized throughput
b = Message packet length
λ = Successful traffic arrival rate
R = Channel capacity

$$G = \frac{b\lambda_t}{R} \tag{5}$$

where, G = Normalized total traffic
b = Message or packet length
λ_t = Total traffic arrival rate per second
R = Channel capacity

Let transmission time of each packet be

$$\tau \text{ given as } \tau = b/R \text{ second/packets}$$

where, τ = Transmission time of each packet
b = Message or packet length
R = Channel capacity

Thus, equations (4) and (5) may be rewritten as

$$\rho = \lambda\tau \tag{6}$$

and

$$G = \lambda_t\tau \tag{7}$$

In order that there may not be any collision between individual packets, a space of 2τ second is needed between each packet. This is because of fact that if another user began a message within the previous τ second; its tail will collide with the current message and similarly, if another user begin a message within next τ second, it will collide with the tail end of current message. The traffic arrival to the satellite repeater may be considered as a Poission process because in most of communication system modelled as queueing system the arrival of traffic best explained by Poission process.

Thus, the probability of having k new messages arrive during a time interval of τ second is represented by the Poission distribution as

$$P(k) = \frac{(\lambda T)^k e^{-\lambda_\tau}}{k!}; \; k \geq 0$$

where λ is the average message arrival rate. Since within time interval of 2τ seconds no message is transmitted that is k = 0 and then

$$P(k = 0) = P_s = \frac{(2\tau\lambda_t)e^{-2\lambda\tau}}{0!} = e^{-2\tau\lambda}t \tag{8}$$

This is the case of a successful packet transmission because then there would be no collision among packets. Here again λ_t represents the total traffic arrival rate. The probability of successful transmission P_s can also be represented in a simple way as:

$$P_s = \frac{\lambda}{\lambda_t} \tag{9}$$

where λ is the successful portion of the total traffic rate λ_t. Thus, from eqs. (8) and (9) one gets:

$$\lambda = \lambda_t e^{-2\tau\lambda_t} \tag{10}$$

Equations (10), (6) and (7) combined together would give

$$\rho = Ge^{-2G} \tag{11}$$

Equation (11) gives the relationship between the normalized throughput ρ and the normalized total traffic from the ALOHA system.

To improve the usage of channel capacity of pure ALOHA system discussed above, a technique termed slotted ALOHA. Here a sequence of synchronization pulses is broadcast to all stations and message are required to be sent in the slot time between synchronization pulses. It must be remembered that the message can be started only at the beginning of a time slot and as with the pure ALOHA packet length are constant. This technique reduces the collision rates of packets by half since only message transmitted; in the same slot can interfere with one another. Thus, with the reduction of collision window from 2τ to τ, the relationship between normalised throughput ρ to the normalized total traffic G. Equation (4) now becomes

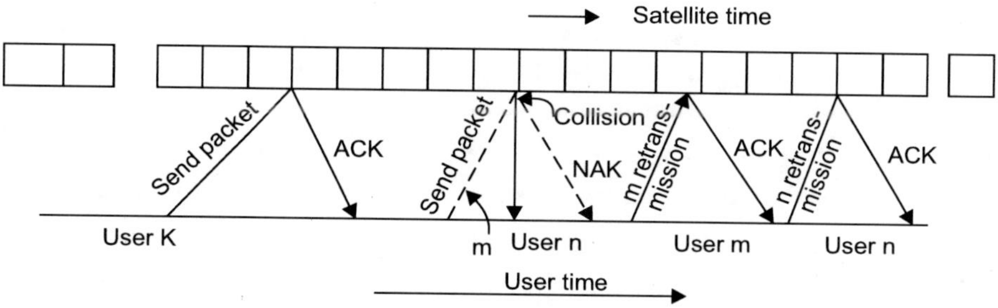

Fig. 12.1: Slotted ALOHA Operation of Random Access Scheme

$$\rho = Ge^{-G} \tag{12}$$

The plot of Equation (12) labelled as slotted ALOHA. Here the maximum value of ρ is 1/e = 0.37 which shows an improvement of two times over pure Aloha system. Thus, in slotted Aloha if a negative acknowledgement (NAK) occurs, the user retransmits after a random delay of an integer number of slot times. Figure 12.1 shows an example of slotted Aloha packet transmission of the random access scheme. Here user k transmits its packets and the successful transmission is indicated by the return of ACK.

When the users m and n simultaneously transmit their packets, they collide with each other and a NAK is returned. After a random delay the users m and n retransmit their packets. Successful transmission if takes place, will be informed by the return of ACK. In case the packets m and n again collide, another retransmission of m and n packets would take place after another random delay. The process may repeat till the successful transmission takes place and is informed by the return of ACKS. Thus, here each using station employs a random number generator to select its retransmission time. It would be worthwhile to mention here that the above pure Aloha and slotted Aloha schemes are also called ALOHA or S-ALOHA protocols.

12.2 TERRESTRIAL PACKET RADIO SYSTEM OR BROADCAST RADIO

(1) Physical Description: In broadcast radio antenna is omnidirectional i.e. it radiate electromagnetic waves equally in all directions. Thus, broadcast radio does not require dish-shaped antennas, and the antennas need not be rigidly mounted to a precise alignment.

(2) Application: Radio is a general term used to encompass frequencies in the range of 3 kHz to 300 GHz. We are using the informal term broadcast radio to cover the VHF and part of UHF band 30 MHz to 1 GHz. This range covers F.M. radio as well as U.H.F. and V.H.F. television. This range is also used for a number of data networking applications.

(3) Transmission Characteristics: The range of 30 MHz to 1 GHz is an effective one for broadcast communication. Unlike the case for lower frequency electromagnetic waves the ionosphere is transparent to radio waves above 30

MHz. Transmission is limited to line of sight and distant transmitters will not interfere with each other due to reflection from the atmosphere. Broadcast radio waves are less sensitive to attenuation from rainfalls.

As a line of sight propagation technique, radio waves obeys equation $d = 7.14\sqrt{kh}$ that is maximum distance between transmitter and receiver is slightly more than the optical line of sight or $7.14\sqrt{kh}$.

A prime source of impairment for broadcast radio waves is multipath interference. Reflection from land, water and natural or human made objects can create multipath between antennas. This effect is frequently evident when T.V. reception displays multiple images as an aeroplane passes by.

12.3 RESOURCE SHARING AND MULTIPLE ACCESS TECHNIQUE

Resource in this case is transponder of satellite. By multiple satellite access system you can access geographically dispered location and share satellite's capacity. In satellite communication major advantage is that single point to multipoint communication is possible. Single transponder access many earth stations. So transponder should have multiple access capacity.

Commonly used multiple access schemes are as follows:

Multiple Access Scheme

- Frequency Division Multiple access (FDMA)
- Time Division Multiple access (TDMA)
- Code Division Multiple access (CDMA)

(1) Frequency Division Multiple Access (FDMA): For satellite communication multiple access technique is commonly used. Each of the earth stations in satellite communication transmits one or more carriers at different centre frequencies. Each carrier is assigned a frequency bandwidth, a small guard band to avoid overlapping of adjacent carriers. The satellite transponder receives all the carriers within its bandwidth does the necessary translation and amplification and retransmits them back to earth.

Figure 12.2 shows basic concepts of FDMA in satellite communication. Different earth stations are capable of selecting carriers containing messages of their interest. Advantage of F.D.M.A. include simplicity of each station equipment and fact that no complex timing and synchronization technique are required. Disadvantage is that intermodulation problem with its adverse effect on S/N ratio.

(2) Time-Division Multiple Access (TDMA): Time division multiple access (T.D.M.A.) is a technique in which different earth stations in the satellite making use of satellite transponder use a single carrier on a time division basis. Different earth station transmit traffic bursts in a periodic time frame called TDMA frame. The traffic bursts from different earth stations are synchronized so that all bursts

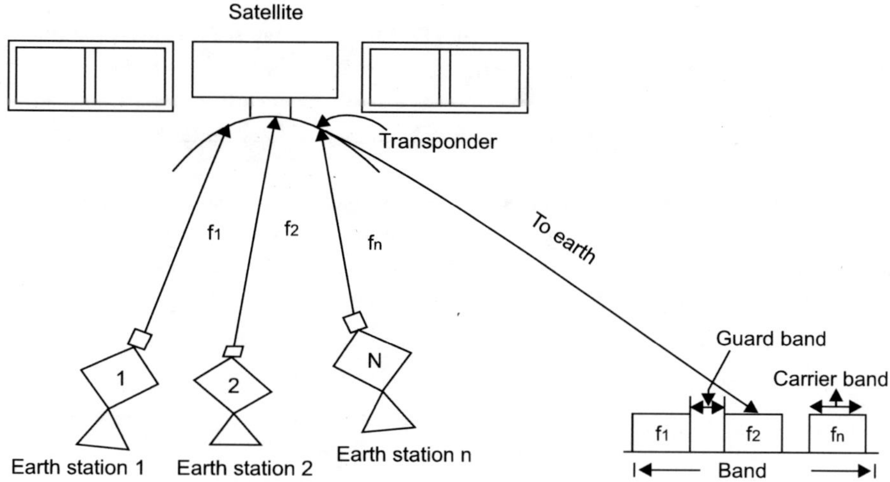

Fig. 12.2: Satellite Earth Station

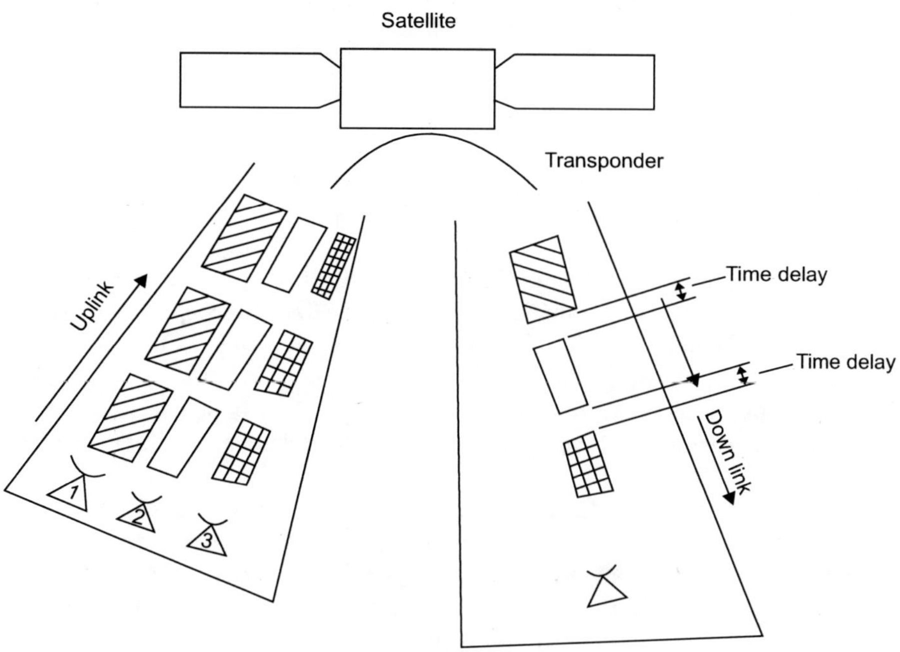

Fig. 12.3: TDMA Satellite Earth Station

arriving at the transponder are closely spaced but do not overlap. The transponder works on a burst at a time and retransmits back to the earth sequence of bursts. All earth stations can receive the entire sequence and extracts the signal of their interest.

Figure 12.3 shows the basic concept of TDMA. Advantages of TDMA include absence of intermodulation problems as the transponder has to deal with one carrier at a time. Disadvantage is that complex and expensive earth station equipment and stringent timing and synchronization requirement. TDMA is suitable for digital transmission only.

(3) Code Division Multiple Access (CDMA): In code division multiple access (CDMA), the entire bandwidth of the transponder is used all the time. Signals from different stations are encoded in such a way that actual transmitter can be recovered only by assigned codes.

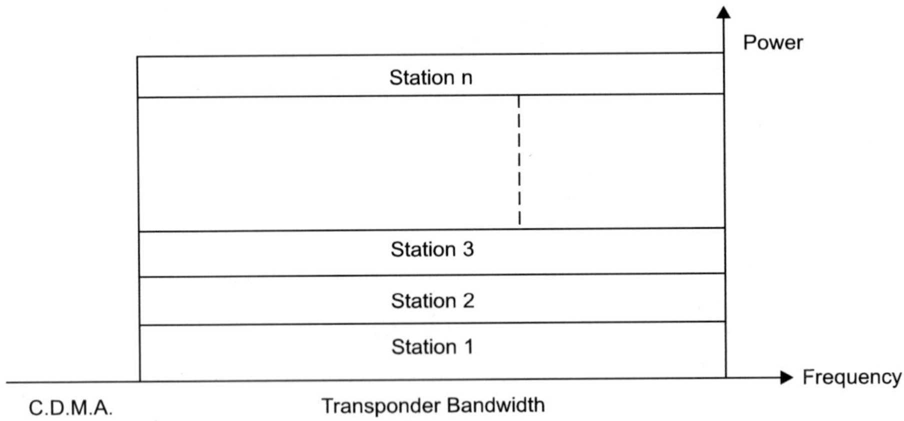

Fig. 12.4: CDMA Access Technique

In code division multiple access (CDMA) all users transmit to the satellite on the same frequency and at the same time. The earth stations transmit orthogonally coded spectrum signals that can be separated at the receiving earth station by correlation with the transmitted code. CDMA is inherently a digital technique. In each of the multiple access technique, some unique property of signal (frequency, time or code) is used to label the transmission such that the wanted signal can be recovered at the receiving terminal in the presence of all other signals.

Code division multiple access is a scheme in which a number of users can occupy all of the transponder bandwidth all times. CDMA signals are encoded such that information from an individual transmitter can be recovered by receiving station that knows the code being used, in the presence of all the other CDMA signals in the same bandwidth. Each receiving station allocated a CDMA code, any transmitting station that wants to send data to that earth station must use the correct code. The CDMA chip sequence modulates the data bits of original message and chip rate is always much greater than the data rate.

CDMA has become popular in cellular telephone system where it is used to enhance cell capacity. However, it has been widely employed in satellite communication systems because it is usually proved to be less efficient, in terms of capacity than FDMA and TDMA. The globarstar LEO satellite system was designed to use CDMA for multiple access by satellite telephone; the advantage of CDMA in this application is soft handoff.

QUESTIONS

12.1. What is difference between multiplexing and multiple access technique?
12.2. What is burst?
12.3. What is guard time? Mention its value in typical TDMA system.
12.4. What is mean by TDMA frame acquisitions and frame synchronization?
12.5. What is satellite switching? Explain difference between statistics and dynamic switching?
12.6. What is CDMA? In what way is it superior to TDMA? Mention the potential applications of CDMA.
12.7. Explain Kepler's law of planetary rotation.
12.8. Explain in detail random multiple access technique.
12.9. What is mean by ALOHA and S-ALOHA?
12.10. What is mean by pure ALOHA and reservation ALOHA?
12.11. Explain in detail queueing system.
12.12. What are the various switching technique and how packet switching is different from them? Discuss various applications of packet switching.
12.13. Why is packet switching being used in random multiple access technique?
12.14. What are polling technique? Explain the polling technique used in binary tree search technique.

Index